CHINA'S PASTORAL REGION

Sheep and Wool,
Minority Nationalities,
Rangeland Degradation and
Sustainable Development

Dedicated
to
Jill and Glenys

CHINA'S PASTORAL REGION

Sheep and Wool,
Minority Nationalities,
Rangeland Degradation and
Sustainable Development

John W. Longworth
Department of Agriculture
The University of Queensland
Australia

and

Gregory J. Williamson
Department of Primary Industries
Queensland
Australia

CAB INTERNATIONAL

In association with:

ACIAR
The Australian Centre for International Agricultural Research

CAB INTERNATIONAL Tel: Wallingford (0491) 832111
Wallingford Telex: 847964 (COMAGG G)
Oxon OX10 8DE Telecom Gold / Dialcom: 84: CAU001
UK Fax: (0491) 833508

© CAB INTERNATIONAL 1993. All rights reserved. No part of this publication
may be reproduced in any form or by any means, electronically,
mechanically, by photocopying, recording or otherwise, without
the prior permission of the copyright owners.

A catalogue record for this book is available from the British Library.

ISBN 0 85198 890 3

Published in association with:

The Australian Centre for International Agricultural Research (ACIAR)
GPO Box 1571
Canberra
ACT 2601
Australia

Tel: (06) 248 8588
Fax: (06) 257 3051
Telex: AA 62491
Telecom Gold/Dialcom: 6007: IAR001

Printed and bound in the UK at the University Press, Cambridge.

Contents

Foreword	xiii
Acknowledgements	xv
List of Tables	xvii
List of Figures	xxii
List of Plates and Maps	xxiii
Glossary	xxiv
About the Authors	xxv

1. **Introduction**   1

PART I - BACKGROUND

2. **Outline of the Research Project**   9
   - 2.1 Research Objectives   10
   - 2.2 Brief Rationale for the Project   10
   - 2.3 Research Methodology   11
   - 2.4 Research Methods   13
     - 2.4.1 Literature Review   13
     - 2.4.2 Survey Questionnaires   13
     - 2.4.3 Survey Design   15
     - 2.4.4 Field Reports   16
     - 2.4.5 Analytical Models   17
   - 2.5 Research Difficulties   18
     - 2.5.1 Vertical Bureaucratic Channels   18
     - 2.5.2 Separating Fact from Fiction   20
     - 2.5.3 Some Special Difficulties Encountered in Surveying Pastoral Areas   21
     - 2.5.4 Definitional and Data Problems   22

3. **Some Characteristics of the Pastoral Region**   25
   - 3.1 Levels of Administration   26
   - 3.2 Pastoral Provinces/Autonomous Regions   28
     - 3.2.1 Population   28
     - 3.2.2 Land Use   30
     - 3.2.3 Herbivorous Livestock   30
     - 3.2.4 Output of Pastoral Products   32
     - 3.2.5 Gross Output Value   32
   - 3.3 Pastoral and Semi-Pastoral Counties   34
   - 3.4 Minority Nationalities   37
     - 3.4.1 Ethnic Autonomy   39
     - 3.4.2 National Representation   40
     - 3.4.3 Development Assistance   41

	3.5	History of Changes to Property Rights in the Pastoral Areas	42
		3.5.1 Land Reform 1947-1950	42
		3.5.2 Reforms to Livestock Ownership 1947-1957	43
		3.5.3 Collectivisation 1952-1958	44
		3.5.4 The Household Responsibility System 1956-1966	45
		3.5.5 Cultural Revolution 1966-1976	46
		3.5.6 Property Rights Reforms Post-1978	46
4.	**The Chinese Sheep and Wool Scene**		48
	4.1	Wool Grading and Sheep Breeds	49
		4.1.1 National Wool Grading Standard	49
		4.1.2 Wool Types and Sheep Breeds	51
	4.2	Sheep Production Systems and Sheep-Raising Localities	52
		4.2.1 Traditional Sheep-Raising Localities	52
		4.2.2 Potential Sheep-Raising Localities in Southern China	54
	4.3	Sheep and Wool in the Pastoral Provinces	55
		4.3.1 Sheep Numbers	56
		4.3.2 Wool Production	60
	4.4	Organisation of Production	63
		4.4.1 Private Sheep Raising	64
		4.4.2 State Farms	65
	4.5	Wool Marketing	68
		4.5.1 Supply and Marketing Cooperatives	68
		4.5.2 Traditional Wool Pricing Arrangements	70

PART II - PROVINCIAL CASE STUDIES

5.	**Inner Mongolia Autonomous Region**		75
	5.1	Identifying the Pastoral Areas of IMAR	76
	5.2	Physical Details	77
		5.2.1 Geography	77
		5.2.2 Population	77
		5.2.3 Land Use	81
		5.2.4 Pasture Land Degradation	81
		5.2.5 Livestock Numbers	84
		5.2.6 Location of Pastoral Production	90
	5.3	Livestock Improvement Programs	91
		5.3.1 Cattle	91
		5.3.2 Fine and Semi-fine Wool Sheep	92
		5.3.3 Meat Sheep	94
		5.3.4 Goats	96
	5.4	Animal Husbandry Research and Extension Organisations	96
		5.4.1 Animal Husbandry Bureau	96
		5.4.2 Economic Management Station/Bureau	97
		5.4.3 Vertical Bureaucratic Connections	97
	5.5	Introduction of the Household Production Responsibility System	98
		5.5.1 Privatising the Livestock	98
		5.5.2 Contracting Out the Pasture Land	99
	5.6	Trends in Incomes and Wool Pricing	100
		5.6.1 Trends in Rural Incomes	100
		5.6.2 Wool Pricing Since 1984	101

		Appendix 5A	106
		Appendix 5B	111

6. Gansu Province — 116
- 6.1 Identifying the Pastoral Areas of Gansu — 117
- 6.2 Physical Details — 118
  - 6.2.1 Geography — 118
  - 6.2.2 Population — 120
  - 6.2.3 Land Use — 120
  - 6.2.4 Types of Grassland — 121
  - 6.2.5 Degradation of Grasslands — 123
  - 6.2.6 Livestock Numbers — 124
  - 6.2.7 Wool Production — 126
- 6.3 Sheep Breeds and Improvement Programs — 126
  - 6.3.1 Breeds and their Location — 127
  - 6.3.2 The "One Million" Fine Wool Sheep Base — 128
  - 6.3.3 Other Sheep and Goat Breeding Programs — 129
- 6.4 Animal Husbandry Research and Extension Organisations — 129
- 6.5 Introduction of the Household Production Responsibility System — 130
- 6.6 Economic Conditions in Pastoral Areas of Gansu — 131
  - 6.6.1 Growth in Incomes since 1978 — 131
  - 6.6.2 Poverty Alleviation Policies — 133
  - 6.6.3 Development of Pastoral Product Processing — 134

7. Xinjiang Uygur Autonomous Region — 135
- 7.1 Identifying the Pastoral Areas of the XUAR — 136
- 7.2 Physical Details — 137
  - 7.2.1 Geography — 137
  - 7.2.2 Population — 137
  - 7.2.3 Land Use — 140
  - 7.2.4 Herbivorous Livestock Numbers — 142
  - 7.2.5 Distribution of Herbivorous Livestock by Prefecture — 144
  - 7.2.6 Location of Pastoral Production — 147
- 7.3 Administration of Agriculture and Animal Husbandry in the XUAR — 148
- 7.4 Xinjiang Production and Construction Corp — 148
  - 7.4.1 A Brief History of the Xinjiang PCC — 148
  - 7.4.2 Sheep and Wool in the PCC System — 150
  - 7.4.3 Other PCC Activities — 151
- 7.5 Livestock Improvement Programs — 153
  - 7.5.1 Fine Wool Sheep Improvement within the XUAR AHB Network — 153
  - 7.5.2 Fine Wool Sheep Improvement within the PCC Network — 155
  - 7.5.3 Breed Improvement Programs for Meat and Coarse Wool Sheep — 156
- 7.6 Xinjiang Animal Husbandry Industrial and Commercial Company — 157
  - 7.6.1 Wool Trading Activities — 158
  - 7.6.2 Livestock Trading Activities — 159
  - 7.6.3 Import/Export of Stud Sheep — 160
- 7.7 Other Economic Issues — 160
  - 7.7.1 Rural Income Trends and Future Sustainable Development — 160
  - 7.7.2 Wool Pricing in the XUAR — 162

PART III - COUNTY CASE STUDIES

8. Balinyou County (IMAR) — 167
   8.1 Resource Endowment — 168
      8.1.1 General Features — 168
      8.1.2 Land Use — 169
      8.1.3 Population — 170
      8.1.4 Administrative Structure — 170
      8.1.5 Livestock Population — 170
   8.2 Improvements to Production Conditions in the Pastoral Systems — 171
      8.2.1 Improved Pastures — 171
      8.2.2 Pasture Management Programs — 171
      8.2.3 Improved Sheep and Wool Production — 173
      8.2.4 Improved Goat Breeds — 178
      8.2.5 Improved Cattle Breeds — 178
   8.3 Socio-Economic Development — 179
      8.3.1 Income Growth — 179
      8.3.2 Changes to Commodity Prices — 180
      8.3.3 Timing of Introduction of the Household Production Responsibility System — 181
      8.3.4 Level of State Assistance — 181
      8.3.5 Development of Pastoral Product Processing — 182
      8.3.6 Growth in Output of Major Pastoral Commodities — 182
   Appendix 8 — 183

9. Wongniute County (IMAR) — 186
   9.1 Resource Endowment — 187
      9.1.1 General Features — 187
      9.1.2 Land Use — 188
      9.1.3 Population — 188
      9.1.4 Administrative Structure — 189
      9.1.5 Livestock Population — 189
   9.2 Improvements to Production Conditions in the Pastoral Systems — 189
      9.2.1 Improved Pastures — 189
      9.2.2 Improved Sheep and Wool Production — 191
      9.2.3 Improved Goat Breeds — 195
      9.2.4 Improved Cattle Breeds — 195
   9.3 Socio-Economic Development — 195
      9.3.1 Income Growth — 195
      9.3.2 Changes to Commodity Prices — 196
      9.3.3 Timing of Introduction of the Household Production Responsibility System — 197
      9.3.4 Level of State Assistance — 197
      9.3.5 Development of Pastoral Product Processing — 197
      9.3.6 Growth in Output of Major Pastoral Commodities — 198

10. Aohan County (IMAR) — 199
   10.1 Resource Endowment — 200
      10.1.1 General Features — 200
      10.1.2 Land Use — 200
      10.1.3 Population — 201
      10.1.4 Administrative Structure — 201

		10.1.5 Livestock Population	202
		10.1.6 Flock Sizes	202
	10.2	Improvements to Production Conditions in the Pastoral Systems	203
		10.2.1 Improved Pastures	203
		10.2.2 Pasture Improvement Programs	203
		10.2.3 Aerial Seeding of Pasture/Tree Planting	203
		10.2.4 Supplementary Feeding	203
		10.2.5 Sheep Production Systems	204
		10.2.6 Improved Sheep and Wool Production	204
		10.2.7 Wool Output	206
	10.3	Socio-Economic Development	206
		10.3.1 Income Growth	206
		10.3.2 Changes to Commodity Prices	207
		10.3.3 Timing of Introduction of the Household Production Responsibility System	209
		10.3.4 Level of State Assistance	209
		10.3.5 Development of Pastoral Product Processing	209
		10.3.6 Growth in Output of Major Pastoral Commodities	210
11.	**Alukeerqin County (IMAR)**		211
	11.1	Resource Endowment	212
		11.1.1 General Features	212
		11.1.2 Land Use	212
		11.1.3 Population	214
		11.1.4 Administrative Structure	214
		11.1.5 Livestock Population	214
	11.2	Improvements to Production Conditions in the Pastoral Systems	216
		11.2.1 Improved pastures	216
		11.2.2 Improved Sheep and Wool Production	217
	11.3	Socio-Economic Development	219
		11.3.1 Income Growth	219
		11.3.2 Commodity Prices	220
		11.3.3 Timing of Introduction of the Household Production Responsibility System	222
		11.3.4 Level of State Assistance	222
		11.3.5 Development of Pastoral Product Processing	223
		11.3.6 Growth in Output of Major Pastoral Commodities	224
	Appendix 11A		227
	Appendix 11B		230
12.	**Wushen County (IMAR)**		237
	12.1	Resource Endowment	238
		12.1.1 General Features	238
		12.1.2 Land Use	239
		12.1.3 Population	239
		12.1.4 Administrative Structure	240
		12.1.5 Livestock	240
	12.2	Improvements to Production Conditions in the Pastoral Systems	240
		12.2.1 Improved Pastures	240
		12.2.2 Improved Sheep and Wool Production	241

	12.3 Socio-Economic Development		244
		12.3.1 Income Growth	244
		12.3.2 Commodity Prices	244
		12.3.3 Timing of Introduction of the Household Production Responsibility System	245
		12.3.4 Level of State Assistance	245
		12.3.5 Development of Pastoral Product Processing	245
		12.3.6 Output of Major Pastoral Commodities	246

13. Sunan County (Gansu) — 247

13.1 Resource Endowment — 248
    13.1.1 General Features — 248
    13.1.2 Land Use — 249
    13.1.3 Population — 249
    13.1.4 Administrative Structure — 249
    13.1.5 Livestock Population — 250
13.2 Improvements to Production Conditions in the Pastoral Systems — 250
    13.2.1 Improved pastures — 250
    13.2.2 Pasture Improvement Programs — 250
    13.2.3 Pasture Management Programs — 251
    13.2.4 Improved Sheep and Wool Production — 251
13.3 Socio-Economic Development — 254
    13.3.1 Income Growth — 254
    13.3.2 Changes to Commodity Prices — 254
    13.3.3 Timing of Introduction of the Household Production Responsibility System — 256
    13.3.4 Level of State Assistance — 257
    13.3.5 Development of Pastoral Product Processing — 257
Appendix 13A — 259
Appendix 13B — 261

14. Dunhuang County (Gansu) — 262

14.1 Resource Endowment — 263
    14.1.1 General Features — 263
    14.1.2 Land Use — 263
    14.1.3 Population — 264
    14.1.4 Administrative Structure — 265
    14.1.5 Livestock Population — 265
14.2 Improvements to Production Conditions in the Pastoral Systems — 266
    14.2.1 Improved Pastures — 266
    14.2.2 Supplementary Feeding — 266
    14.2.3 Improved Sheep Breeds — 266
    14.2.4 Specific Sheep Breeding Programs — 267
    14.2.5 Specific Animal Health Programs — 267
14.3 Socio-Economic Development — 268
    14.3.1 Income Growth — 268
    14.3.2 Changes to Commodity Prices — 268
    14.3.3 Timing of Introduction of the Household Production Responsibility System — 269
    14.3.4 Level of State Assistance — 270
    14.3.5 Growth in the Output of Major Pastoral Commodities — 270

Contents

15.	**Tianzhu County (Gansu)**	272
	15.1 Resource Endowment	273
	15.1.1 General Features	273
	15.1.2 Land Use	273
	15.1.3 Population	273
	15.1.4 Livestock Population	274
	15.2 Improvements to Production Conditions in the Pastoral Systems	274
	15.2.1 Improved Sheep and Wool Production	274
	15.2.2 Improved Cattle Breeds	275
	15.3 Socio-Economic Development	275
16.	**Cabucaer County (Xinjiang)**	276
	16.1 Resource Endowment	277
	16.1.1 General Features	277
	16.1.2 Land Use	277
	16.1.3 Population	278
	16.1.4 Administrative Structure	278
	16.1.5 Livestock Population	279
	16.2 Improvements to Production Conditions in the Pastoral Systems	279
	16.2.1 Improved pastures	279
	16.2.2 Supplementary Feeding	279
	16.2.3 Improved Sheep and Wool Production	279
	16.2.4 Improved Goat Breeds	282
	16.2.5 Improved Cattle Breeds	282
	16.3 Socio-Economic Development	282
	16.3.1 Income Growth	282
	16.3.2 Changes to Commodity Prices	282
	16.3.3 Timing of Introduction of the Household Production Responsibility System	283
	16.3.4 Level of State Assistance	284
	16.3.5 Development of Pastoral Product Processing	284
	16.3.6 Growth in Output of Major Pastoral Commodities	284
17.	**Hebukesaier County (Xinjiang)**	285
	17.1 Resource Endowment	286
	17.1.1 General Features	286
	17.1.2 Land Use	286
	17.1.3 Population	287
	17.1.4 Administrative Structure	287
	17.1.5 Livestock Population	287
	17.2 Improvements to Production Conditions in the Pastoral Systems	288
	17.2.1 Improved pastures	288
	17.2.2 Supplementary Feeding	291
	17.2.3 Improved Sheep and Wool Production	291
	17.3 Socio-Economic Development	293
	17.3.1 Income Growth	293
	17.3.2 Commodity Prices	294
	17.3.3 Development of Pastoral Product Processing	295
	17.3.4 Growth in Output of Major Pastoral Commodities	295

PART IV - THE BIG PICTURE

18. Constraints to Development in Pastoral Areas   299
    18.1 Population Pressures   300
        18.1.1 Link Between Population Pressure, Poverty
             and Rangeland Degradation   300
        18.1.2 Policies Contributing to Population Pressures   304
    18.2 Market Distortions   308
        18.2.1 Product Market Distortions   309
        18.2.2 Factor Market Distortions   312
    18.3 Institutional Uncertainties   320
        18.3.1 Property Right Uncertainties   320
        18.3.2 The "Policy Mirage" Syndrome   321
    18.4 Technical Improvements   322
        18.4.1 Pasture Improvement   322
        18.4.2 Crop Production   325
        18.4.3 Fencing   326
        18.4.4 Animal Health and Water Wells   326
        18.4.5 Livestock Improvement   326

19. The Way Ahead   328
    19.1 Sheep and Wool   328
    19.2 Minority Nationalities   331
    19.3 Rangeland Degradation   332
    19.4 Sustainable Development   334

References   337

Index   340

About the Authors   351

Foreword

The Australian Centre for International Agricultural Research (ACIAR) promotes collaborative research into improving sustainable agricultural production in developing countries. Its activities are funded from the Australian official development assistance program. This book represents part of the output from an ACIAR research project joining researchers from The University of Queensland with two Chinese institutions - the Institute of Rural Development (Academy of Social Sciences) and the Institute of Agricultural Economics (Academy of Agricultural Sciences). The project is an excellent model of an ACIAR project. Collaboration was strong, with all parties contributing both human and financial resources. ACIAR resources have played a catalytic role in linking the various groups.

The Australian research team brought a unique combination of practical rural skills and western economic thinking to apply to the problems facing suppliers and marketers of raw wool in China, and more broadly to the problems of development in the largely unknown pastoral areas of China. This has complemented the range of knowledge and skills resident in the Chinese teams. The project has profited by the direct involvement of the Directors of both Chinese Institutes.

The focus of agricultural research throughout the world, whether it be economic or technological, has moved to embrace environmental issues. Most concern has been voiced with regard to rainforest areas and agricultural production areas on steeply sloping lands. The rangelands have been relatively neglected. Yet degraded rangelands cover huge tracts of the Earth's land surface especially in Central Asia and Northern China. In these areas rangeland management is intertwined with the fate of the ethnic minorities who have traditionally inhabited these localities and many of whom now face difficult economic circumstances.

A total of five books (including one in Chinese), a technical report and many professional papers have already emerged (or will soon emerge) from the study. However, this volume represents the major publication and it is pleasing to see such a substantial and definitive work resulting from the project. The book makes an especially important contribution to the literature by highlighting the need for a more holistic view of the impact of development assistance. In the past, assistance has usually been confined to financing technical modernisation and has generally focused on increasing productivity rather than ensuring sustainability. This book demonstrates the need for greater attention to be given to policy as a means of ensuring sustainability and thereby increasing the long-term effectiveness of development assistance in environmentally sensitive areas.

Kenneth M Menz
Research Program Coordinator (Economics and Farming Systems)
Australian Centre for International Agricultural Research, Canberra.

Acknowledgements

This book summarises a major team effort made possible by institutional and financial support from the Australian Centre for International Agricultural Research. The research project, which generated the information in this book, benefited from a long gestation or planning period. Initially conceived in mid-1986, it did not formally commence until March 1989. From the beginning, collaboration between the Australian and the Chinese scientists working on the project has been excellent. One of the major reasons for this and for the success of the whole research program has been the commitment to the project of the Directors of the two collaborating Chinese Institutes, Professor Niu Ruofeng and Professor Chen Jiyuan. Without their support and personal involvement, it would have been impossible to conduct the fieldwork. In this connection it is also important to acknowledge the support of the Chinese Ministry of Agriculture and in particular the Department of Animal Husbandry and Veterinary Science. It is hoped that the foreign professional perspective, as reported in this book, may assist the Ministry in its endeavours to improve the lot of the people living in China's pastoral region.

Of course, a great deal of the understanding and insight obtained by the Australian team during the project came from their Chinese colleagues. The professionalism of the Chinese members of the project team was always to be admired. The two senior Chinese scientists who worked on the project, Zhang Cungen and Lin Xiangjin, brought a great depth of experience and an invaluable network of Chinese colleagues to the project. They were ably supported by Zhou Li, Du Yintang, Liu Yuman and Xu Ying. All of these people have contributed immensely to the technical content of this book.

In addition to the technical input of our Chinese colleagues, they were also unstinting in providing administrative and organisational support without which the Australian team could not have functioned. In this regard the most crucial practical contribution of the Chinese team members related to language. None of the Australians working on the project had any Chinese language capabilities before the research began in 1989. On the other hand, almost all the Chinese scientists allocated to the project had some English language capability. The interpreting burden during fieldwork, therefore, fell on the shoulders of our Chinese colleagues especially Du Yintang and Liu Yuman and to a lesser degree Zhang Cungen and Xu Ying. Other members of the Chinese team also made major contributions in relation to translation on some occasions, especially Lin Xiangjin. The bilingual capabilities of these people are to be admired and the contributions their language skills made to the project (and to this book in particular) are gratefully acknowledged.

The authors also want to thank the other Australians working on the project. Colin Brown and Ross Drynan generously allowed us to use unpublished reports prepared by them on topics relevant to this book. They have also read earlier drafts of the manuscript and made some valuable suggestions. Similarly, Peter Lynch from ACIAR has contributed to the volume by making editorial suggestions and providing the resources to have the maps computer enhanced. The literary standard of the text has been greatly improved by Colleen Moss who has made innumerable suggestions on points of English expression, punctuation and grammar. Cherelle Mungomery has also made major contributions to the polishing of the manuscript. Two research assistants, Warren Campbell and Kathy Jones, have contributed to aspects of this volume. However, the person to whom our biggest vote of appreciation must go is Deborah Noon. She has patiently and meticulously typed all the field reports and other papers and documents for the project. Single-handedly, she has managed all the files and professional paperwork associated with the project. Finally and most importantly, she has prepared draft after draft of this manuscript and typeset the final version. The original drawings for the maps and the figures are also her work. Deb, your skill, dedication, patience and perseverance have been the backbone of the project.

Others in the Department of Agriculture at The University of Queensland have generously assisted both the Australian team members and visiting Chinese collaborating scientists. In particular, Sue Young and other staff in the Overseas Projects Office have taken care of the accounts, travel arrangements and other administrative matters associated with the project.

The junior author would like to acknowledge the generous terms on which his employer, the Queensland Department of Primary Industries has allowed him to participate in the project.

Finally, the authors want to express their gratitude to their partners, Jill and Glenys, for accepting both the long absences the fieldwork in China has occasioned over the last five years and the nights and weekends which have been devoted to preparing this document.

While gratefully acknowledging all of the above for their assistance, the authors accept full responsibility for the finished product, warts and all.

John W. Longworth
Gregory J. Williamson

September, 1993

List of Tables

Table 2.1	Three Research Institutions and Staff Members Involved in ACIAR Project No. 8811	9
Table 2.2	Defining "Farm Household Net Income" in Pastoral Areas	23
Table 3.1	Outline of the Administrative Structure in China	26
Table 3.2	Some Population and Land Use Characteristics of the Pastoral Provinces/Autonomous Regions of China, 1990	29
Table 3.3	Herbivorous Livestock Numbers and Output of Pastoral Products in the Pastoral Provinces/Autonomous Regions of China, 1990	31
Table 3.4	Gross Output Values for the Pastoral Provinces/Autonomous Regions of China, 1990	33
Table 3.5	Provincial/Autonomous Region Distribution of the 266 Counties Defined by the Central Government as being Pastoral and Semi-Pastoral in 1990	34
Table 3.6	Number of Livestock in the Pastoral and Semi-Pastoral Counties Compared with All China, 1949 to 1990 (selected years)	36
Table 3.7	Wool Production in the Pastoral and Semi-Pastoral Counties Compared with All China, 1985-1990	37
Table 3.8	Provincial Distribution of the Minority Nationalities in China	38
Table 3.9	Transformation of the Pastoral System from Independent Households to Mutual Aid Teams, to Cooperatives and Finally to Communes in the IMAR, 1952-1959	45
Table 4.1	Number of Sheep by Wool Type in each of the 12 Pastoral Provinces/Autonomous Regions, 1981 and 1991	57
Table 4.2	Proportion of Chinese Sheep Flock by Wool Type in each of the 12 Pastoral Provinces/Autonomous Regions, 1991	58
Table 4.3	Wool Production by Wool Type in each of the 12 Pastoral Provinces/Autonomous Regions, 1981 and 1991	61
Table 4.4	Proportion of Chinese Wool Production by Wool Type in each of the 12 Pastoral Provinces/Autonomous Regions, 1991	62
Table 4.5	Official State Purchase Prices for Raw Wool in the Four Years 1981 to 1984	70
Table 5.1	Number of County/Banner Level Administrative Units in the IMAR by League/City Recognised by Governments as being Pastoral or Semi-Pastoral	76
Table 5.2	Some Characteristics of the Central Government Recognised Pastoral and Semi-Pastoral Counties/Banners in IMAR, 1990	78

Table 5.3	Population and Land Use in the IMAR by Prefectural Administrative Unit, 1988	81
Table 5.4	Extent of Pasture Degradation in the IMAR by Prefectural Administrative Unit, 1988	82
Table 5.5	Production Characteristics of the *Gramineae* Group of Grasslands in Various Stages of Degradation in Chifeng City Prefecture, 1981-1985	84
Table 5.6	Yields of Different Grades of Forage Grasses for the *Gramineae* Group of Grasslands in Various Stages of Degradation in Chifeng City Prefecture, 1981-1985	84
Table 5.7	Long-Term Changes in Herbivorous Livestock Numbers in the IMAR and All China, 1949 and 1990	85
Table 5.8	Number and Proportion of Total Sheep Represented by Each Major Category of Sheep in the IMAR, 1981 to 1991	86
Table 5.9	Livestock Numbers in the IMAR by Prefectural Administrative Unit, 1988	87
Table 5.10	Output of Major Pastoral Products in the IMAR by Prefectural Administrative Unit, 1990	90
Table 5.11	Number of Sheep by Wool Type in Chifeng City Prefecture as a Whole and in Balinyou and Alukeerqin Banners, 1980 to 1991	95
Table 5.12	Maximum and Minimum Prices for the Various Types of Wool Established by the IMAR Price Bureau for the 1989 Wool Purchasing Season	105
Table 6.1	Number of Sheep and Goats in Prefectures and Cities of Gansu, 1988	117
Table 6.2	Number of Sheep in the Seven Pastoral and Two Semi-Pastoral Counties Recognised at the Provincial Level in Gansu, 1987	118
Table 6.3	Some Characteristics of the Central Government Recognised Pastoral and Semi-Pastoral Counties in Gansu, 1990	119
Table 6.4	Minority Nationalities with Population Greater than One Thousand in Gansu, 1990	120
Table 6.5	Long-Term Changes in Herbivorous Livestock Numbers in Gansu and All China, 1949 and 1990	124
Table 6.6	Number and Proportion of Total Sheep Represented by Each Major Category of Sheep in Gansu, 1980 to 1991	126
Table 6.7	Production of Greasy Wool, Goat Hair and Cashmere in Gansu, 1980 to 1990	127
Table 7.1	Number of County Level Administrative Units in the XUAR by Prefecture/City Together with Number of Counties Recognised as being Pastoral or Semi-Pastoral, 1989	136
Table 7.2	Some Characteristics of the Central Government Recognised Pastoral and Semi-Pastoral Counties in XUAR, 1990	138
Table 7.3	Population of the XUAR by Ethnic Group, 1949, 1970 and 1990	140

List of Tables

Table 7.4	Population and Land Use in the XUAR by Prefectural Administrative Unit, 1989	141
Table 7.5	Long-Term Changes in Herbivorous Livestock Numbers in XUAR and All China, 1949 and 1990	143
Table 7.6	Number and Proportion of Total Sheep Represented by Each Major Category of Sheep in XUAR, 1981 to 1991	144
Table 7.7	Distribution of Sheep and Goats in XUAR by Prefecture/City, 1989	145
Table 7.8	Output of Major Pastoral Products in the XUAR by Prefectural Administrative Unit, 1990	147
Table 7.9	Some Major Indices of the Importance of the Xinjiang PCC in Relation to the XUAR Economy, 1990	150
Table 7.10	Distribution of Sheep and Goats in the Xinjiang PCC System by Division/Bureau and Prefecture/City, 1989	151
Table 7.11	Distribution of Wool Production in the Xinjiang PCC System by Division/Bureau and Prefecture/City, 1989	152
Table 7.12	Administratively Determined Prices for the Base Grades of Wool in the XUAR, 1970 to 1991	163
Table 8.1	Annual Rainfall at Daban the Capital of Balinyou County, 1959 to 1990	168
Table 8.2	Administrative Units in Agricultural, Pastoral and Semi-Pastoral Areas of Balinyou County, 1989	170
Table 8.3	Percentage of Ewes Artificially Inseminated in Balinyou County, 1980 to 1990	175
Table 8.4	Proportion of Each Grade of Wool Purchased by the Balinyou SMC in 1989	177
Table 8.5	Average Net Income Per Capita in Balinyou County, 1970 to 1989	179
Table 8.6	Prices Paid by the SMC for Wool Purchased from Farmers in Balinyou County, 1975 to 1991	180
Table 8.7	Production of Raw Wool in Balinyou County, 1978 to 1992	182
Table 9.1	Annual Rainfall at Wudan the Capital of Wongniute County, 1969 to 1990	187
Table 9.2	Administrative Units in Agricultural, Pastoral and Semi-Pastoral Areas of Wongniute County, 1989	189
Table 9.3	Annual Increase in Area of Improved Pasture and Fenced Pasture in Wongniute County, 1985 to 1990	190
Table 9.4	Average Net Income Per Capita in Wongniute County, 1978 to 1989	195
Table 9.5	Income, Expenditure, and Investment Statistics for Pastoral Households in Wongniute County, 1983 to 1989	196
Table 9.6	Prices Paid for Raw Wool (by grades) in Wongniute County, 1987 to 1990	196
Table 9.7	Prices Paid for Beef, Mutton and Cashmere in Wongniute County, 1985 to 1990	197

List of Tables

Table 9.8	Wool and Cashmere Production in Wongniute County, 1978 to 1988	198
Table 10.1	Annual Rainfall at Xinhui the Capital of Aohan County, 1957 to 1990	201
Table 10.2	Administrative Units in Agricultural, Pastoral and Semi-Pastoral Areas of Aohan County, 1989	201
Table 10.3	Distribution of Sheep-Raising Households in Aohan County by Flock Size, 1990	202
Table 10.4	Average Net Income Per Capita for Aohan County, 1970 to 1989	207
Table 10.5	Average Prices Paid by the SMC for Wool and Cashmere in Aohan County, 1985 to 1990	208
Table 10.6	Prices Paid for Beef, Mutton and Goat Meat in Aohan County, 1978 to 1990	208
Table 10.7	Wool and Cashmere Production in Aohan County, 1983 to 1987	210
Table 11.1	Annual Rainfall at Tianshan the Capital of Alukeerqin County, 1968 to 1991	212
Table 11.2	Administrative Units in Agricultural, Pastoral and Semi-Pastoral Areas of Alukeerqin County, 1989	214
Table 11.3	Average Net Income Per Capita for Alukeerqin County, 1978 to 1990	220
Table 11.4	Maximum and Minimum Wool Prices Offered by the SMC in Alukeerqin County, 1988 and 1989	221
Table 11.5	Throughput and Scouring Costs in Alukeerqin County, 1987 to 1990	224
Table 11.6	Wool and Cashmere Production in Alukeerqin County, 1983 to 1990	225
Table 11.7	Amount of Wool Purchased by the Alukeerqin County SMC, 1973 to 1988	225
Table 11.8	Meat Production in Alukeerqin County, 1987 to 1990	226
Table 12.1	Fleece Weights for the Best Erdos Fine Wool Sheep, 1985 to 1991	242
Table 12.2	Average Prices Paid for Raw Wool (by grades) in Wushen County, 1991 and 1992	245
Table 13.1	Annual Rainfall at Hongwan the Capital of Sunan County, 1960 to 1989	248
Table 13.2	Average Net Income Per Capita for Sunan County, 1978 to 1990	254
Table 13.3	Prices Paid for Quota Wool in Sunan County, 1985 to 1989	255
Table 13.4	Prices Paid for Cashmere in Sunan County, 1985 to 1989	255
Table 13.5	Prices Paid for Livestock Destined for Slaughter, Sunan County, 1986 to 1989	256
Table 14.1	Monthly and Annual Rainfall in the Capital of Dunhuang County, 1960 to 1989	264
Table 14.2	Ethnic Composition of Dunhuang County Population, 1990	265

Table 14.3	Average Net Per Capita Rural Income for Dunhuang County, 1970 to 1989	268
Table 14.4	Prices Paid for Animal Fibres in Dunhuang County, 1982 to 1990	269
Table 14.5	Procurement Prices Paid for Sheep Meat, Goat Meat and Beef in Dunhuang County, 1978 to 1990	270
Table 14.6	Total Wool, Fine Wool and Semi-fine Wool Production in Dunhuang County, 1985 to 1989	271
Table 14.7	Wool, Cashmere and Goat Hair Production in Dunhuang County, 1975 to 1989	271
Table 15.1	Wool Production in Tianzhu County, 1988 and 1989	275
Table 16.1	Ethnic Groups in Cabucaer County, 1990	278
Table 16.2	Average Net Income per Capita in Cabucaer County, 1979 to 1991	283
Table 16.3	Wool, Cashmere and Meat Production in Cabucaer County, 1970 to 1991	284
Table 17.1	Annual Rainfall in the Capital of Hebukesaier County, 1970 to 1991	286
Table 17.2	Distribution of Sheep-Raising Households in Hebukesaier County by Flock Size, 1991	291
Table 17.3	Long-Term Changes in the Composition of the Hebukesaier County Sheep Flock, 1967 to 1992	292
Table 17.4	Average Net Rural Incomes Per Capita in Hebukesaier County, 1970 to 1991	293
Table 17.5	Wool, Cashmere and Meat Production in Hebukesaier County, 1970 to 1991	295
Table 18.1	Incomes, Poverty, Population Density and Rangeland Degradation in Pastoral and Semi-Pastoral Counties of the IMAR	301
Table 18.2	Changes in Selected Minority Populations in All China and the 12 Pastoral Provinces, 1982 and 1990	307

List of Figures

Figure 2.1	Methodological Framework	12
Figure 4.1	Categories of Wool Established by the Chinese National Wool Grading Standard	50
Figure 4.2	Total Sheep Numbers and Wool Production in China, 1949 to 1991	56
Figure 5.1(a)	End of Year Total Sheep Equivalents and Sheep Numbers in Balinyou and Wongniute Counties of Chifeng City Prefecture, IMAR	88
Figure 5.1(b)	End of Year Total Sheep Equivalents and Sheep Numbers in Aohan and Alukeerqin Counties of Chifeng City Prefecture, IMAR	89
Figure 5.2(a)	Nominal and Real Average Net Rural Income Per Capita for Balinyou and Wongniute Counties in IMAR	102
Figure 5.2(b)	Nominal and Real Average Net Rural Income Per Capita for Aohan and Alukeerqin Counties in IMAR	103
Figure 6.1	Livestock Numbers in Gansu Province, 1970 to 1991	125
Figure 6.2	Nominal and Real Average Net Rural Income Per Capita for Two Case Study Counties in Gansu	132
Figure 7.1	Long-Term Trends in the Sheep Population and Total Sheep Equivalents in the XUAR	143
Figure 7.2	Long-Term Trends in the Sheep Population and Total Sheep Equivalents in Yili and Tacheng Prefectures of the XUAR	146
Figure 7.3	Nominal and Real Average Net Rural Income Per Capita for Two Case Study Counties in XUAR	161
Figure 11.1	End of Year Sheep and Goat Numbers in Alukeerqin County, 1970 to 1991	215
Figure 11.2	State Purchase Prices for Wool and Cashmere in Chifeng City Prefecture, 1978 to 1991	222
Figure 17.1	Total Stocking Rate in Sheep Equivalents and Sheep Numbers in Hebukesaier County, 1949 to 1990	288
Figure 18.1	Major Factors Influencing Development Outcomes in the Pastoral Areas of China	300
Figure 18.2	Long-Term Changes in Rural Population Densities and Livestock Densities for Two Pastoral Counties	302
Figure 18.3	Long-Term Changes in Rural Poulation Densities and Livestock Densities in an "Agricultural" County	303
Figure 18.4	Impact of Technical Improvements on Rangeland Degradation when Policy, Institutional and Incentive Structures are Inappropriate	323
Figure 18.5	Herbivorous Livestock Population and Amount of Hay Stored in Chifeng City Prefecture, 1970 to 1988	324

List of Plates and Maps

Plates

Plate 1	Geographic Location of the 12 Pastoral Provinces of China
Plate 2	Geographic Location of the Pastoral and Semi-Pastoral Counties of China
Plate 3	Proportion of the Population Belonging to Minority Nationalities

Maps

Map 5.1	Prefectural-Level Administrative Units of the IMAR	75
Map 6.1	Prefectural-Level Administrative Units in Gansu Province	116
Map 7.1	Prefectural-Level Administrative Units of the XUAR	135
Map 8.1	Balinyou County in Chifeng City Prefecture of IMAR	167
Map 9.1	Wongniute County in Chifeng City Prefecture of IMAR	186
Map 10.1	Aohan County in Chifeng City Prefecture of IMAR	199
Map 11.1	Alukeerqin County in Chifeng City Prefecture of IMAR	211
Map 12.1	Wushen County in Yikezhao Prefecture of IMAR	237
Map 13.1	Sunan County in Zhangye Prefecture of Gansu Province	247
Map 14.1	Dunhuang County in Jiuquan Prefecture of Gansu Province	262
Map 15.1	Tianzhu County in Wu Wei Prefecture of Gansu Province	272
Map 16.1	Cabucaer County in Yili Prefecture of XUAR	276
Map 17.1	Hebukesaier County in Tacheng Prefecture of XUAR	285

Glossary

Abbreviations

ABC	=	Agricultural Bank of China
AHB	=	Animal Husbandry Bureau
SMC	=	Supply and Marketing Cooperative
DM	=	Dry Matter
HPRS	=	household production responsibility system
HRS	=	household registration system
FRS	=	fiscal responsibility system
NSP	=	nationally supported poverty counties
PSP	=	provincially supported poverty counties
IFAD	=	International Fund for Agricultural Development
ACIAR	=	Australian Centre for International Agricultural Research
AIDAB	=	Australian International Development Assistance Bureau
IMAR	=	Inner Mongolia Autonomous Region
XUAR	=	Xinjiang Uygur Autonomous Region
PRC	=	People's Republic of China
CCP	=	Chinese Communist Party

Units

yuan	=	unit of money (see below)
mu	=	unit of land (15mu = 1ha)
μm	=	micron ($1\mu m = 10^{-6}m$)
spinning count	=	traditional wool trade term to describe the fineness of wool with lower values indicating coarser wool and higher values signifying finer wool

Currency

The Chinese currency is called the Renminbi (RMB) but the unit of money is the yuan. The exchange rate yuan per US dollar over the last decade has been as follows:

1982 and 1983	1.89	1987 and 1988	3.72
1984	2.33	1989	3.76
1985	2.94	1990	4.78
1986	3.45	1991	5.32

Source: Bank of China

About the Authors

John Longworth is Professor of Agricultural Economics and Pro-Vice-Chancellor (Social Sciences) at The University of Queensland. With a life-long involvement in the family sheep/wheat/cattle farm and having been employed as a wool classer prior to commencing an academic career, John brought a unique combination of practical rural skills and research experience to ACIAR Project No. 8811. As President of the International Association of Agricultural Economists 1988/91, as a frequent visitor to China after 1986, and as editor of *China's Rural Development Miracle: With International Comparisons* (1989), he was well known in China prior to the commencement of the research on which this book is based. He has also published *The Wool Industry in China: Some Chinese Perspectives* (1990).

Prior to his involvement with China's pastoral region, John Longworth researched the socio-economic background to Japanese domestic rural policies. His publications on this topic include the definitive volume *Beef in Japan: Politics, Production, Marketing and Trade* (1983). John's extensive fieldwork experiences in Japan, where language and cultural barriers create similar barriers to those encountered by foreign researchers in China, contributed greatly to the success of ACIAR Project No. 8811.

John travelled widely in the pastoral region of China between 1989 and 1992 conducting interviews at all levels. He has been primarily responsible for the preparation of this book in which he demonstrates a comprehensive understanding of the subject matter, from the finer points of wool grading to the broadest principles of public policy.

Greg Williamson is Principal Officer in charge of International Projects for the Queensland Department of Primary Industries (QDPI). In 1989, he was seconded to The University of Queensland as the Senior Research Assistant on ACIAR Project No. 8811. Greg lived and worked in the pastoral region of China for up to five months in each of the four years of the project. As a result, he has an in-depth understanding of the grass roots situation in some of China's most remote areas. In particular, Greg has concentrated on defining the nature and the extent of rangeland degradation in these ecologically sensitive parts of the country.

Prior to commencing his university studies, Greg obtained extensive practical experience with dairy and beef cattle as well as horticulture. He graduated in Agricultural Economics with Honours I and a University Medal from The University of Queensland in 1987 and was employed by the Australian Bureau of Agricultural and Resource Economics in Canberra before joining the QDPI.

Kenneth M Menz
Research Program Coordinator (Economics and Farming Systems)
Australian Centre for International Agricultural Reserarch, Canberra

Chapter One

Introduction

Pastoralism is not an activity readily associated with China. Yet the 12 pastoral provinces which broadly define China's pastoral region occupy almost three-quarters of the country (Plate 1). Within these provinces, there are 266 counties which are formally classified as pastoral or semi-pastoral (Plate 2). These counties cover about one-third of China and are the home of the overwhelming majority of the descendants of the traditional pastoralists who roamed these parts of northern and north western China for millennia. This book addresses key issues associated with the long-term sustainable development of these pastoral communities.

Sheep raising was always a dominant feature of the culture and subsistence economy of the ethnic groups or minority nationalities who lived in these parts of China. Since 1949 and especially after 1978, the traditional subsistence way of life of these peoples has been replaced with a more modern economy. The raising of sheep for meat, hides and wool has now become a major determinant of the commercial livelihood of the residents of these remote pastoral areas. At the same time, degradation of the natural rangelands on which pastoralism depends has emerged as an increasingly serious threat to the long-term viability of economic activities based on grazing animals.

Broadly, this book analyses the major constraints to sustainable development of pastoralism in China's pastoral region. More specifically, the focus is on the problems facing further development of the sheep and wool industry and the closely related issues surrounding rangeland degradation. Special attention is devoted to the minority nationalities both because most sheep-raising households belong to these ethnic groups and because they add a critically important strategic and political element to the significance of China's pastoral region.

From a strategic viewpoint, the pastoral region is of major importance. A large proportion of China's mineral, fuel and rangeland resources are located in this part of the country. The region also borders on countries such as Mongolia, Russia, Kazakhstan, Kirghizstan, Tadzhikistan, Afghanistan and India which, in recent times, have been subject to varying degrees of political turmoil largely as a result of tensions between different ethnic minorities. The government of the People's Republic of China (PRC) is concerned that these troubles may spill over into the relatively large ethnic populations resident in China's pastoral region. This is potentially a major political problem given that a relatively large proportion of the people living in the pastoral region belong to minority nationalities.

Owing to the strategic nature of the location of many of the minority nationalities in China, the People's Government has introduced a system of autonomous rule under which minorities are afforded a high degree of administrative and cultural autonomy. Perhaps the best known example of this system is the Inner Mongolia Autonomous Region which is a provincial-level administrative unit but there are also a significant number of autonomous prefectures and counties. The policy of allowing autonomous rule is significant in that the Chinese government views it as being a key element in their effort to maintain social cohesion and stability in otherwise remote and relatively poor areas. The Chinese government is anxious to ensure the success of this policy, and in so doing recognises the importance of sustaining real economic growth in these parts of the pastoral region.

Raw wool production is one of the few major industries upon which further economic development of the strategically important pastoral areas of China can be built. However, in the context of the Chinese agricultural sector, animal husbandry ranks a poor second in importance to grain production in particular and cropping in general. Furthermore, within the animal husbandry sub-sector, small ruminants such as sheep and goats have not in the past received much emphasis compared with poultry, pigs, dairy cattle and draught animals. Consequently, at the national level and even in most pastoral provinces, relatively few research or administrative resources have been traditionally devoted to sheep and wool industry problems. Nevertheless, there is growing awareness among Chinese policy-makers at all levels that the rangelands and the industries which are based on these pastures are under serious threat. Perhaps the most dramatic evidence of this was the enactment of the National Rangeland Law in 1985.

In addition to the emerging strategic and political significance of the pastoral areas, the changing food consumption patterns in China have awakened new interest in grazing animals at the highest levels of government. The growing consumer preference for high protein foods such as eggs, milk and all kinds of meat is forcing a reassessment of priorities within the Chinese animal husbandry sub-sector. As poultry and pig production expands to meet the rapidly increasing consumer demand for animal protein, the concomitant increase in the demand for grain-based animal feedstuffs is posing a growing threat to human food grain supplies. Raising grazing animals for meat is an alternative means of increasing animal protein supplies which is now attracting much greater attention than in the past. In addition, dietary considerations suggest that lean meat from grazing animals is more nutritionally advantageous than meat from animals raised and fattened on grain.

Another critical element contributing to the increasing importance of sheep and wool and hence the pastoral areas is the growing market for wool-based fabrics in China. As incomes rise and consumers have greater freedom of choice, not only do they choose to eat more meat but they also elect to purchase better quality garments, blankets, etc. made from wool or wool blends. Therefore, the demand of textile manufacturers for raw wool, especially better quality fine wool, is expanding rapidly in China. From the viewpoint of policy-makers in Beijing, the domestic fine wool growing industry should be expanded not only to take advantage of these marketing opportunities but also to reduce the need to import wool from overseas. Under these circumstances, the Chinese government is faced with a happy coincidence in terms of outcomes between domestic policies aimed at developing the pastoral areas and foreign trade policies aimed at restricting wool imports to conserve foreign exchange.

Despite the growing awareness of and interest in pastoralism in Chinese policy-making circles, remarkably little research has been undertaken on an industry-wide basis. For example, while considerable effort has been devoted to surveying the extent of rangeland degradation, there have been almost no studies of the policy/institutional framework within which the degradation problem has emerged. Indeed, within China, rangeland degradation is widely perceived as a technical problem for which there are technological solutions. One major contribution of this book is to challenge this concept and to identify the policy/institutional factors causing degradation.

Perhaps an equally important aspect of the book is that it presents an account of the "grass roots" situation. Throughout the book, first-hand observations and primary data collected at the county, township, village and household level are analysed to arrive at conclusions about the real situation in relation to the matters of interest. Even the most careful "desk studies" frequently fail to appreciate the actual situation in China. While this will generally be the case, it is likely to be especially true in regard to investigations concerned with the pastoral region. The enormous distances and remoteness of the pastoral areas, together with the cultural, language and other barriers to reliable communications between the traditional pastoralists and Han-dominated administrative hierarchies, mean that there is no substitute for visiting the grass roots to obtain an accurate picture of the real situation.

The research on which this book is based has been a unique undertaking. It has involved a highly successful collaborative effort by two Chinese research groups and an Australian team. The motivation for the project, the procedures and some of the problems are outlined in the next chapter. The fieldwork spanned four years 1989 to 1992, and the researchers travelled widely in northern and north western China. The original focus of the investigations was sheep raising and wool production and marketing, but as the research progressed the issues being investigated broadened. Nevertheless, the emphasis remained on sheep and wool.

In Part I of the book, therefore, once the research process itself has been briefly explained (in Chapter 2) and the pastoral region described in some detail (Chapter 3), a major chapter (Chapter 4) has been devoted to the wool scene in China. Much of the remainder of the book assumes an understanding of sheep and wool industry terminology in general and of the Chinese industry in particular. In addition to briefly introducing some sheep and wool terminology, Chapter 4 demonstrates the dominant part that the pastoral region plays in the Chinese raw wool industry and explains how the production and marketing of wool is presently organised in China. It also discusses the relative importance of the various provinces in the pastoral region in relation to sheep and wool. The Inner Mongolia Autonomous Region (IMAR), Gansu Province and Xinjiang Uygur Autonomous Region (XUAR) are identified as three key provinces (Plate 1). Besides being most important in regard to the production of raw wool (especially fine wool), these three provinces span the full range of ecological and socio-cultural environments under which sheep are raised in northern and north western China.

While the background discussion in Part I is at the national and regional level, the case studies of the three major wool-growing provinces presented in Part II examine matters at the provincial level. Different issues are dealt with in greater depth in each case study. For example, livestock improvement programs and animal husbandry research and development organisations are reported in detail for the IMAR, while the Gansu case study has a major section on the types of grassland which form the basis for pastoral activities in that province. In the case of the XUAR, the Xinjiang Production and Construction Corp (PCC) is extremely important and consequently it is one of the topics covered at some length in the

XUAR chapter. An important contribution of all three provincial case studies is the identification and description of pastoral and semi-pastoral counties within each province. In particular, the case study counties are placed in context and the extent to which they may be considered "representative" is examined.

The ten county case studies are reported in Part III of the book, with each study organised under three major sub-headings: resource endowment; improvements to production conditions in the pastoral systems; and socio-economic development. However, while the basic facts are presented for all ten counties, different features of the pastoral region are illustrated in each county and it is these special characteristics which receive the emphasis in each chapter.

China's pastoral region encompasses a wide range of ecosystems. Three of the case study counties (Balinyou, Wongniute and Alukeerqin) are located in the somewhat more climatically favoured grassland areas of eastern IMAR. Another important case study county, Aohan, lies just to the south of the area traditionally known as the Eastern Grasslands and it is representative of the more agriculturally-oriented sheep-raising areas in the eastern part of the pastoral region. Wushen County (Chapter 12) is located in the once famous Eerduosi Grasslands of the IMAR south of the Yellow River. Nowadays, more than half the land area in Wushen is covered by sand dunes which have replaced the once lush rangeland. In this and many other ways, Wushen presents an interesting contrast with the four case study counties which are situated in eastern IMAR.

In Gansu Province, three counties were surveyed. Sunan and Tianzhu are essentially alpine sheep-raising areas in which the production systems and many other features are dramatically different to those found in the five counties investigated in the IMAR. Dunhuang, the third case study county in Gansu, is a large desert county where the residents are entirely dependent on a few oases. Perhaps surprisingly, given their restricted resource base, the people of Dunhuang County enjoy relatively high incomes, generated by high-value agricultural production rather than by grazing animals.

Although the XUAR covers a vast area, a large part of it is inhospitable and uninhabited desert. Most of the sheep in this autonomous region are raised on a semi-nomadic basis because: the summer pastures are high up in the mountain ranges; the winter pastures are around the agricultural areas located along the rivers and on the edges of the deserts; and the spring/autumn pastures are on the loess plains between the agricultural settlements and the mountain ranges. The sheep are moved from pasture to pasture with the seasons. There are many variations on this "four season" grazing system and the two case study counties in the XUAR illustrate two extremes. Cabucaer County is a relatively small, rich area located in Yili Prefecture, which is considered by many to be the best fine wool growing prefecture in all China. Hebukesaier, on the other hand, is more than seven times larger than Cabucaer in area but it is semi-desert. The environment in Hebukesaier is extremely harsh and not well suited to the production of fine wool.

Not only do the ten county case studies illustrate the range of ecological conditions in the pastoral region, they also span the socio-cultural spectrum. All the counties except Aohan (Chapter 10) and Dunhuang (Chapter 14) have a relatively high proportion of residents who belong to minority nationalities. And five of the ten counties studied are considered among the poorest group of counties in all China and are, therefore, provided with special assistance under the poverty alleviation program funded by the Central government.

The provincial and county case studies presented in Parts II and III provide a first-hand, detailed picture of conditions in the pastoral areas of China today. Part IV draws together the main themes which emerge from these case studies and presents the "big picture". The major aim in Chapter 18 is to identify the constraints to further development and to suggest policy reforms which will contribute to future progress. In particular, it is argued that rangeland degradation is a symptom rather than the root cause of the problems facing the pastoral areas of China. Chapter 19, the final chapter, draws the four threads of the book listed in the sub-title together and places the key issues in a forward-looking global context.

Rangeland destruction by cultivation and degradation by overgrazing, alkalisation and sand erosion are not new phenomena in China. For example, as discussed in the Wushen County case study (Chapter 12), serious damage was inflicted on large areas of natural grassland in the middle of last century. Substantial areas of the best pastures have been destroyed by cultivation since 1949 especially during the "Great Leap Forward" era (1958 to 1961). More recently, policies implemented after 1978 have created both incentives and opportunities while new technology has provided the means for private households to "mine" the rangelands by adopting non-sustainable grazing management practices.

About one-third of China is officially described as pasture land and virtually all of it is degraded to some extent, while perhaps as much as 35% is beyond repair. The case studies presented in this book provide detailed information on the nature and the extent of the problem. Clearly, in many parts of the pastoral region, pasture degradation is continuing at an alarming rate. In relation to the whole Chinese socio-economic system, the pastoral industries may be of little significance. But in that strategically important and political-powder-keg part of the country known as China's pastoral region, the survival of pastoralism as a viable economic activity is a crucial ingredient in regional stability. Pastoralism cannot survive without rangelands. Hence the importance of identifying the root causes of rangeland degradation and the urgent need for action to arrest the insidious processes which threaten the very existence of many minority nationality communities living in this part of China.

PART I

BACKGROUND

Chapter Two

Outline of the Research Project

China's pastoral region, defined broadly as 12 provinces, accounts for the most remote and backward three-quarters of the country. Foreign researchers who aim to investigate this vast area by fieldwork will need not only the collaboration of Chinese colleagues but also the official support of the relevant government officials. The research project on which this book is based was made possible because both these prerequisites for success were arranged as part of the China-Australia Agricultural Cooperation Agreement. Under this bilateral agreement, the Australian Centre for International Agricultural Research (ACIAR) has responsibility for responding to requests from the Government of the People's Republic of China, through its Ministry of Agriculture, for assistance with agricultural research endeavours in which both countries have an interest.

In late 1986, a request was issued by the Ministry of Agriculture to ACIAR for assistance with an industry-wide study of Chinese raw wool production and marketing. Out of this request developed ACIAR Project No. 8811 entitled "Economic Aspects of Raw Wool Production and Marketing in China". This major four-and-a-half-year project, which was completed in June 1993, was a collaborative research effort involving Chinese and Australian agricultural economists from three research institutions. The names of these three institutions and a full list of researchers officially involved in the project may be found in Table 2.1. ACIAR provided funding to complement the contributions made by the Chinese side.

Table 2.1 Three Research Institutions and Staff Members Involved in ACIAR Project No. 8811

Institute of Rural Development within the Chinese Academy of Social Science	Institute of Agricultural Economics within the Chinese Academy of Agricultural Science	Department of Agriculture at The University of Queensland, Australia
Chen Jiyuan (Director)	Niu Ruofeng (Director)	John W. Longworth
Lin Xiangjin	He Changmao	Ross G. Drynan*
Du Yintang	Zhou Li	Colin G. Brown
Liu Yuman	Zhang Cungen	Gregory J. Williamson
Wu Jinghua	Shi Zhaolin	Deborah D. Noon
Liu Zheng	Xu Ying	Warren J. Campbell
	Xie Xiaocun	
	Guo Jianjun	

*Participated in the project from the University of Sydney since January 1990.

2.1 Research Objectives

In preliminary discussions concerning the research task, it was recognised that: wool production was only one aspect of sheep raising in China; sheep are only one of several important types of grazing animal raised by pastoralists; traditional pastoralists are not the only people raising sheep in the pastoral region; while pastoralism is a relatively minor economic activity in relation to the total economy of the pastoral region, it is of crucial importance to the traditional pastoralists; the overwhelming majority of the traditional pastoralists are members of minority nationalities; and rangeland degradation is an increasingly important issue. That is, in the planning of the research, cognisance was taken of the wider ramifications of an investigation which, initially at least, was to be focused narrowly on wool production and marketing. This book is the final manifestation of this broader approach.

However, both to facilitate the development of a collaborative effort and to secure official approval for fieldwork in remote, inaccessible parts of the country, it was essential to have relatively sharply focused research objectives. Formally, therefore, the major objectives of ACIAR Project No. 8811 were:
- to identify and, where possible, to quantify the technical, economic and institutional constraints to the production and marketing of raw wool in China; and
- to establish a strong basis for longer-term collaborative research between Chinese and Australian scholars on wool economics in China.

2.2 Brief Rationale for the Project

Despite the traditional low profile afforded small ruminants (i.e. sheep and goats) in China, in recent years as pointed out in Chapter 1, the Chinese government has begun to place a higher relative priority on preserving the rangelands and on developing fine wool production. It is considered that a prosperous wool industry would make a major contribution to the socio-economic stability of strategically important parts of the country. At the same time, increased domestic output of wool would reduce China's dependence on imports and conserve foreign exchange. The Central government, therefore, is seeking technologies and policies which will increase the productivity of the wool industry.

Prior to the commencement of ACIAR Project No. 8811, the major research emphasis in China in relation to sheep and wool had been on new technology (including genetic material). Little attention had been given to evaluating the effectiveness of existing policies or to developing more appropriate new policies and institutional arrangements. Although not explicitly stated in the formal objectives of Project No. 8811, one of the aims of this project has been to redress this imbalance between the attention given to new technology on the one hand and the need for policy/institutional reforms on the other.

Raw wool production does not have a long history as a major commercial industry in China. It is concentrated in the twelve pastoral provinces which are collectively referred to as the pastoral region. These provinces accounted for 89% of the sheep raised and 84% of the greasy wool produced in China in 1991. The three provincial-level administrative units surveyed as part of ACIAR Project No. 8811 (i.e. the IMAR, Gansu Province and the XUAR) contributed 124,000 tonne of greasy wool or 52% of the 1991 Chinese clip. Even more importantly, these three provinces produced 66,000 tonne or 61% of the fine and improved fine wool grown in China in that year. (Fine and improved fine wool in China refers to wool of ≤ 25μm average fibre diameter. See Section 4.1 for more details.)

During the 1980s, the People's Republic of China emerged as a major export destination for Australian wool. In terms of value, only 5.5% of Australian wool exports went to China in 1980/81 but by 1986/87 exports of wool to China had reached 11.5% of the total. However, Chinese purchases declined sharply from late in 1988, and during the 1989/90 financial year only 3.6% of wool exports by value went to China. Since 1991, there has once again been a steady increase in the value of wool exports to China, and in the later part of the 1992/93 wool buying year China, for the first time, emerged as the principal buyer of Australian wool.

The wide fluctuations in Chinese purchases during the 1980s had a major destabilising influence on the Australian wool industry. Indeed, it could be argued that the sudden withdrawal of the Chinese buyers from the market in 1989 was one of the major factors which led to the disastrous collapse of the Australian wool reserve price and buffer stock scheme. Although a number of obvious and simplistic explanations for the variations in Chinese demand for Australian wool were widely circulated in the 1980s, the real situation in China was not well understood in Australia.

Chinese demand for imported wool is both a derived and a residual demand. It is derived from the demand, both in China and from overseas, for Chinese wool textile products. It is a residual demand because Chinese textile manufacturers are "encouraged" to use raw wool produced domestically before turning to imports. For reasons well understood by economists, derived demand is less stable than final consumer demand and a residual-derived demand will be even more unstable.

Considerable research effort by the International Wool Secretariat (IWS) and others has been devoted to analysing the changes occurring in the domestic market for wool textiles in China. Likewise, the role of China in the world textile market has also attracted the attention of IWS researchers and others. On the other hand, prior to the initiation of ACIAR Project No. 8811, there had been little substantive research undertaken on the supply side of the Chinese wool industry. Even in China and especially in Beijing where most of China's important policy decisions are made, little detailed information was available about the effects of the rapid changes which occurred in the pastoral areas of China during the 1980s.

From both the Chinese and the Australian viewpoint, therefore, ACIAR Project No. 8811 had the potential to fill important information gaps. For the Chinese government, the project aimed to provide data and ideas about the constraints facing the raw wool industry in China and how these limitations on the future economic development of the pastoral region could be moderated and, perhaps, even overcome. From the Australian perspective, the project had the potential to contribute to the formation of more realistic expectations about the Chinese market for Australian wool.

2.3 Research Methodology

Research methodology is concerned with the broad philosophical approach implicit in the research process. ACIAR Project No. 8811 cannot easily be described as having been moulded by a particular research philosophy. The range of issues being examined, the number and background of the researchers involved, and the time and space dimensions of the project, all militate against any simplistic description of the research process.

Nevertheless, the essence of the research process in this project is captured in Fig. 2.1. The three key aspects of the project, namely normative research, positive research and policy recommendations, are identified and placed in context.

Throughout the project, the research team used well established normative paradigms or theoretical frameworks to formulate ideas and hypotheses and to provide a focus for discussion. Normative theories about such things as: the distribution of power, food, work, economic rewards, etc. in the household; the incentive structures created for households by certain property rights and other policy settings; the political and economic market places which exist at village, county, prefectural, provincial and national levels; the valuation of non-market resources; natural resource management; and many other broad topics; all have shed light on various aspects of the project.

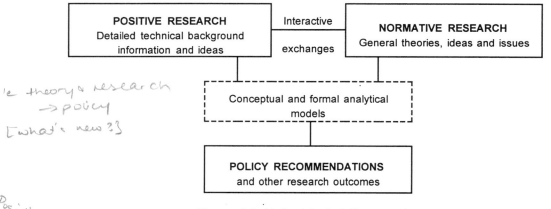

Figure 2.1 Methodological Framework

At the same time, the fieldwork was positivistic. The purpose was to discover first-hand "what is" happening. The reality at the "grass roots" in a country such as China may not always accord with the "official" view at higher levels. The collection of data and ideas on technical, social, economic, cultural, and geographic phenomena was aimed at getting the facts right. The provincial and county case studies in this book principally present detailed background information of this kind.

These normative and positive approaches to research need to be combined in a modelling framework to analyse "scientifically" the various aspects of greatest concern. Irrespective of whether a formal mathematical model or simply a conceptual model derived from theory is used, the ultimate aim is to generate useful policy recommendations. These recommendations may cover the full range from the micro-technical level (e.g. concerned with ways to improve the winter feeding regime for sheep) to the macro-political level (e.g. suggestions relating to national fiscal and population policies). Only when these policy recommendations can be substantiated by a thorough analysis of both "what is" and "what ought to be" *and* "why", are the recommendations likely to have any real credence.

2.4 Research Methods

In principle, the research methods employed were extremely conventional. However, in practice the project had a number of unique characteristics and some aspects of the procedures adopted and experience gained may prove useful to other researchers undertaking fieldwork in China.

2.4.1 Literature Review

As would be expected, a considerable amount of "desk research" was conducted both in China and in Australia prior to entering the field. Selection, organisation and, where necessary, the translation both of the documents collected in this initial phase of the project and of the literature discovered subsequently, has been an important on-going aspect of the project.

The main aim of the initial phase of documentary research was to generate hypotheses for testing in the field. It was also important to ascertain both the nature and extent of secondary information available and the kind of primary data needed to test the hypotheses.

Relevant English language materials were identified by Australian researchers with the help of a number of bibliographic databases. Initial literature searches were aimed at gathering "general" information that would serve as an introduction to the Chinese economy. As understanding and knowledge of the research issues increased, document searches focused on specific relevant aspects of the agricultural economy in China such as "policy reforms" and "the household production responsibility system". Finally, the modest amount of English language literature on the pastoral region of China and on sheep and wool in China was collected.

Once research began in China, important Chinese research papers were collected and translated into English for inclusion in the document collection. This aspect of the project often presented considerable difficulties for the collaborating Chinese scientists because some of the best material was on the "restricted list" and could not officially be made available to foreigners.

An important part of the initial desktop research in China was the preparation of a series of benchmark papers which summarised the information and ideas available at the beginning of the project. These papers, originally written in Chinese in 1989, were translated and then further developed by the editor before being published in English (Longworth, 1990).

The primary purpose of the literature search and document collection was to assemble a systematically organised set of reference materials for ready access by those working on the project. A project bibliographical database (using *Papyrus* software) has now been constructed with over 2,000 entries. The English language materials are stored at The University of Queensland. Most Chinese language materials are kept at the two collaborating Institutes in Beijing. However, certain key Chinese language data sources and other references together with many translations are also held at The University of Queensland.

2.4.2 Survey Questionnaires

The survey questionnaires represent one of the principal research tools used by the researchers. Separate questionnaires were prepared for household, village, township, county, prefecture, and provincial/autonomous region administrative levels, as well as for supply and marketing cooperatives, local wool processors and the Agricultural Bank of China. The questionnaires were drafted in Australia using information obtained from the initial literature review and subsequently refined before fieldwork commenced in 1989 in collaboration with Chinese project scientists using secondary information available in China. Further revisions were made during fieldwork to incorporate new developments.

In designing the questionnaires, one of the major factors taken into account was the time allowed for fieldwork by the resources available. Given the comparatively broad nature of the research task and the relatively small secondary information base upon which the project could draw, it appeared that a great deal of data needed to be collected. On the other hand, there was a need to resist the temptation to ask too many questions. The questionnaires were structured as tightly as possible to allow for maximum information from a minimum number of questions. Questions were kept as simple in concept and terminology as possible to facilitate translation and for ready comprehension and comparability of information between respondents. The questionnaires also included a significant number of open-ended questions, to permit the necessary flexibility to explore subjects of increasing interest, or subjects of previously unforseen relevance as and when the opportunity arose.

Another measure, introduced in the second year of fieldwork (i.e. in 1990), and designed to alleviate further the problems associated with the lack of time, entailed the division of each of the questionnaires into "comprehensive" and "data only" questionnaires. The latter questionnaires were forwarded to the relevant interviewees prior to the arrival of the project team. The concept of "comprehensive" and "data only" questionnaires was generally well received by local officials in the second and subsequent years of the project. In the first year, many officials had complained of difficulties associated with having insufficient time to assemble and supply the data requested. In addition, the use of "data only" questionnaires enabled local officials to transcribe the data in their own time and present this data to project staff without any further need for transcription, thus avoiding the loss of valuable survey time and minimising translation and transcription errors.

In the construction of the questionnaires, as many safeguards as possible were incorporated to avoid misunderstandings and other sources of error since the data collected usually could not be checked once the research team had left the field. In this respect, the use of dual-language questionnaires was of great assistance.

Each English language questionnaire was translated into Chinese before the fieldwork commenced. The interviews were then conducted with the aid of interpreters from the collaborating Chinese Institutes. There were many advantages in having pre-translated formal questionnaires. The most obvious was that a great deal of time was saved at each interview because the interpreters (not always the same people) did not have to decide how to translate each question since this problem had been solved in advance. The second advantage concerned the precise choice of words. Even though the interview was conducted in a foreign language, the Australian researcher could be reasonably confident that the question they had intended to be asked was actually the question put to the interviewee. Thirdly, the Australian researcher could retain control over the flow of the interview. For example, once a certain question had been answered, the Australian researcher could change the thrust of the interview by requesting the Chinese researcher to ask a specific question by referring to the desired question by number.

The Chinese language questionnaires also enabled local officials to follow the interview and to contribute both to the questions and the answers. This "local official" input sometimes added an "overlay" to the interview which needed to be carefully handled. On the other hand, local knowledge often helped in unravelling the inconsistencies and other problems which emerged during some interviews.

The comprehensive questionnaires were primarily used as a "check list" to ensure nothing was overlooked at each interview. No interviewee was ever subjected to every question in the relevant questionnaire. Having a comprehensive list of questions which was cross-coded to a Chinese language version of the questionnaire enabled the researchers to adopt a flexible and informal approach to most interviews. Interviewees could be encouraged to elaborate on topics of interest to them even when this disrupted the "intended" flow of the interview, without the risk of something being overlooked owing to the "apparently" disorganised nature of the interview.

2.4.3 Survey Design

The 12 pastoral provinces occupy almost three-quarters of China. Selecting the most appropriate parts of this vast geographical expanse in which to undertake fieldwork presented major difficulties. There was really no practical alternative to selecting purposefully certain pastoral provinces and then choosing counties within these provinces which, taken together, would provide a good coverage of the various sheep-raising conditions (and hence production systems). The counties were also selected so as to sample the range of socio-economic circumstances found in the pastoral areas of China. While there was some attempt to minimise the logistical and administrative difficulties associated with organising the fieldwork by choosing more accessible counties, especially in the first year, in the final analysis these considerations did not play a major part in determining which counties were surveyed.

For reasons which will be made clear in the next two chapters, the three provincial-level units within the pastoral region selected for study were the IMAR, Gansu Province and the XUAR (Plate 1). In the case studies of these three provinces/autonomous regions presented in Part II of this book, the counties surveyed are discussed and information is provided which demonstrates the degree to which these counties are "representative". Each of the provincial/autonomous region case studies in Part II and the county case studies in Part III contains a map showing the precise geographic location of the administrative units with which the chapter is concerned.

As already indicated, practical considerations suggested that fieldwork in the first year (1989) should be concentrated in an eastern prefecture of the IMAR which was comparatively easy to access from Beijing. Consequently, four counties in Chifeng City Prefecture were selected for investigation in 1989. Two of these are described as having typical pastoral production systems (Balinyou and Alukeerqin), one has pastoral/semi-pastoral/agricultural production systems (Wongniute), and one is a more or less agricultural area (Aohan).

For 1990, it was decided that the team should return to the Chifeng City Prefecture and re-visit many of the agencies (but not the households) visited in 1989. In addition, the team spent some time in Huhehot (the capital of IMAR) before travelling on to Gansu Province.

In Gansu, two counties were selected for intensive study in 1990. One county (Sunan) has an alpine production environment while the second (Dunhuang) is more typical of the semi-desert/irrigated agricultural production systems. These two counties provide sharp and differing contrasts with the areas investigated in eastern IMAR.

In 1991, the third year of the original project, no new counties were surveyed. However, further follow-up interviews were conducted in Chifeng City Prefecture (especially in Balinyou and Alukeerqin), in Huhehot and in Lanzhou (the capital of Gansu) as well as in Beijing.

After the project was reviewed in September 1991, ACIAR extended Project No. 8811 for another 15 months to allow surveys to be undertaken in important wool-growing areas not included in the earlier investigations.

Four more counties were studied in 1992. Two of these (Cabucaer and Hebukesaier) were in the XUAR, one was in the Eerduosi Grasslands area of southern IMAR (Wushen), and the fourth was a Tibetan Autonomous County (Tianzhu) in Gansu. The research in the XUAR in 1992 followed the same general pattern as that adopted in the IMAR in 1989 and in Gansu in 1990. Interviews were conducted at the provincial, prefectural, county, township, village and household levels. However, the surveys undertaken in Wushen and Tianzhu were much less formal, being tightly constrained by the lack of time. The main purpose for visiting these two "extra" counties was to broaden the geographical coverage of the overall research effort by making first-hand observations in two additional counties.

Within each of the eight counties surveyed in detail, two townships were selected for intensive investigation. The township selection process was checked as far as possible to ensure that the areas being studied in depth in each county were indeed typical of that county.

Within each township area, villages were selected on the basis of average statistics on per capita incomes. The aim was to sample villages with average per capita incomes close to the average income for the township. At the village level, whenever possible, households were selected on the basis of sheep numbers. That is, the distribution of households in the village by flock size was examined and households chosen for interview which were representative of the range of flock sizes found in the village.

For many reasons (including limited resources and the inability of the research team to obtain a suitable sampling frame prior to actually arriving in the village), it was decided to interview relatively few households but each household surveyed was questioned in great detail. A case study approach rather than a properly structured sample survey was adopted in regard to the survey of households.

While the initial intent of the data collection program was to focus on households, much more time was actually devoted to interviewing other relevant agencies at the county level, in the prefectural and provincial capitals, and in Beijing. Although the household interviews provided invaluable opportunities to observe conditions at the grass roots and to investigate the extent to which various policies were actually being implemented, the sample was too small to permit any detailed statistical analysis of the data collected.

2.4.4 Field Reports

In addition to the data collected via the formal questionnaires, the Australian members of the research team adopted the policy of preparing field reports as a means of gathering and assembling relevant ideas and information. A field report may have been based on a single interview or represent the integration of several interviews and perhaps other written sources as well. The field reports contain factual information along with views and opinions of both the interviewees and interviewer.

The roles of the field reports in the research process were two-fold. First, they provided a mechanism by which information obtained during fieldwork could be rapidly communicated between project team members. Second, the reports provided written records of the interviews which could be assessed and reassessed as the project progressed. The real meaning and significance of material collected during an interview often did not emerge until some time later when the project team had "climbed further up the learning curve".

In many instances, the interviewer's ideas and suggestions are expressed in the field reports. It is important to note, however, that these ideas and suggestions were expressed in relation to the particular interview or discussion being reported in the field report in question and did not necessarily reflect a more comprehensive overview of the topic. Nevertheless, the ideas and suggestions expressed in the reports were regarded by the project team as being a valuable input into the research process, often providing "starting points" for more comprehensive research efforts.

Similarly, the field reports contain views which are not always those of the author of the report. In many cases, the viewpoint of the interviewees has been recorded irrespective of the opinion of the author. The recording of this primary data is considered to be an essential part of the team research effort, as professional interpretation of these viewpoints may differ between team members and with other professionals. More importantly perhaps, as already mentioned, the value and meaning of some of the ideas and information collected only emerged some considerable time after the original interview.

The provincial and county case studies in this book draw heavily upon many of these field reports.

2.4.5 Analytical Models

The valuing of inputs and the pricing of outputs to reflect their social marginal values in the context of the Chinese economy presents major problems. All markets in China are distorted by policy constraints and other rigidities. In addition, as with all governments, the Chinese government undoubtedly has policy objectives other than economic welfare maximisation. For these and many other reasons, the application of simple neo-classical economic models in the Chinese context must be undertaken with great caution. In the case of ACIAR Project No. 8811, the approach has been to adopt an institutional perspective which places the emphasis on analysing the political economy of the situation rather than on empirical models based on simple neo-classical theory. Within this framework, most of the analyses undertaken by the project team could be described as descriptive/conceptual rather than quantitative. The researchers have attempted to use the ideas and data collected to present meaningful explanations of what is happening in the pastoral areas of China.

Apart from the methodological problems with positivistic empirical economic models, the nature of most of the data (limited length and accuracy of economic time series and non-random samples in the case of cross-sectional data) largely precludes sophisticated statistical analysis. Even when reliable longer-term economic data series are available, the structural changes which have occurred in China over the last four decades, and especially since 1978, create obstacles to an econometric approach.

One useful means of "massaging" the available data is to adopt a linear programming approach. The advantages and disadvantages of mathematical modelling in a situation such as that facing the research team investigating the problems of the Chinese raw wool industry are well known and will not be further elaborated here. The project survey data (especially the household survey information) provide a basis from which to construct synthetic activity budgets, define resource constraints, etc. Considerable progress has been made in the construction of linear programming and dynamic programming models of the pastoral production systems. These models represent analytical tools which may, in time, be developed to the point where they could be used to evaluate the various policy options identified as a result of the descriptive/conceptual analysis referred to earlier. Furthermore, the Animal Husbandry Bureaus in Chifeng City Prefecture and in Sunan County have both invited Chinese members of the research team to develop linear programming models specifically designed to assist them plan the development of the animal husbandry sectors under their jurisdiction.

2.5 Research Difficulties

With the "opening" of China, Western researchers from many different disciplines have "invaded" the country. Most foreign investigators studying the rural sector have concentrated on agricultural areas. Relatively few studies have focused on pastoral areas. Indeed, it would seem that the research project on which this book is based is the first attempt to gain a comprehensive appreciation of grass roots conditions in China's pastoral region.

Undertaking fieldwork and primary data gathering in any part of China is likely to be a difficult exercise. However, the problems which confront non-Chinese speaking Western social scientists aiming to conduct research work in remote pastoral areas are probably several orders of magnitude greater than those encountered in more densely settled agricultural areas closer to large urban centres. This section briefly discusses some of the key issues which create obstacles to research in pastoral areas of China.

2.5.1 Vertical Bureaucratic Channels

Even the most casual visitor to China will quickly begin to appreciate the importance of being able to cope with Chinese bureaucracy. Foreign researchers will, of course, usually be entirely dependent on one or more arms of the Chinese bureaucracy. Therefore, it is critically important to understand as much as possible about the relevant bureaucracy in advance.

A particular Chinese bureaucracy unit such as the Animal Husbandry Bureau will exist at various levels of government under various names. For example, the Animal Husbandry and Veterinary Science Department at the Central government level is part of the Ministry of Agriculture; the Gansu Animal Husbandry Bureau is part of the Gansu Agricultural Commission (i.e. the provincial department of agriculture); the Wu Wei Animal Husbandry Bureau belongs to the Wu Wei prefectural Agriculture and Animal Husbandry Commission; and the Tianzhu Animal Husbandry Bureau is part of the Tianzhu County government. (Tianzhu County in Wu Wei Prefecture of Gansu Province was one of the counties surveyed.)

The bureaucratic unit is divided at each level into a number of branches or stations. Thus at the county level, the Animal Husbandry Bureau will include stations such as the Animal Improvement Station and the Pasture

Improvement/Management Station. These stations are not usually based on a physical site or experiment station. Rather they are extension-oriented branches of the county Animal Husbandry Bureau. There will be corresponding branches (or divisions) in higher-level Animal Husbandry Bureaus, with a vertical responsibility. For example, the Pasture Improvement Station in Balinyou County is vertically responsible to the Pasture Improvement Station at the Chifeng City prefectural level, which in turn is responsible to the corresponding provincial/autonomous region level station in Huhehot, the capital of IMAR. The Huhehot station is under the "guidance" of the relevant branch/division in Beijing.

The vertical linkages and lines of command are very strong and powerful. Information flows up and down these vertical channels. On the other hand, horizontal communications between the vertical channels are weak and often non-existent even within the same general bureaucracy. For example, at any level of government each station within the Animal Husbandry Bureau communicates primarily with the corresponding station above or below its position in the vertical hierarchy. Communication and influence between the stations concerned with different professional matters at the same level of government are extremely limited.

One factor moderating the professional isolation of the vertical hierarchies is the overriding political influence of the Chinese Communist Party (CCP). Within each major bureaucratic unit such as a provincial Animal Husbandry Bureau exists a separate CCP structure headed by a Secretary who is usually more powerful than the Director of the unit when these two key posts are not held by the same person. While the Party structure will have separate sub-units in each vertically-organised branch/station of the parent bureaucracy, the various Party committees form horizontal linkages which, at least in regard to political and administrative matters, tend to reduce the communication gap between the branches/stations.

The existence of vertical bureaucratic channels which are horizontally isolated, especially in relation to professional issues, has major implications for wide-ranging and comprehensive research projects such as ACIAR Project No. 8811. Unless the necessary support, introductions and lines of authority have first been established at the higher levels of government, it is almost impossible to obtain cooperation at the lower level. Furthermore, even within one major bureaucracy such as the Animal Husbandry Bureau, separate connections and lines of approval must be developed down to each of the stations (branches or divisions) of interest to the research project at the "grass roots". The only effective way to achieve the necessary authority is to start at or near the top, that is at the Central government level, and then to meet sequentially with people all the way down the chain of command through the provincial, prefectural, county and eventually to the township and village level. While this is a time- and patience-consuming exercise, the experience of the researchers involved in ACIAR Project No. 8811 demonstrated again and again the problems (in fact, blank walls) which can be encountered if such an approach is not methodically adopted.

The surveys reported in this book involved interviews and other interactions with a significant number of separate vertical bureaucracies such as the Animal Husbandry Bureau and its various Stations; the Supply and Marketing Cooperatives; the Agricultural Bank of China; the Textile Corporations and associated textile mills; the Fibre Testing Bureau; the Xinjiang Production and Construction Corps, etc. It was not always possible to obtain the full support and cooperation of these agencies at the various levels.

One important aspect of the vertical hierarchy problem in China for collaborative research relates to the status of the collaborating Chinese research institutions. ACIAR Project No. 8811 involved two senior or national Chinese research institutes. The high status not only of these institutes but also of the senior scientists from these institutes who participated actively in the fieldwork played a major part in establishing the necessary lines of authority referred to above.

Another issue is the command over resources. While not a matter of critical concern for ACIAR Project No. 8811, the injection of aid funds (or collaborative research resources) into one vertical bureaucracy in the anticipation that this bureaucracy will have access to the resources (e.g. experimental sites) under the control of another vertical bureaucracy, is a recipe for disaster (or, at best, for years of frustration). Funds, especially foreign aid funds, do not flow horizontally from one vertical structure to another. Research groups in China do not share information let alone physical resources. Both are too scarce and too valuable. It is vital, therefore, to select the most appropriate vertical hierarchy with which to undertake collaborative work and to enter this hierarchy at the highest possible point if the collaboration is to have any chance of a successful outcome.

2.5.2 Separating Fact from Fiction

Considerable effort must be devoted to sorting fact from fiction in China. There are often two entirely different views of the world - the official understanding of what is happening and the real situation at the grass roots level. Deng Xiaoping, for example, has alluded to this situation on many occasions by exhorting public and party officials to "seek truth from facts". The "policy mirage" syndrome is widespread in China. Officials "see" and report what they believe is the successful implementation of policy but at the grass roots level reality is very different. Perhaps the most infamous example of the dangers of the "policy mirage" syndrome is the widespread starvation which resulted from overly optimistic reports of agricultural production during the "Great Leap Forward" era (1958 to 1961). Indeed, it is possible that the magnitude of this disaster was an important catalyst in Deng's efforts after 1978 to encourage the Party (and the public generally) towards maintaining a healthy scepticism about "facts" and "figures" offered up by party cadres and government bureaucrats to demonstrate the "success" of Party and government policy.

All too often, Western scholars in studying the Chinese system choose to ignore what the Chinese themselves now implicitly acknowledge as being extremely important in formulating policy; that is, the need to understand precisely the nature of what is happening at the grass roots. It is simply not possible to achieve a level of knowledge consistent with making a useful contribution to policy formulation through desktop research which ultimately must attempt to fit an understanding around official statistics. Nor is it sensible to qualify desktop research findings by simply talking to one or two "key" people at the Central level in the hope that these individuals have the capability to "sort the wheat from the chaff" or, in Deng's words, to determine the "truth from the facts". In many cases, such people may have only a limited or narrow understanding of the real situation at the grass roots level. This arises in part because of the distinct vertical "self-interested" nature of each of the government agencies within the Chinese bureaucracy referred to in the previous section. It also arises because bureaucrats at the Central level often rely heavily on information and policy feedback from the next lower level of government administration (i.e. the provincial or autonomous region level of government) rather than seeking advice directly from officials much closer to the grass roots.

Collaborative field survey research of the type undertaken by ACIAR Project No. 8811 has the potential to make a unique contribution to Chinese policy debates. Not only do foreign scientists have an opportunity to assess and report on the real situation first hand, but also their Chinese colleagues can take advantage of the interest created by the novelty of the foreign visitors to probe local officials in much greater depth than might otherwise be possible.

2.5.3 Some Special Difficulties Encountered in Surveying Pastoral Areas

There are a number of difficulties unique to fieldwork research in pastoral areas in China. Perhaps the most important of these is how to secure cooperation at the local level. Within China, governments at the local level have traditionally had a strong degree of autonomy and control. In the past, this can be traced back to the widespread influence of local warlords. With the introduction of Mao Zedong's version of Communism after 1949, particular emphasis was placed on self-reliance at the local level in terms of food and other essential items of daily life. In recent times, with the introduction of the economic reforms and the development of a more open economy, local governments have been forced to accept a greater degree of fiscal self-reliance.

It is often even more difficult to obtain cooperation at the local level in pastoral areas than would normally be the case for agricultural areas. This is because local governments in pastoral areas tend to be rather conservative and generally suspicious of outside intervention and "official" probes into their local activities and way of life. Furthermore, many of the pastoral counties are "closed" which means special permission is required from national, provincial, prefectural and county administrations before foreigners are permitted to travel to these areas. Four of the counties surveyed (Wushen, Sunan, Tianzhu and Hebukesaier) are closed counties rarely, if ever, previously visited by a foreign survey team.

Hence, in order to conduct surveys in pastoral areas, it is of paramount importance to have as a counterpart organisation a Chinese research institute which is highly influential in government circles and preferably in Communist Party circles as well. In general, this will mean that the foreign researcher needs to be working with an institution based at the Central government level and one with members who have a large network of influential contacts throughout the country. This network of influential contacts is of critical importance in breaking down the barriers between the various vertical bureaucratic hierarchies which, as discussed above, often constrain any meaningful cooperation on an official basis.

Household surveys in pastoral areas entail a number of difficulties not encountered in surveys of agricultural areas. Furthermore, these problems require a high degree of local cooperation, which as already stressed, may not be readily forthcoming if they are to be overcome. Perhaps the most serious of these difficulties is the widely scattered nature of the pastoral population. That is, compared to an agricultural community, a pastoral population at any one point in time may be dispersed over a large geographical area, and the location of any one particular household at any particular time is often not known to the local administering officials. Hence, any attempt to ensure a reliable statistical survey of a given pastoral population is likely to be costly and may even be physically impossible when the population has moved to areas which are inaccessible to conventional means of transport.

A second major difficulty is the relatively short period each year during which it is feasible to survey pastoral households. That is, snow and extreme cold prevent access to the great majority of pastoral households in winter, while in summer and autumn, many pastoral households move to largely inaccessible areas in the mountains to allow their animals to graze on summer pastures. Thus the only real window of opportunity to survey households is in the early spring period or late autumn period. Clearly, therefore, a high degree of organisation and local cooperation is required in order to minimise delays in what is already a relatively short period of time in which to conduct the planned surveys.

A third difficulty unique to pastoral areas is the considerable communication problems that arise as a result of the differences in the languages of the Han Chinese and the minority people which dominate the pastoral populations. In regions close to Beijing and the eastern coastal areas in general, the overwhelming majority of people are able to speak and read Chinese. However, in the more remote pastoral areas, minority nationality people are more likely to be less well educated and Putonghua (the national language) is typically not widely spoken. Under these circumstances, it is often necessary to enlist the assistance of an additional interpreter to enable communication between the minority household and the accompanying Han Chinese. Given that the answers then have to be translated into English, the risks of information being lost or distorted are increased. Of course, the time taken to complete the interview is also greatly increased. As is explained in Section 2.4.2, these difficulties were largely overcome in the case of ACIAR Project No. 8811 through the use of dual-language questionnaires and, where necessary, through the use of multi-lingual translators.

A fourth problem is that while researchers in agricultural areas will also encounter the local view that "everything has to be paid for", the costs of transport and almost everything else are likely to be much higher for surveys in pastoral areas. In poor pastoral areas, rarely if ever visited by foreigners, a Western research team is obviously going to be seen as a good "one-off" source of extra income. This is likely to be the view of local officials, hoteliers, restauranters and farm households alike. The research team will be required to pay handsomely for everything including the hiring of local cars, local officials, and compensating for any lost time incurred by the households being interviewed. In addition, if the research team is operating from the Central level, the team must work systematically down through each administrative level of government and, at each point in the chain, a charge will be levied for use of official time and resources.

2.5.4 Definitional and Data Problems

Obtaining a "standard" definition of key physical and economic terminology is often a major difficulty in China. For example, the term "net income" is defined in various ways by different agencies. The Statistical Bureau definition used to calculate the official estimates of average net income per capita presented later in this book is set out in Table 2.2. A major difficulty with this definition in terms of the Western definition of net income is that it includes taxes. It is important, therefore, to recognise that the term "net income" used throughout this book refers to after-tax net income. It is also significant that production costs do not include any imputed charge for family labour.

Table 2.2 Defining "Farm Household Net Income" in Pastoral Areas

	Net income = gross income - production costs - depreciation - taxes		
Gross income	• live animal sales • animal product sales (wool, meat, skins, etc.) • animals consumed by the household • gifts received from relatives and friends • money received from urban relatives and friends • household sideline income • income from work outside the farm • other income		
Production costs	• pasture contract fee • labour employed • disease prevention • concentrate and roughage feed costs • interest payments on debts related to production • other production-related expenses		
Depreciation on	• pasture constructions (fences, wells, etc.) • stud animals • shed building • production machinery • other assets used in production		
Taxes	• government taxes	- animal tax - slaughter tax on animals slaughtered and marketed	
	• collective taxes	- collective administration costs - collective project costs - teachers' subsidies - collective welfare fund - entertainment and other costs	

[handwritten annotations: "in Chinese figures these are not subtracted" (next to taxes); "(exclude family labour)" (next to production costs)]

Another important term in the context of this book is "pasture land". In China, native grasslands or rangelands are referred to as pasture land, while improved grasslands are referred to as being either artificial, semi-artificial, or aerial-sown. As discussed in more detail in some of the case studies, the definitions of these three types of improved pasture are essentially based on the degree of land cultivation entailed in improving the native grasslands. If the land is thoroughly ploughed and introduced plant species are sown, it is referred to as artificial pasture. In many parts of China, any plants grown entirely for animal feed may be classified as "artificial pasture". That is, grazing oats, for example, will be described as artificial npasture. Where the natural grassland is not completely cultivated prior to planting the introduced species, the resulting pasture is described as semi-artificial pasture. The term "aerial-sown pasture" is self-explanatory.

Measurements used to describe herbivorous livestock populations also vary. Generally the term "sheep units" refers to a weighted total of all the various kinds of grazing livestock (see Appendix 5B at the end of Chapter 5 for details of the weights used in China). Often, however, livestock populations may be described in terms of the number of large and small livestock without any prior standardisation for different types of animals. In this book, an independent method developed by Minson and Whiteman (1989) was used to standardise livestock numbers. The conversion coefficients adopted included sheep – 1; cattle – 7.12; and goats – 0.82.

There were also serious difficulties relating to what actually constituted a village in pastoral areas. In pastoral areas, a village may be either a physical village or an administrative village; the difference being that an administrative village encompasses several physical villages. The concept of an administrative village evolved essentially from the old commune system where the administrative village was equivalent to the old commune brigade level of administration, whereas the physical village was equivalent to the production team or the lowest level of commune administration. The township which encompasses both administrative and physical villages formed the pinnacle of the commune structure. The township is now the lowest level of official government in China.

Particular difficulties relate to obtaining data for pastoral areas at the national level. Unlike agricultural areas, there is a great paucity of information about pastoral areas at the Central government level. This is largely a product of the greater emphasis afforded to agricultural production *vis-à-vis* pastoral production, as well as the inherent difficulties in obtaining information from a relatively small population scattered over a large and inhospitable area. The difficulty for the researcher is that there is relatively little secondary information available. Much of the information which can be obtained relates to only a few key variables, and in many cases these variables are aggregated in such a way that their usefulness is greatly reduced. For example, fine and improved fine grades of wool are not separated out in the production statistics, while in some data sets it is even difficult to obtain separate population statistics for sheep and goats. Indeed, in almost all cases, it is impossible to obtain a figure for the amount of mutton produced since the only available data refer to the aggregate "mutton and goat meat". Historical information on prices is extremely difficult to obtain because price series are frequently not recorded. On the other hand, excellent records are maintained for key physical data such as daily rainfall and temperature. However, this kind of information is sometimes withheld because it is regarded as being highly classified.

Fieldwork in pastoral areas, therefore, involves collecting primary data as well as locally available secondary data not passed on to the relevant authorities for publishing at the national level. Because the local secondary data are not generally available at higher administrative levels, these records may need to be "purchased" by the researchers. That is, village, township and even county authorities often require payment for information.

Chapter Three

Some Characteristics of the Pastoral Region

The pastoral region of China can be defined in different ways. The broadest interpretation refers to the 12 provincial-level administrative units officially designated as pastoral provinces (Plate 1). These provinces/autonomous regions encompass almost all the pastoral activities in China, and pastoralism occupies more land area than any other activity in most of these provincial-level units. Nevertheless, the raising of grazing animals is just one form of animal husbandry practised in these 12 provinces. Furthermore, animal husbandry is only part of the agricultural sector which in turn is one sector in a rapidly diversifying economy. Therefore, while it is convenient to consider the pastoral industries of China as being located in the 12 pastoral provinces, it is important to appreciate that these provinces vary greatly in the extent to which they are economically dependent on industries based on grazing animals.

Within the 12 pastoral provinces, 266 of the 1,224 county-level administrative units are officially designated by the Central government as pastoral or semi-pastoral counties (Plate 2). Since aggregate statistics are available for this group of counties, it is sometimes said that these 266 counties constitute the pastoral region of China. However, many of these counties include significant agricultural areas. In addition, there are many counties not in this group of pastoral and semi-pastoral counties which include large pastoral areas. Two such counties (Dunhuang and Cabucaer) are discussed as case studies in this book. While for some purposes it may be appropriate to refer to the 266 pastoral and semi-pastoral counties as the pastoral region, in many cases such an approach would be too restrictive.

It is not easy, therefore, in any aggregate statistical sense to define unambiguously China's pastoral region. While this book is generally concerned with the pastoral region broadly defined as 12 pastoral provinces/autonomous regions, the narrower concept of the region as consisting of 266 pastoral and semi-pastoral counties is adopted when appropriate. The term "pastoral areas" is used in a qualitative sense to refer to all of the genuinely pastoral parts of the country irrespective of administrative/statistical groupings.

This chapter reviews some important features of the pastoral region. One unusual and potentially confusing aspect concerns the terms used to describe certain administrative units in some pastoral areas. The first section briefly describes Chinese administrative structures and introduces these special terms. The next two sections review some of the key statistical characteristics of the pastoral region, initially in terms of the 12 pastoral provinces/autonomous regions and then in relation to the 266 pastoral and semi-pastoral counties. Of course, one of the most important features of much of the pastoral region is the relatively high proportion of the population which belongs to minority nationalities. These ethnic minorities are discussed in the fourth section. The chapter concludes with a brief historical sketch of how property rights in relation to livestock and pasture land have changed in the pastoral areas since 1949.

3.1 Levels of Administration

The government of the People's Republic of China, which is referred to as the "Central government" in this book, and its various ministries, commissions, bureaus, etc. represents the first level of administration in China. Below the Central government, there are three large cities (Beijing-shi, Tianjin-shi, and Shanghai-shi), twenty-three provinces and five autonomous regions (Table 3.1). The three large cities directly under the Central government are divided into several major counties and districts which are ranked higher than ordinary counties but lower than prefectures. For example, Beijing-shi has nine major counties, six suburban districts, and four urban districts. While the distinction between provinces and autonomous regions is discussed in detail below in Section 3.4.1, the most important difference is that the governments of autonomous regions have considerable freedom to legislate in relation to minority groups living within the region while provincial governments do not have this authority.

Table 3.1 Outline of the Administrative Structure in China

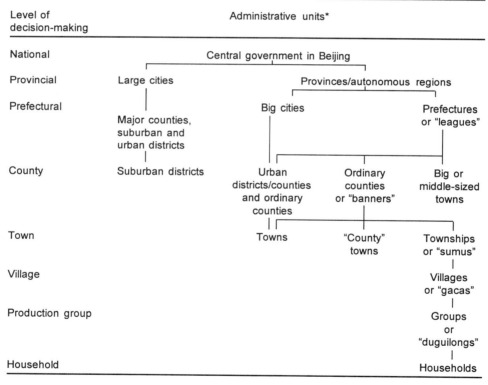

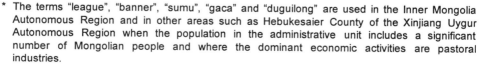

* The terms "league", "banner", "sumu", "gaca" and "duguilong" are used in the Inner Mongolia Autonomous Region and in other areas such as Hebukesaier County of the Xinjiang Uygur Autonomous Region when the population in the administrative unit includes a significant number of Mongolian people and where the dominant economic activities are pastoral industries.

Provinces/autonomous regions are divided into prefectures and big cities. For example, the Inner Mongolia Autonomous Region (IMAR) is divided into eight prefectures and four big cities. The eight prefectures in the IMAR are called "leagues" because there are a significant number of Mongolian Nationality people living in these areas and because pastoral activities constitute a major part of the prefectural economy. The prefectural level cities are Huhehot (the capital), Baotou (the steel city on the yellow river), Wuhai and Chifeng City. These cities have a number of urban districts, suburban districts and/or counties under their control.

As mentioned in Section 2.4.3, four of the counties surveyed as part of ACIAR Project No. 8811 and presented as case studies in this book, were located in Chifeng City Prefecture. Chifeng City, which was formerly known as "Zhaowudamon" (i.e. Zhaowuda League), is not a "city" in the usual sense of the word. Indeed, it embraces a larger geographical and administrative area than some of the leagues/prefectures in the IMAR. To emphasise this fact, it is usually referred to in this book and elsewhere as "Chifeng City Prefecture". For a general overview of this prefecture, see Lin (1990).

Prefectures are divided into urban or suburban districts, counties and sometimes towns at the county level. For example, there are 12 administrative units under the control of the Chifeng City prefecture-level government. Three of these administrative units are referred to as urban and suburban districts, two are always called "counties" while the remaining seven are often referred to as "banners" rather than "counties". The word "banner" is used for a county when there are a significant number of Mongolian people living in the county and where the pastoral industries are the dominant economic activity. While some counties (e.g. Ningcheng County in Chifeng City Prefecture) may have a large number of livestock (especially sheep), agricultural activities are much more important in counties than in banners.

Counties/banners are sub-divided into towns (which are large administrative centres and there is usually only one per county); "county" towns (which are the more economically developed townships with some industry); and the much more numerous ordinary townships. Aohan County in Chifeng City Prefecture, for example, has one town (Xinhui, the county capital), no "county" towns and 30 townships. One of these township-level administrative units is actually a State farm (the Aohan State Farm and Fine Wool Sheep Stud) and another is a Mongolian township which is referred to as a "sumu". The various types of State farm are discussed below in Section 4.4.2.

Townships are usually sub-divided into administrative villages. The administrative villages may consist of a single village or a number of physically separate natural villages which in turn are made up of one or more production groups (or teams). These groups can vary in size from about 10 households up to 150 households or more. Pastoral administrative villages in which there are a significant number of Mongolian people are called "gacas". Mongolian production teams are referred to as "duguilongs".

The basic decision-making unit is the household, which may consist of several generations of the one family.

3.2 Pastoral Provinces /Autonomous Regions

The geographic location of the 12 pastoral provinces is shown in Plate 1. Historically, much of this part of China represented a "buffer" between the heartland of the Han peoples in central and south east China and their Asian neighbours to the north and west. The present national borders of China probably include a larger pastoral region than at any previous time in the long history of the Chinese people. Since 1949, the Central government has resettled significant numbers of Han people in these traditionally non-Han areas. Nevertheless, the population density in most of the pastoral provinces (especially the border provinces) remains very low and non-Han ethnic groups still constitute a major proportion of the population in most of the pastoral provinces. While the long-term trends in the size and composition of the population in some of these provinces are discussed in the provincial case studies (see, for example, Sections 5.2.2 and 7.2.2), the general situation as regards the population in the 12 pastoral provinces as at the end of 1990 is discussed in this section.

As already stressed, pastoralism is not the dominant economic activity in all 12 pastoral provinces/autonomous regions. The nature and economic importance of pastoral activities varies a great deal within this group of provincial-level administrative units. This section also briefly examines some 1990 statistics on land use, livestock numbers, output of pastoral products and gross output values. The objective is both to provide an overview of the 12 pastoral provinces and to place the three case study provinces (IMAR, Gansu and XUAR) in perspective. To assist in the latter respect, Tables 3.2 to 3.4 (and Table 3.5 in the next section) have the data for the three case study provinces highlighted.

3.2.1 Population

Residents of the 12 pastoral provinces now constitute about one-third of the population of China (Table 3.2). With just over 15 million people, the Xinjiang Uygur Autonomous Region (XUAR) has a population a little smaller than that of Australia while all of the other pastoral provinces except Tibet, Qinghai and Ningxia, have much larger populations. Indeed, Sichuan has almost 10% and Hebei more than 5% of the national total.

Three out of every four people living in the 12 pastoral provinces are classified as belonging to the agricultural (and pastoral) population. Interestingly, this is below the national average (Table 3.2). Only in four of the pastoral provinces (Hebei, Sichuan, Tibet and Gansu) is the population more "agricultural" than for the nation as a whole. In particular, the two most important pastoral provinces/autonomous regions (XUAR and IMAR) have significantly lower fractions than the national average of their population classified as agricultural and they are also both below the pastoral region average in this regard.

Most of the minority nationality people would be classified as part of the agricultural population and in some of the pastoral provinces, they represent a large fraction of this population (Table 3.2). Even in regard to the total provincial population, ethnic minorities constitute more than 10% in seven of the 12 provinces and about one-third or more in four provinces. Three of the four large sparsely populated "border" provinces (IMAR, XUAR and Tibet) have especially high proportions of both their agricultural and provincial populations in the ethnic minority classification. In the case of Tibet, the ethnic minority population exceeds the total agricultural population. Clearly, in Tibet many of the Tibetans are classified as urban dwellers.

Table 3.2 Some Population and Land Use Characteristics of the Pastoral Provinces/Autonomous Regions of China, 1990

Pastoral province/ autonomous region	Population						Land use			
	Total provincial population		Agricultural population as a prop. of provincial population[2]	Minority nationality population		Population density	Total size of province		Cultivated area as prop. of total provincial area	Pastoral area as prop. of total provincial area
	No. of people	Prop. of All China		As prop. of prov. pop.	As prop. of agric. pop.		Area	As prop. of All China		
	(million)	(%)	(%)	(%)	(%)	(people/km^2)	('000 km^2)	(%)	(%)	(%)
Hebei	61.1	5.41	86.09	3.93	4.56	324.72	188.16	1.96	34.84	25.36
Shanxi	28.8	2.55	77.08	0.35	0.45	183.65	156.82	1.63	23.55	28.18
IMAR	**21.5**	**1.90**	**69.77**	**19.53**	**28.00**	**18.78**	**1,144.65**	**11.92**	**4.34**	**67.00**
Liaoning	39.5	3.49	58.48	15.70	26.84	270.68	145.93	1.52	23.76	23.22
Jilin	24.7	2.18	62.35	10.12	16.23	129.27	191.07	1.99	20.62	15.48
Heilongjiang	35.2	3.11	59.38	5.65	9.52	77.43	454.63	4.74	19.42	16.57
Sichuan	107.2	9.48	85.82	4.57	5.33	186.07	576.13	6.00	10.93	40.31
Tibet	2.2	0.19	86.36	95.45	110.53	1.83	1,204.25	12.54	0.18	56.26
Gansu	**22.4**	**1.98**	**83.93**	**8.48**	**10.11**	**49.07**	**456.46**	**4.75**	**7.62**	**39.96**
Qinghai	4.5	0.40	71.11	42.22	59.38	6.11	736.32	7.67	0.78	52.41
Ningxia	4.6	0.41	76.09	32.61	42.86	88.82	51.79	0.54	15.37	57.62
XUAR	**15.2**	**1.34**	**71.71**	**62.50**	**87.16**	**9.17**	**1,657.59**	**17.27**	**1.86**	**48.26**
Sub-total	366.9	32.45	76.18[3]	10.68[3]	14.02[3]	52.69	6,963.80	72.54	6.99	51.95
All China[1]	1,130.5	100.00	80.55[4]	8.08[4]	10.03[4]	117.76	9,600.00	-	-	-

[1] Excluding servicemen and the population of Taiwan, Hong Kong and Macao.
[2] The definition of "agricultural population" includes people dependent on pastoral activities.
[3] As a proportion of the sub-total of 12 pastoral provinces/autonomous regions.
[4] As a proportion of the grand total for All China.

Sources: *China Agricultural Yearbook, 1991*, Agricultural Publishing House, Beijing
Statistics of Ethnic Minorities of China, 1949-1990, Statistical Press of China, Beijing

3.2.2 Land Use

The 12 pastoral provinces cover more than 72% of China (Table 3.2). Heilongjiang, IMAR, XUAR and Tibet (the four border provinces) together represent almost 40% of the nation's land area. Climatic conditions in these remote parts of China are extremely harsh. Nevertheless, significant areas of these four border provinces have been "reclaimed" for agricultural purposes, especially in Heilongjiang which has much more land under cultivation for crop production than IMAR, XUAR and Tibet combined.

Of the remaining eight pastoral provinces, Sichuan, Gansu and especially Qinghai have a relatively low proportion of their total land used for cropping, while the other five provinces, with the exception of Ningxia with 15%, have over 20% under cultivation. Indeed, Hebei grows crops on more than one-third of its territory.

Land not used for cultivation may be under pastures, forests or be classified as unusable because of its altitude, lack of water, soil salinity, or for other reasons. The proportion of each pastoral province classified as under pasture is shown in Table 3.2. Seven of the 12 provinces have 40% or more of their area classified as pastoral with more than three-quarters of the IMAR being described as pasture land. Over half the total area included in the 12 pastoral provinces, or more than one-third of China, is classified as being used for pastoral pursuits.

3.2.3 Herbivorous Livestock

In terms of total feed requirements, the most important type of herbivorous animal raised in the pastoral region is cattle. The 41 million cattle require almost three times as much dry matter per year as the 100 million sheep. However, a large proportion of the cattle are raised in provinces such as Heilongjiang, Sichuan, Hebei and Shanxi. In these provinces it is common for cattle to be tethered or housed most of the time and fed by cut-and-carry methods. On the other hand, a large proportion of the 100 million sheep in the 12 pastoral provinces obtain most of their roughage by grazing, especially in the spring/summer/autumn seasons.

In terms of grazing animals, therefore, sheep are the dominant type of herbivore, especially in pastoral areas. As shown in Table 3.3, the 100 million sheep in the 12 pastoral provinces/autonomous regions represented around 89% of all sheep in China in 1990. The data in Table 3.3 also demonstrate that the IMAR and the XUAR are the major sheep raising provinces but Qinghai, Tibet and Gansu also have large sheep flocks.

In terms of feed requirements, horses are the next most important herbivorous type of animal after cattle and sheep. About 72% of all horses in China are kept in the 12 pastoral provinces, especially in the IMAR and the XUAR.

Goats rank last of the four major types of herbivorous animals raised in the pastoral region in terms of total feed requirements. There were 41 million goats in this part of China in 1990, which was 42% of the national herd. Goats outnumber sheep in Hebei and Sichuan and they play a major role in most of the harsher grazing environments in all the pastoral provinces. Even in some of the more environmentally favoured pastoral areas, goats and sheep are often raised together. Farmers can readily increase the number of one of these types of animal at the expense of the other in response to changes in the relative profitability of sheep and goat raising.

Table 3.3 Herbivorous Livestock Numbers and Output of Pastoral Products in the Pastoral Provinces/Autonomous Regions of China, 1990

Pastoral province/ autonomous region	Major herbivorous animals[1]				Major pastoral products					
					Meat output			Animal fibre output		
	Number of sheep	Number of goats	Number of cattle	Number of horses	Total[2]	Mutton and goat meat	Beef	Goat wool	Sheep wool	Cashmere
	(million)	(million)	(million)	(million)	('000t)	('000t)	('000t)	(tonne)	(tonne)	(tonne)
Hebei	5.12	5.63	2.08	0.57	1,301	80	56	2,389	12,634	400
Shanxi	4.06	3.04	1.79	0.11	319	40	28	936	5,608	344
IMAR	**20.75**	**9.49**	**3.85**	**1.57**	**536**	**127**	**86**	**2,292**	**59,203**	**2,076**
Liaoning	1.94	0.73	1.51	0.47	904	13	34	291	7,803	135
Jilin	2.19	0.15	1.88	0.81	518	8	36	22	8,487	2
Heilongjiang	2.49	0.34	2.37	0.99	559	13	52	93	9,614	4
Sichuan	3.47	5.99	10.08	0.52	4,428	37	69	167	2,739	4
Tibet	11.11	5.66	5.06	0.33	88	39	44	893	8,264	448
Gansu	**8.79**	**2.31**	**3.38**	**0.42**	**395**	**38**	**35**	**1,132**	**15,543**	**259**
Qinghai	14.05	2.04	5.39	0.45	153	56	52	420	17,155	157
Ningxia	2.27	0.90	0.28	0.03	69	17	5	290	3,780	158
XUAR	**23.81**	**4.49**	**3.38**	**1.05**	**305**	**158**	**71**	**1,635**	**49,297**	**607**
Sub-total	100.04	40.77	41.05	7.32	9,575	626	568	10,560	200,127	4,594
Sub-total as percentage of All China	88.68	41.94	39.90	71.98	33.52	58.61	45.22	63.98	83.58	79.88
All China	112.82	97.21	102.88	10.17	28,567	1,068	1,256	16,506	239,457	5,751

[1] Year-end livestock numbers.
[2] Includes pork, beef, mutton, goat meat, poultry meat and rabbit meat.
Source: *China Agricultural Yearbook, 1991,* Agricultural Publishing House, Beijing

3.2.4 Output of Pastoral Products

As would be expected, given that almost nine out of 10 sheep are raised in this part of the country, the pastoral provinces account for most of the wool grown in China (84% in 1990, see Table 3.3). In addition, four-fifths of the cashmere produced in China also comes from the pastoral provinces. The IMAR and the XUAR together produce over half of both the wool and the cashmere grown in the pastoral region.

Pork dominates the meat sector of the national Chinese food production system and this is true of the pastoral region as well. The 12 pastoral provinces produce almost 60% of the national output of mutton and goat meat, but this represents only about 6.5% of the total meat produced in the pastoral region because of the large pork output. Similarly, although more than 45% of all beef produced in China in 1990 came from the 12 pastoral provinces, this output represented only about 6% of the total meat produced in these provinces. As with animal fibres, the IMAR and the XUAR are the dominant producers of mutton and goat meat and of beef.

3.2.5 Gross Output Value

Unfortunately, it is not possible to obtain aggregate output-value statistics for pastoral products as distinct from other animal husbandry products. In particular, the dominant role of pork in the output of meat from the animal husbandry sector in most pastoral provinces makes the statistic "gross animal husbandry output value" of doubtful value as an indicator of the relative economic significance of pastoral products. Nevertheless, since it is the only aggregate statistic which is available for this purpose, it has been presented in Table 3.4 both in value terms and as a proportion of gross agricultural output value.

Seven of the 12 pastoral provinces have animal husbandry sectors which, in gross output value terms, contribute a smaller fraction to the output of the whole provincial agricultural sector than the national average of 26%. On the other hand, Tibet, Qinghai, Sichuan, and IMAR have animal husbandry sectors which contribute well above the national average proportion to gross agricultural output value. With the exception of Sichuan, the pastoral industries dominate the animal husbandry sector in these provinces. Overall, animal husbandry in the 12 pastoral provinces contributes only marginally more to gross agricultural output value than the national average.

The agricultural sector in the pastoral provinces as a whole, however, is a significantly greater contributor to gross rural society output value than the average for all China. In fact, this is true for nine of the 12 pastoral provinces, the exceptions being Hebei, Shanxi and Liaoning. This situation is of considerable significance since it demonstrates the relative lack of economic activity in rural areas other than in the conventional agricultural sector (including animal husbandry). In other parts of China, economic development in rural areas has shifted the emphasis away from traditional agricultural pursuits as a source of output value to a much greater extent than in most pastoral provinces. For China as a whole, less than half (46%) of gross rural society output value was generated by agriculture in 1990. The corresponding figure in Table 3.4 for the 12 pastoral provinces is 53%, but Tibet (94%), XUAR (87%), Qinghai (81%) and IMAR (77%) are all still remarkably dependent on traditional agriculture (and animal husbandry) to generate rural income. Clearly, steps need to be taken to create more non-agricultural income-generating activities in rural areas in these provinces in particular, and in the pastoral region in general. In reality, however, there are major limitations on the opportunities for non-farm enterprises in the remote, sparsely populated pastoral areas within these provinces. In these areas, improved productivity within the pastoral industries will remain the major source of real income growth for the foreseeable future.

Table 3.4 Gross Output Values for the Pastoral Provinces/Autonomous Regions of China, 1990
(Calculated at 1990 current prices and expressed in 100 million yuan)

Pastoral province/ autonomous region	Total gross rural society output value[1] (TGRSOV)	Gross agricultural output value[2] (GAOV)		Gross animal husbandry output value (GAHOV)	
		Amount	Prop. of TGRSOV	Amount	Prop. of GAOV
			(%)		(%)
Hebei	873	358	41.01	83	23.18
Shanxi	322	125	38.82	28	22.40
IMAR	**204**	**157**	**76.96**	**46**	**29.30**
Liaoning	704	274	38.92	76	27.74
Jilin	322	189	58.70	41	21.69
Heilongjiang	386	245	63.47	49	20.00
Sichuan	1,069	637	59.59	210	32.97
Tibet	18	17	94.44	8	47.06
Gansu	**170**	**103**	**60.59**	**26**	**25.24**
Qinghai	31	25	80.65	11	44.00
Ningxia	36	25	69.44	5	20.00
XUAR	**167**	**145**	**86.83**	**29**	**20.00**
Sub-total	4,302	2,300	53.46	612	26.61
Sub-total as percentage of All China	25.89	30.02	-	31.16	-
All China	16,619	7,662	46.10	1,964	25.63

[1]Total gross rural society output value (TGRSOV) refers to the total output of products, expressed in value terms, produced by enterprises of township industry (previously commune industry), village industry (previously brigade industry) and industry below village level (previously production team industry, joint sponsored industry by farmers, and individual industry). It includes both the value of finished products and the value of industrial operation services provided to other enterprises. Also included is the value of semi-finished products.
TGRSOV has five major components: agriculture (GAOV); rural industry; rural construction; rural transport; and rural domestic trade, catering and service trade.
[2]The agriculture component of TGRSOV includes the contributions made by crop farming, forestry, animal husbandry (GAHOV), sideline production and fishery.
Source: *China Agricultural Yearbook, 1991.* Agricultural Publishing House, Beijing

Unfortunately, however, as the case studies in this book demonstrate, overall productivity in most pastoral areas is actually on the decline owing to the continued degradation of the natural pastures. Unless the current downward trend in overall productivity can be reversed, it is difficult to see how current real income levels can be maintained let alone improved. If, at the same time, the number of people to be supported continues to increase, then real income per capita, already relatively low in most pastoral areas, must inevitably fall.

3.3 Pastoral and Semi-Pastoral Counties

Obviously, significant portions of some of the 12 pastoral provinces/autonomous regions are not really pastoral areas. Furthermore, in most of these 12 large administrative units, the majority of the population are not dependent on pastoral activities for their livelihood. Consequently, in some respects it could be misleading to consider these 12 geographic areas as constituting the pastoral region of China.

At present, there are 2,833 county-level administrative units in China. All such units are classified by the Central government as urban or rural, and the rural counties are further sub-divided into agricultural, semi-pastoral, or pastoral.

The Central government recognised 97 semi-pastoral and 90 pastoral counties in 1978 but the number of counties in each category was increased in 1979, 1984 and 1985 so that in that year there were 147 semi-pastoral and 119 pastoral counties. In 1987, the number of semi-pastoral counties was reduced by two but in 1988 one more pastoral county and another semi-pastoral county were added to the list which has since remained unchanged. At present, therefore, according to the Central government there are 146 semi-pastoral and 120 pastoral counties in China, all of which, as shown in Table 3.5, are located in the 12 pastoral provinces/autonomous regions.

Table 3.5 Provincial/Autonomous Region Distribution of the 266 Counties Defined by the Central Government as being Pastoral or Semi-Pastoral in 1990

Province or autonomous region	Prefecture or cities	Counties and cities or districts at county level			Agricultural counties or urban cities and districts
		Total	Pastoral	Semi-pastoral	
	(no.)	(no.)	(no.)	(no.)	(no.)
Hebei	18	172		6	166
Shanxi	12	118		1	117
IMAR	**12**	**100**	**33**	**21**	**46**
Liaoning	14	100		6	94
Jilin	8	59	1	9	49
Heilongjiang	14	132	7	8	117
Sichuan	21	217	10	38	169
Tibet	7	78	13	24	41
Gansu	**14**	**85**	**7**	**12**	**66**
Qinghai	8	43	26	4	13
Ningxia	4	24	1	2	21
XUAR	**15**	**96**	**22**	**15**	**59**
Sub-total	147	1,224	120	146	958
All China*	336	2,833	120	146	2567

*Excluding Taiwan Province
Sources: *Statistical Yearbook of China, 1991*, State Statistical Bureau of the PRC, Beijing
Statistics of Ethnic Minorities of China, 1949-1990, Statistical Press of China, Beijing

Characteristics of the Pastoral Region

The precise definition of what constitutes a "pastoral" or "semi-pastoral" county does not appear to be public knowledge in China but four factors are known to be important. To qualify as a pastoral or semi-pastoral county, pastoralism must be a traditional and important economic activity in the county; a significant proportion of the rural population must be undertaking pastoral activities; and the ratio of pasture land area to cultivated land area must be high. Counties which satisfy these rather vague criteria are divided into pastoral and semi-pastoral counties on the basis of the proportion which gross animal husbandry output value represents of gross agricultural output value. (See notes below Table 3.4 for an explanation of these terms.) This proportion is normally greater than 50% for pastoral counties and between 25 and 50% for semi-pastoral counties.

Not only counties but all administrative units in China are classified roughly according to these criteria. Hence, there are 12 pastoral (and semi-pastoral) provinces, certain prefectures are described as pastoral or semi-pastoral, and townships, villages and even households are identified as pastoral or semi-pastoral. Most rural administrative units in China not described as pastoral or semi-pastoral are called agricultural households, villages, townships, counties, etc. Of course, even in most of the 12 pastoral provinces, many more rural counties are classified as agricultural than pastoral or semi-pastoral although the latter occupy a much larger proportion of the total land area.

Another important aspect of the pastoral and semi-pastoral county classification system is that not all provincial governments recognise counties in the same way as the Central government. See, for example, the discussion in this regard in the Gansu Province case study in Section 6.1. In addition, some of the counties traditionally regarded as being pastoral have become semi-pastoral and some former semi-pastoral counties have become more or less agricultural counties. Of the case study counties, Tianzhu (Gansu) is an example of the former while Aohan (IMAR) is still considered semi-pastoral by the Central government but locally it is regarded as an essentially agricultural county.

The geographic location of the 266 pastoral and semi-pastoral counties currently recognised by the Central government is shown in Plate 2. Some pastoral provinces include few pastoral or semi-pastoral counties. For example, only one of the 118 county-level administrative units in Shanxi is classified as a semi-pastoral county and there are no pastoral counties in this province (Table 3.5). On the other hand, 54 of the 100 county-level units in the IMAR are either pastoral or semi-pastoral counties.

The 266 pastoral counties and semi-pastoral counties shown on Plate 2 include a major share of the usable pasture land in the 12 pastoral provinces/autonomous regions. They are also the home of a high proportion of the households whose incomes depend heavily upon pastoral pursuits. Nevertheless, as already pointed out, there are also many other counties in the pastoral provinces/autonomous regions which are not included in the Central government's list of 266 counties, in which large numbers of sheep are raised and in which pastoral activities are important. This point, together with the vagueness with which the counties are classified as "pastoral" or "semi-pastoral" and the changes in the membership of this group of counties over time, all suggest that aggregate data for "the pastoral region" defined as consisting of pastoral and semi-pastoral counties should be carefully interpreted.

In any event, data for the pastoral and semi-pastoral counties are only available for a limited number of characteristics. Table 3.6 presents some long-term data on the number of head and the proportion of the national total in the pastoral and semi-pastoral counties for large livestock and for sheep and goats. The figures for large livestock (cattle, horses, mules, donkeys, yaks and camels) suggest that the pastoral and semi-pastoral counties accounted for little more than a fifth in 1990, down from the peak of almost 27% in 1982. The proportion of the national small ruminant flock (sheep and goats) to be found in the pastoral and semi-pastoral counties reached 46.5% in 1985 but declined to around 37% in 1990.

Table 3.6 Number of Livestock in the Pastoral and Semi-Pastoral Counties Compared with All China, 1949 to 1990 (selected years)

Year	Large livestock			Sheep and goats		
	All China	Pastoral and semi-pastoral counties		All China	Pastoral and semi-pastoral counties	
		Number	Proportion of all China		Number	Proportion of all China
	('000)	('000)	(%)	('000)	('000)	(%)
1949	60,020	10,142	16.9	42,350	19,023	44.9
1952	76,460	12,269	16.0	61,780	24,615	39.8
1957	83,820	14,627	17.5	98,580	36,231	36.8
1965	84,210	18,029	21.4	139,030	55,911	40.2
1978	93,890	21,660	23.1	169,940	62,205	36.6
1979	94,590	22,706	24.0	183,140	68,821	37.6
1980	95,250	23,580	24.8	187,310	70,856	37.8
1981	97,640	24,955	25.6	187,730	76,232	40.6
1982	101,130	27,038	26.7	181,790	74,980	41.2
1983	103,500	25,492	24.6	166,950	70,197	42.0
1984	108,690	26,245	24.2	158,400	70,544	44.5
1985	113,820	27,633	24.3	155,880	72,454	46.5
1986	118,960	27,810	23.4	166,230	71,248	42.9
1987	121,910	27,956	22.9	180,340	73,004	40.5
1988	125,380	28,170	22.5	201,530	77,144	38.3
1989	128,050	27,805	21.7	211,640	77,444	36.6
1990	130,210	28,108	21.6	210,020	78,199	37.2

Note: As explained in the text, the Central government increased the number of counties recognised as pastoral and semi-pastoral after 1978. However, as far as possible, the data for all years shown in this table are the aggregates for the 266 counties recognised by the Central government in 1990. The figures shown for years prior to 1978 should be treated with caution.

Sources: *Statistical Yearbook of China, 1991*, State Statistical Bureau of the PRC, Beijing
Statistics of Ethnic Minorities of China, 1949-1990, Statistical Press of China, Beijing

Wool production data are available for the pastoral and semi-pastoral counties from 1985. This information, together with the corresponding data for all China, is presented in Table 3.7. Although the time series in Table 3.7 is rather short, the figures suggest that the pastoral and semi-pastoral counties have not kept pace with the general expansion in wool production in other parts of China. The proportion of the national clip grown in these counties declined from 57% to 49% over the 1985 to 1990 period.

Table 3.7 Wool Production in the Pastoral and Semi-Pastoral Counties Compared with All China, 1985-1990

Year	All China	Wool production in the 266 pastoral and semi-pastoral counties	
		Amount	Proportion of all China
	('000 t)	('000 t)	(%)
1985	178.0	102.0	57.3
1986	185.2	101.2	54.6
1987	208.9	105.7	50.6
1988	221.7	110.6	49.9
1989	237.3	114.0	48.0
1990	239.5	116.6	48.7

Sources: *Statistical Yearbook of China, 1991*, State Statistical Bureau of the PRC, Beijing
Statistics of Ethnic Minorities of China, 1949-1990, Statistical Press of China, Beijing

Table 3.3 indicates that in 1990, the 12 pastoral provinces/autonomous regions contributed about 84% of the national wool clip. Since the 266 counties produced only around 49% of the wool grown in China in 1990, over one-third (35%) of the national clip must have been grown in the 12 pastoral provinces/autonomous regions outside the 266 pastoral or semi-pastoral counties in that year. Indeed, agricultural counties located in the same general areas as the pastoral and semi-pastoral counties often have some townships and villages which are classified as pastoral or semi-pastoral. For example, as already mentioned, two of the ten case study counties included in ACIAR Project No. 8811 (Dunhuang in Gansu and Cabucaer in XUAR) are excluded from the official Central government list of 266 pastoral and semi-pastoral counties, yet they include townships and villages which are defined as being "purely" pastoral.

Owing to the diversity and mixing of pastoral and agricultural activity within and between localities, there is no simple way of identifying and aggregating all the pastoral areas as such in China. The 266 pastoral and semi-pastoral counties officially recognised by the Central government taken together represent a convenient base upon which to build. On the other hand, the 12 pastoral provinces/autonomous regions, which include all of the 266 pastoral/semi-pastoral counties as well as almost all the agricultural counties in which pastoral activities are also important, delineate a useful set of outer bounds for China's pastoral region.

3.4 Minority Nationalities

China is a country with a great diversity of ethnic minority people and the Central government recognises 55 different groups or minority nationalities (see Table 3.8). According to the 1990 census, the minority population was 91.2 million (8% of the mainland's total) compared with 67.2 million (6.7% of the total) at the time of the previous national census in 1982. Eighteen minority groups in 1990 had a population in excess of 1 million compared with 15 such groups in 1982 and only 10 in 1964. These 18 minority nationalities had a total population of 85.41 million in 1990. Many of the other 37 ethnic minorities which made up the remaining 5.79 million of the minority population include relatively small numbers of people.

Table 3.8 Provincial Distribution of the Minority Nationalities in China

Minority nationalities resident in the pastoral provinces/autonomous regions		Minority nationalities resident in other provinces/autonomous regions	
Nationality	Province/autonomous region[1]	Nationality	Province/autonomous region
Mongolian*	Inner Mongolia, Liaoning, Xinjiang, Jilin, Heilongjiang, Qinghai, Hebei and Gansu (Henan and Yunnan)	Zhuang*	Guangxi, Yunnan, Guangdong and Guizhou
Hui*	Ningxia, Gansu, Xinjiang, Qinghai, Hebei, Liaoning, Inner Mongolia, Heilongjiang and Jilin (Henan, Yunnan, Shandong, Anhui, Beijing, Tianjin and Shaanxi)	Yao*	Guangxi, Hunan, Yunnan, Guangdong and Guizhou
Tibetan*	Tibet, Sichuan, Qinghai and Gansu (Yunnan)	Dong*	Guizhou, Hunan, Guangxi and Hubei
Korean*	Jilin, Heilongjiang, Liaoning and Inner Mongolia	Dai*	Yunnan
Manchu*	Liaoning, Heilongjiang, Jilin, Hebei and Inner Mongolia (Beijing)	She	Fujian, Zhejiang, Jiangxi and Guangdong
Daur	Inner Mongolia, Heilongjiang and Xinjiang	Blang	Yunnan
Miao*	Sichuan (Guizhou, Yunnan, Hunan, Guangxi, Hainan and Hubei)	Bai*	Yunnan and Hunan
Yi*	Sichuan (Yunan, Guizhou and Guangxi)	Hani*	Yunnan
Tujia*	Sichuan (Hunan, Hubei and Guizhou)	Va	Yunnan
Uygur*	Xinjiang (Hunan)	Buyi*	Guizhou
Kazak*	Xinjiang and Gansu	Li*	Hainan and Guangdong
Lisu	Sichuan (Yunnan)	Gaoshan	Taiwan and Fujian
Dongxiang	Gansu and Xinjiang	Lahu	Yunnan
Naxi	Sichuan (Yunnan)	Shui	Guizhou and Guangxi
Kirgiz	Xinjiang	Jingpo	Yunnan
Tu	Qinghai and Gansu	Mulam	Guangxi
Qiang	Sichuan	Maonan	Guangxi
Sala	Qinghai and Gansu	Gelo	Guizhou and Guangxi
Xibe	Xinjiang, Liaoning and Jilin	Achang	Yunnan
Tajik	Xinjiang	Pumi	Yunnan
Uzbek	Xinjiang	Nu	Yunnan
Russian	Xinjiang	De'ang	Yunnan
Ewenki	Inner Mongolia and Heilongjiang	Jing	Guangxi
Bao'an	Gansu	Dulong	Yunnan
Yugur	Gansu	Jino	Yunnan
Tatar	Xinjiang		
Oroqen	Inner Mongolia and Heilongjiang		
Hezhe	Heilongjiang		
Moinba	Tibet		
Lhoba	Tibet		

Note: The 18 minorities marked with an asterisk (*) had populations in excess of 1 million in 1990.
[1]Provinces/autonomous regions in brackets are non-pastoral provincial-level units in which the relevant minority also reside.
Source: *China Statistical Yearbook, 1991*, State Statistical Bureau of PRC, Beijing

Although minority nationalities account for only a small proportion of the total Chinese population, these people are distributed over more than half of China. As Plate 3 demonstrates, the areas in which the ethnic groups represent a significant proportion of the population tend to be located in the more remote northern and western border regions. In particular, a major fraction of the minority nationality people live in the 12 pastoral provinces (Table 3.2). The ethnic minorities which live in the 12 pastoral provinces are listed in Table 3.8.

For thousands of years, the minority nationalities have been of major strategic importance to the Han Chinese. Spence (1990, p.556) notes for example that during the Qing Dynasty (1644-1911) the minority nationalities "served a buffer function between the Han Chinese and the inhabitants of other lands". In more recent times during the war against Japan and the ensuing civil war, ethnic minorities formed an important base from which to fight the Japanese and nationalist armies.

Relations between the dominant Han Chinese and the ethnic minorities, however, have not always been conducted in a spirit of mutual self-interest and understanding. Following the long history of discrimination and oppression of national minorities, a deep sense of mistrust had developed between the Han and national minority peoples. Spence (1990, p.556), for example, illustrates the depth of this mistrust citing local sayings such as, "a rock does not make a good pillow, nor does a Han Chinese a friend", and "if we read (Chinese) our stomachs will ache, our crops won't grow, and our women will become barren". This deep sense of mistrust between the Han and the national minorities has, on many occasions since 1949, spilt over into open hostility. Spence (1990, p.556) cites incidents such as those involving the Hui people of the Muslim faith in Gansu Province in the early 1950s, being "forbidden to enter certain cities", and even being "fired upon by Han Chinese settlers" and "Tibetans who collaborated with Han Chinese on road building projects, being killed or mutilated by their fellow villagers".

The Han-dominated government of New China recognised from the beginning the need to heal the historical division between the Han and the ethnic minorities. Since 1949, therefore, the Central government has placed great emphasis on three key areas of policy: promoting autonomous rule; ensuring a sufficient level of national representation; and accelerating development in minority autonomous areas.

Despite the implementation of these policies and other less subtle strategies, such as resettling large numbers of Han people in remote border regions and maintaining a significant military presence in the more troublesome areas, there remains the potential for serious social unrest in some parts of China's pastoral region. For example, there were serious riots in Tibet in March 1989 to mark the thirtieth anniversary of the March 1959 rebellion. The western border areas of the XUAR were also subjected to an outbreak of violence in April 1990. This incident, and the car bomb which killed six people and injured another 20 in Urumqi the capital of the XUAR in February 1992, were ascribed to "Muslim separatists" (Mackerras, 1994).

3.4.1 Ethnic Autonomy

From very early in the life of The People's Republic of China, providing ethnic minorities with a limited amount of autonomous self-government within the limits prescribed by the Constitution and State Laws was seen as an important first step towards "upholding *national unity* and promoting *border stability* and unity amongst ethnic groups" (*China Daily*, 1992a, emphasis added).

The policy of ethnic autonomy was first implemented on 1 May 1947, with the establishment of The Mongolia Autonomous Region (later renamed The Inner Mongolia Autonomous Region). The policy was subsequently extended to other areas of China after being formalised on the 21 September 1949 with the endorsement of the "Common Program" or the Provisional Constitution by the First Chinese People's Political Consultative Conference, which set forth the constitutional rights of minority nationalities in China (Anon., 1991). The "Common Program" stated that, 'Greater Nationalism' and 'Local Nationalism' should be opposed. Acts of discrimination, oppression, and dividing of the various nationalities should be prohibited ... and ... regional autonomy should be exercised in areas where national minorities are concentrated. The "Common Program" was later formally endorsed at the First Session of the First National People's Congress on 20 September 1954, and officially incorporated into the National Constitution as the Program of the People's Republic of China for Implementing Regional National Autonomy (Ma, 1989, p.21).

By the end of 1991, a total of 159 national autonomous areas had been established throughout China. Of these, five are autonomous regions, 30 are autonomous prefectures, and 124 are autonomous counties or banners. In addition, where minority populations are small and scattered, "national townships" were established. Currently there are more than 1,500 national townships registered throughout China (*China Daily*, 1991).

Regional autonomous self-governments are established in areas where minority nationalities live in what are referred to as "compact" communities. As Ma (1989, p.24) explains, "these autonomous governments, apart from exercising the functions and powers of normal State bureaucracies, also exercise the functions and powers of autonomous self-government within the limits prescribed by the Constitution and State laws". However, as Ma elaborates, "regional national autonomy is designed to ensure the right of autonomy, not only for ethnic people living in considerable numbers in dense communities, but also for those living in small communities". That is, a minority people may have several autonomous areas in accordance with their distribution.

3.4.2 National Representation

The minority people are also guaranteed representation in national government. According to the "Electoral Law" adopted by the People's Congress in 1952, "all minority nationalities are entitled to appropriate representation in the National People's Congress" (*China Daily*, 1991). This Law was later amended by the Fifth National People's Congress in 1979 to include a provision stating that "even a minority nationality with an exceptionally small population shall have at least one Deputy in the National People's Congress" (Ma, 1989, p.23).

Of the members of the National People's Congress held in 1954, 178 (or 14.5%) were from minority nationalities. While this proportion dropped to a low of 9.3% for the 1975 Congress (owing to the Cultural Revolution), it subsequently increased again reaching 15% at the 1988 Congress. That is, the minorities have always been over-represented in these forums relative to their share of the national population. On the other hand, minority membership of the CCP has never exceeded 6% and there has been a tendency for the under-representation in the CCP to worsen since the 1950s (Mackerras, 1994).

Few ethnic minority leaders have risen to national prominence. Perhaps the best known is the Mongol Ulanfu who was a long-serving member of the Central Committee of the CCP and The Politburo. He was a Vice Premier in the 1950s and became Vice President of China in 1983. Ulanfu's son, Buhe, was appointed governor of the IMAR in 1982 and promoted to Vice President of the Standing Committee of the National People's Congress in the early 1990s.

3.4.3 Development Assistance

Promoting development in ethnic minority areas is seen as being another important means of reducing tension in troubled border areas. In this regard, special development funds are allocated annually by the Central government to support minority areas. These funds include a wide range of relief funding, loans and subsidies, to facilitate the rehabilitation and development of production in the national autonomous areas (Ma, 1989). Currently, under the Eighth Five-Year Plan (1991–95) the State is making available ¥1.1 billion to minority areas. This represents an increase in nominal terms of ¥300 million or 37.5% on the previous allocation in the Seventh Five-Year Plan (1985–1990). Other support being provided in the Eighth Five-Year Plan includes "encouraging State-owned, large and medium-sized firms, colleges and institutes to train more technical persons and provide more technological support to minority regions while encouraging scientists and technicians to make contributions in the (minority) regions" (*China Daily*, 1992, p.1).

An important subset of the development funds provided to minority areas is funding provided by the Central government for poverty alleviation. Within the pastoral region, for example, of the 153 pastoral and semi-pastoral counties with minority populations greater than 30% of the total population, 38 or one-quarter of the total number of counties are recognised by the Central or provincial levels of government as being poverty counties.

The criteria for defining poor areas are described in considerable detail in Anon. (1989). Briefly, counties receiving State aid can be divided into two broad categories: those receiving direct Central government assistance (nationally supported poverty or NSP counties); and those receiving direct provincial or autonomous region government assistance (provincially supported poverty or PSP counties).

Within the NSP category of counties, there are three sub-categories: poor counties; poor counties in pastoral regions; and poor counties in "Sanxi" regions. As explained in more detail in the Aohan and Alukeerqin County case studies (see Sections 10.3.4 and 11.3.4), counties included in the "poor county" sub-category would include "counties that saw (*sic*) the annual average per capita net income in local rural areas being lower than ¥150 in 1985; old revolutionary bases and autonomous counties of minority nationalities that saw (*sic*) the annual average per capita net income in local rural areas being lower than ¥200 in 1985; and a few minority autonomous counties with special difficulties in Inner Mongolia, Xinjiang and Qinghai, as well as some counties in major old revolutionary bases that had made great contributions to the Chinese revolution" (Anon., 1989, p.52).

Counties included in the "poor counties in the pastoral regions" sub-category include "pastoral counties (or banners) that saw (*sic*) the annual average per capita net income in their rural areas during the three years from 1984 to 1986 being lower than ¥300; and semi-pastoral counties (or banners) that saw (*sic*) the annual average per capita net income in their rural areas during the 1984-1986 period being lower than ¥200" (Anon., 1989, p.53).

The third sub-category of NSP counties known as the "poor counties in Sanxi regions" refers to the most impoverished counties in China, including those counties in "the central arid region represented by Dingxi (Prefecture) of Gansu Province, the Hexi (Corridor) Region, and the Xihaigu area in Ningxia Hui Autonomous Region" (Anon., 1989, p.48).

Financial assistance for poverty counties was first provided to the Sanxi group of poor counties in the early 1980s. In 1986, however, following the Fourth Session of the Sixth National People's Congress, the assistance originally only given to the Sanxi areas was extended to include "old revolutionary bases, areas inhabited by minority people as well as remote and poor regions" (Anon., 1989, p.49). The assistance to be provided to poor areas was considered to be "an important part of the country's Seventh Five-Year Plan of National Economic Development (1986-1990)" (Anon., 1989, p.49). This resolution still forms the basis for the Central government poverty assistance programs.

In total, there are 664 poor counties in China. Of these, 294 are classified as being NSP counties, and the remaining 370 are classified as being PSP counties. As discussed in the county case studies, the assistance provided to the NSP counties is in the form of special loans with preferential interest rates.

The PSP category of counties is not sub-divided into sub-categories. However, the type of assistance provided to PSP counties varies considerably according to the "concrete conditions in different localities" within each province or autonomous region (Anon., 1989, p.54).

According to the guidelines set down by the Fourth Session of the Sixth National People's Congress referred to earlier, poverty is to be tackled on two basic fronts: first, absolute poverty is to be redressed by ensuring that people have sufficient food and clothing to survive; and second, that relative poverty is to be addressed through efforts aimed at furthering regional economic development.

The government's motivation for attempting to address both absolute and relative poverty is based primarily on concerns that "the situation has direct impacts on China's reform, political stability, national unity, social equilibrium and the long-term balanced and coordinated development of the country's national economy. Therefore special policies and effective measures should be adopted to solve this problem in a serious way" (Anon., 1989, p.48).

3.5 History of Changes to Property Rights in the Pastoral Areas

In many respects, the introduction of the household production responsibility system (HPRS) in rural China after 1978 was not as major a break with the past as first impressions suggest. The following brief review of how property rights have changed in the pastoral areas since 1947 demonstrates that this is especially true in regard to these parts of the country.

3.5.1 Land Reform 1947–1950

Following the victory of the Chinese Communist Party forces in Inner Mongolia and north western China in 1947, an Ethnic Reform Law was proclaimed which established guidelines for the redistribution of rights to pastoral property. The general principle in the Ethnic Reform Law was that ownership rights to rangeland be confiscated in favour of village or State ownership. In the Inner Mongolia Autonomous Region (IMAR) and Gansu Province, the ownership rights to a large proportion of the rangeland passed to the village, whereas in the Xinjiang Uygur Autonomous Region (XUAR) the rights mostly passed to the State.

Before the Ethnic Reform Law was introduced, the rangeland was, in theory, publicly owned. That is, it belonged to the people of the minority nationalities living on the rangelands. These rights are said to have been first established in the Qing Dynasty around the year 1740. In practical terms, however, the rights to the rangeland were overwhelmingly controlled by the so-called "animal landlords" who were rich herders and buddhist monks (in more densely populated parts of the pastoral region) and tribal leaders (in less densely populated parts of the pastoral region).

The landlords/tribal leaders used their control over the rangelands to limit the number of livestock owned by "landless" herders and to boost the size of their own flocks/herds. This facilitated a supply of low-cost labour to care for the landlords'/tribal leaders' disproportionate share of livestock numbers. For example, a typical herder's salary just prior to Liberation in Inner Mongolia in 1947 was said to be half a sheep per month or six sheep per year which was equivalent to around 6% of the total annual output value of the livestock cared for by the herder.

3.5.2 Reforms to Livestock Ownership 1947–1957

The 1947 Ethnic Reform Law was intended to reverse the inequities of the pre-Liberation period and provide all herders with free access to the rangeland. However, the ethnic reforms did not specify how the other major pastoral asset, livestock, should be distributed.

The Outline Law for Agrarian Land Reform in 1947 formed the legal basis for redistribution of livestock from landlords and other rich livestock owners to poor herder households. Specifically, the proportion of livestock confiscated was directly related to the so-called exploitation index assigned to landlords and rich herders. The index was determined on the basis of the proportion of hired labourers to working family members. For example, if a family with four family labourers had hired only one labourer, then the family was assigned an exploitation index of 25%. A hired labourer was defined as someone who had worked for the family for a period of greater than three months. Families with exploitation indexes of between 25% and 50% were described as being rich herders/farmers, while those families above 50% were described as being landlords. Therefore, it was possible for an individual who had owned large amounts of land but had hired very little labour to have avoided the critical labels of "landlord" or "rich herder/farmer".

The implementation of the Outline Law was left to the various provincial/autonomous region administrations. The agriculturally-oriented provinces adjoining the pastoral areas tended to implement the redistribution fully in accordance with the definitions and provisions of the Outline Law. Liaoning Province, for example, which in the 1950s included Chifeng City Prefecture, redistributed 90% of the livestock in Alukeerqin County and all of the animals in Balinyou County. (These two counties are discussed as case studies in Chapters 8 and 11.)

By contrast, in the major pastoral provinces and autonomous regions (i.e. IMAR, Gansu Province, etc.) it was decided that livestock would not be redistributed from landlords/tribal leaders (and buddhist monks and richer herders) to landless or tenant herders. This decision was taken as a result of problems experienced with the redistribution of livestock in Chifeng City Prefecture (then belonging to Liaoning Province). In this essentially pastoral prefecture, a large proportion of the livestock were either sold or killed following redistribution. Herders at the time were apparently concerned that a counter-revolutionary shift in power would deprive them of their new-found assets. The high real prices which were available for livestock products following two decades of war with Japan and the more recent civil war, also provided a further powerful incentive for herders to sell a sizeable proportion of their newly acquired assets.

The depth of these concerns surrounding the fragility of the pastoral economy was such that governments of major "pastoral" provinces such as the IMAR and Gansu Province opted to refrain from a general redistribution of livestock, choosing instead to increase the wages paid to the herders responsible for tending livestock owned by the landlords/tribal leaders and rich peasants. In the IMAR, for example, the monthly salary for a herder was increased four times from a half a sheep per month to two sheep per month. In some provinces such as Gansu, in addition to a general wage increase for herders, a law titled "The Three Nos and Two Favourables" was introduced to protect livestock owned by former landlords/tribal leaders and rich peasants. The general principles of the Law as it related to the "Three Nos" component were: no reallocation of existing livestock; no disputation of ownership of existing livestock; and no class distinction arising out of livestock ownership. In terms of the "Two Favourables" component, the general principles were said to be: a favourable increase in the wage for poor herders; and favourable treatment of landlords/tribal leaders.

3.5.3 Collectivisation 1952–1958

The move towards collectivisation in China first occurred in 1952 with the establishment of mutual aid teams and cooperatives (Wen, 1989). The formation of mutual aid teams was essentially a formalisation of the traditional Chinese practice of pooling labour or exchanging equipment during peak seasons of production (Prosterman and Hanstad, 1990). Cooperatives were much more advanced organisationally and required herders to share income on the basis of contributed labour and livestock.

The formation of cooperatives progressed through two stages: primary cooperatives followed by advanced cooperatives. Primary cooperatives entailed herders receiving a share of cooperative income on the basis of contributed labour and livestock. A predetermined level of cooperative income (usually 60%) was distributed according to the amount of labour contributed, while the remainder was distributed according to the number of livestock contributed to generate cooperative income. The advanced cooperative stage entailed the cooperative purchasing the livestock from cooperative members at an agreed price and all cooperative income being distributed according to the level of labour contributed.

The development from mutual aid teams (and cooperatives) to communes began in 1958 and was completed by 1959 (Wen, 1989). Communes differed from mutual aid teams and cooperatives in the system of livestock ownership and income-sharing arrangements. Livestock were sold to the collective, and income was shared on the basis of contributed labour in the form of work points. In a situation analogous to similar reforms in agricultural areas occurring around the same time, households were permitted to retain ownership of a fixed number of livestock.

However, the number of private livestock permitted in pastoral areas was often much greater than in agricultural areas. In the IMAR, for example, a household was permitted to retain 3 cattle, 10 to 20 sheep, and 1 to 2 horses. Within a few years these numbers were increased to 50 sheep, 20 cattle, and 5 to 10 horses. The timing and speed of the shift from mutual aid teams to communes is demonstrated in Table 3.9.

In the commune era (1958 to 1978), the term "collective" could apply to the whole commune or to a brigade or to a production team or to an administrative village within the commune. Nowadays, the term collective is used almost exclusively to refer to a village-level organisation which owns or controls the land contracted out to the households.

Table 3.9 Transformation of the Pastoral System from Independent Households to Mutual Aid Teams, to Cooperatives and Finally to Communes in the IMAR, 1952-1959

Year	Proportion of pastoral households participating in mutual aid teams	Proportion of pastoral households participating in cooperatives	Proportion of pastoral households participating in communes
	---------------------------- (%) ----------------------------		
1952	6.84	0.05	-
1953	7.28	0.04	-
1954	42.71	0.18	-
1955	40.94	0.43	-
1956	47.17	19.17	-
1957	56.50	27.09	-
1958	16.73	-	80.16
1959	-	-	100

Note: The data in this table refers to the geographic area currently called the IMAR. That is, although in the 1950s Hulunbeier, Xingan, Zhelimu and Chifeng City prefectural-level administrative units were not included in the IMAR, data for these areas are included in this table.
Source: IMAR Chinese Academy of Social Sciences, 1991

3.5.4 The Household Responsibility System 1956–1966

The household responsibility system (HRS) was a major forerunner to the present day household production responsibility system (HPRS). The HRS was first introduced in the poor and remote parts of Shanxi and IMAR in 1956. It entailed households being required to meet targets or quotas for production, work points, and production costs in order to receive a reward for over-quota production. The system was particularly important in terms of its emphasis on the devolution of certain property rights from the collective to the individual household. In essence, the introduction of the HRS represented the first major reversal of policy in regard to the socialisation of property rights.

Following the "great struggle between socialism and capitalism" in 1957, the HRS was criticised as "a mistake of principle departing from the socialist road" (Khan, 1984, p.85). After the establishment of the communes (which occurred in 1958), the HRS was abolished. However, within a year or so many communes in the poorer areas of the country revived the HRS. Khan (1984) points out that this was done despite a then current campaign to oppose "Right deviations" in which the HRS was ruthlessly criticised as "extremely backward, retrogressive and reactionary".

The HRS began to be more widely applied in 1962, and by 1963 had been extended to many parts of the IMAR. However, the HRS was once again discontinued in 1965 during the "four clean-ups" campaign in favour of a system in which work points were assessed on the basis of "a correct attitude to work" and "a devotion to the cause of serving the people". Physical strength and skill were correspondingly de-emphasised (Khan, 1984).

3.5.5 Cultural Revolution 1966–1976

The Cultural Revolution period saw further marked changes in the distribution of rights to pastoral property as private ownership of livestock and private production came under attack for being capitalistic practices. During this time, it was argued that China's countryside was threatened by the resurgence of the landlord and bourgeois class, and there existed a sharp struggle between "two classes, two roads and two lines" in the rural economy. In particular, principles linking work points to one's contribution to production or in other words "to each according to his work" were regarded as contributing to the preservation of the bourgeois right. In order to overcome this threat posed by the bourgeois right, the Party exhorted the virtues of "keeping political instead of economic accounts" and "better to grow socialist weeds rather than capitalist shoots" (Chen and Buckwell, 1991).

In the IMAR, the HRS was eliminated and in the latter half of 1968 revolutionary committees led by Red Guards declared that all private ownership of livestock should be eliminated. This declaration resulted in herders releasing large numbers of livestock into the open range where a great many perished owing to unseasonably bad weather (i.e. heavy snows, drought and strong winds). Between 1968 and 1969, livestock numbers in the IMAR are reported to have decreased by a massive 40%. It is interesting that throughout this period, pastoral households retained private ownership of housing.

Once the more extreme parts of the Cultural Revolution were over, a small proportion of livestock was again returned to private ownership. The timing of this policy reversal, however, was extremely variable. In the IMAR, for example, households were once again permitted to own livestock in 1971 while in Sunan County, which is one of the case study counties in Gansu Province, herders were not permitted to own livestock until the introduction of the present HPRS in 1983.

3.5.6 Property Rights Reforms Post-1978

As is well known, the Third Plenary Session of the Central Committee of the Chinese Communist Party, which followed the Eleventh National Congress of the Party in late 1978, authorised the implementation of major reforms in rural areas in connection with the control over production decisions. These changes are now collectively referred to as the introduction of the household production responsibility system (HPRS) to replace the communes.

Liu (1990) describes the three steps which were commonly followed in pastoral areas. Initially, the livestock were contracted (or leased) to the individual households by the commune management. The next step involved the sale of the livestock ownership rights to the households which in return contracted with the commune (or collective as the relevant authority had become known) to pay certain taxes and to make agreed contributions to various communal welfare funds. The third stage involved the households being assigned use rights to the pasture land previously controlled by the commune. (As pointed out in Section 3.5.1, the ownership rights to the land were sometimes vested in the village collective and sometimes held entirely by the State.) The obligations and rights of the household in regard to the use of pasture land were spelt out in a pasture land contract. Hence the so-called "double contract" HPRS under which private households contract with the village collective to accept certain obligations in return for the privilege both of owning the livestock and of having the right to use certain pastures.

While the three steps to the double contract HPRS described by Liu more or less occurred throughout the pastoral region during the 1980s, the case studies in this book demonstrate that there have been a great many variations to accommodate "local factors". Furthermore, in most areas the system is still evolving.

The post-1978 property rights reforms together with other subsequent economic reforms such as the move to greater fiscal responsibility for county governments after 1983, and the introduction of free markets for wool and cashmere after 1985, induced major structural changes in pastoral areas during the 1980s. However, as discussed in Part III of this book, the impact of these changes on such things as sheep numbers, wool output, rangeland degradation and household incomes, varied widely from county to county.

Chapter Four

The Chinese Sheep and Wool Scene

The emphasis which the traditional pastoralists of China's pastoral region placed on sheep has been stressed in earlier chapters. With the founding of New China in 1949, breed improvement and upgrading of sheep flocks was recognised as one potential means of raising the productivity of animal husbandry in pastoral areas. During the 1950s, a number of breeding programs were initiated to develop new breeds of sheep capable of growing more wool of a much finer fibre diameter than the coarse wool types of sheep traditionally raised in these parts of China. With the aid of artificial insemination technology, the new genotypes created by these breeding programs were widely disseminated and have dramatically changed the amount and type of wool grown in China's pastoral region. Originally, the new genes were imported from the Soviet Union and East Germany but the availability of Australian merinos since 1972 has had a profound impact on the development of fine wool sheep breeds in China (Longworth and Williamson, 1993).

Many of the county case studies in this book describe, in the local context, how the breeding and sheep improvement programs have evolved, how they are operating in the 1990s, and the impact of the new genotypes on both the quantity and quality of the wool grown. The case studies also demonstrate that many problems still remain to be overcome. Indeed, the swing towards finer woolled sheep initiated during the 1960s and 1970s under the commune system and continued during the 1980s despite the introduction of the household production responsibility system (HPRS), may now be faltering in some areas. With the greater freedom to determine their own sheep-breeding programs under the HPRS, private sheep-raising farmers in some localities have become less than enthusiastic about upgrading their sheep to fine wools. The reasons for this are discussed in Parts II and III of this book.

In parallel with developments on the production side, the marketing of raw wool has also undergone major changes since the 1950s. In particular, the economic reforms since 1978 have had an enormous impact on wool marketing, especially after 1985 when the Central government devolved control over the marketing of wool to the provinces. The sudden move towards a free market for wool led to chaos in some pastoral areas during the famous "wool war" period of 1986 to 1988 (Watson and Findlay, 1992). The collapse of the world wool market in the second half of 1989 and into 1990 was mirrored in China and resulted in a major build-up in stocks of domestic raw wool. As with production, therefore, serious problems associated with the marketing of wool emerged in China during the 1980s.

Chinese policy-makers concerned about formulating sustainable development strategies for China's pastoral region have become increasingly conscious of the problems facing the Chinese sheep and wool industry. As stressed in Chapter 2, one of the major objectives of the research on which this book is based was to identify and analyse these important constraints to progress in pastoral areas. However, understanding the Chinese sheep and wool scene requires some background information on such things as: Chinese wool grading and sheep breeds, sheep production systems and the location of sheep and wool production; the relative importance of private sheep-raising households compared with State farms; the role of the Supply and Marketing Cooperatives in wool marketing; and traditional wool pricing arrangements. While these topics and related issues are taken up in this chapter, each subject is dealt with only briefly and in a selective manner. That is, there is no attempt to explore the technical details of these topics exhaustively. Instead, aspects of general interest are raised and, where appropriate, reference is made to more detailed discussions both later in this book and elsewhere.

4.1 Wool Grading and Sheep Breeds

There are three recognised grading systems for raw or greasy wool currently in use in China: the simple National Wool Grading (or Purchasing) Standard by which wool is bought from farmers; the Industrial Wool-Sorts Standard by which scouring plants and textile mills regrade raw wool prior to processing; and the Quality Standard for Auctioned Wool which sets out the grading and other procedures required before wool can be sold by auction.

An outline of the first two grading systems may be found in Longworth (1993b). Full details of all three systems together with information about the National Top Standard and recent attempts to revise the Top Standard are presented in Longworth and Brown (1994). However, a few of the terms used in regard to the grading of raw wool in China need to be explained before some of the material in the remainder of this book can be fully understood. Similarly, a brief introduction to wool types and sheep breeds is also required.

4.1.1 National Wool Grading Standard

Greasy wool is purchased from farmers on the basis of the National Wool Grading Standard. This set of standards was initially developed in 1957 and revised in 1976. A new purchasing standard is being developed by the various interested parties but it is proving difficult to reach agreement on any major changes.

In practical terms, the National Standard currently in use is a very simple grading system. It essentially establishes a 2×2 matrix of wool categories with each of the resulting four possible categories in the matrix further subdivided into Grade I and Grade II (see Fig. 4.1). The basic four-cell matrix is determined on the basis of fibre diameter and degree of wool homogeneity (especially the presence or absence of hair fibres in the wool).

For fine and improved fine wool, the basic difference between Grade I and Grade II is that Grade I wool must have a staple length of between 6.0 and 7.9cm while Grade II requires that the wool have a staple length of between 4.0 and 5.9cm. The National Standard also specifies other differences between Grades I and II but staple length is the dominant differentiating factor.

Fine wool (≤ 25μm)	Semi-fine wool (> 25 to ≤ 40μm)	
Fine wool / \ Grade I Grade II	Semi-fine wool / \ Grade I Grade II	Homogeneous wool
Improved fine wool / \ Grade I Grade II	Improved semi-fine wool / \ Grade I Grade II	Mostly homogeneous with some heterogeneous fibres

Fig. 4.1 Categories of Wool Established by the Chinese National Wool Grading Standard

Similarly, for semi-fine and improved semi-fine wool, Grade I essentially means wool of a staple length of between 7.0 and 9.9cm and Grade II requires that the staple length be between 4cm and 6.9cm.

Although not depicted in Fig. 4.1, the Standard also provides for a "special grade" fine wool (must be of outstanding quality and have a staple length of 8cm or more) and a "special grade" semi-fine wool (must be of outstanding quality and have a staple length of 10cm or more).

Chinese statistics relating to wool which subdivide "wool" into categories invariably refer to:

(a) fine and improved fine wool;
(b) semi-fine and improved semi-fine wool; and
(c) "other" wool (i.e. coarse wool, local wool, etc.).

Furthermore, "fine and improved fine wool" will often be referred to simply as "fine wool" while "semi-fine and improved semi-fine wool" may be referred to as "semi-fine wool". It is usually extremely difficult to obtain statistics for "fine" as opposed to "improved fine wool" or for "semi-fine" compared with "improved semi-fine wool".

In China, "improved wool" refers to a wide range of wool types grown by crossbred sheep. In theory, if the crossbred sheep are part of a program of upgrading local sheep by mating them to purebred fine wool rams, the wool from the first three crosses (F_1, F_2 and F_3) using the fine wool rams is referred to as "improved fine wool". Wool from F_4 crosses is usually homogeneous enough (and sufficiently free of hair fibres) to be graded as "fine wool". Of course, when the upgrading program is aimed towards a semi-fine wool breed of sheep, the wool of the F_1, F_2 and F_3 is graded as "improved semi-fine wool" etc.

In practice, wool buyers do not know whether a particular fleece came from an F_1, F_2, F_3, F_4, (etc.) crossbred sheep. Consequently, the distinction between fine and improved fine (and between semi-fine and improved semi-fine) wool is not clear-cut. As discussed in some of the county case studies, while the major buying agency (the Supply and Marketing Cooperative) at the county level may be prepared to *estimate* the distribution of the county clip between fine and improved fine wool (or between semi-fine and improved semi-fine wool), this distinction is not made in the official production statistics recorded by county governments. At administrative levels above the county, it seems impossible to obtain even reliable unofficial *estimates* of the amount of fine as distinct from improved fine wool which is being grown.

The failure to distinguish between, for example, fine wool and improved fine wool, makes Chinese statistics in relation to raw wool production of limited value. The upgrading programs will have changed the proportion of the national clip coming from F_1, F_2, F_3, F_4 (etc.) sheep and hence the composition of the statistical aggregate "fine and improved fine wool". Since the kind of textile products which can be made from the wool is likely to vary greatly depending upon whether it comes from an F_1 or an F_4, for example, the changing composition of the statistical aggregate over time may be extremely important to anyone interested in investigating the domestic supply side of the Chinese wool industry in any degree of detail.

4.1.2 Wool Types and Sheep Breeds

As already mentioned, three basic domestic wool types are generally recognised in China: fine wool (and improved fine wool); semi-fine wool (and improved semi-fine wool); and "other" wool (coarse, local, etc.). Furthermore, as might be expected, sheep breeds are classified on the basis of the wool type they produce.

There are a relatively large number of recognised fine wool sheep breeds such as the Xinjiang Fine Wool, the Gansu Alpine Fine Wool, the North East Fine Wool, the Erdos Fine Wool, the Aohan Fine Wool, etc. All of these breeds of sheep would be broadly classified as being of the merino type and they are sometimes called the Xinjiang merino, Gansu Alpine merino, etc. However, the term "merino" does not occur in the official breed standards for Chinese fine wool breeds.

The term "Chinese Merino" was introduced in December 1985. It refers not to a distinct gene pool or unique breed but to a breed standard. Different breeding programs at the four nationally recognised fine wool sheep studs (Gongnaisi, Ziniquan, Gadasu and Chaganhua) have developed sheep which meet this standard (Longworth and Williamson, 1993).

There are also many recognised breeds which produce semi-fine wool. Similarly, there are many local breeds or strains of sheep which produce coarse or local wool.

Just as the distinction between fine wool and improved fine wool would seem to be extremely vague, there is also no clear-cut, unambiguous definition of a "fine wool sheep" as opposed to an "improved fine wool sheep". One explanation, not particularly helpful, is that a fine wool sheep is one that produces fine wool while an improved fine wool sheep is one which grows improved fine wool (and similarly in the case of semi-fine wool sheep).

Another much more restricted definition of a fine wool sheep is a sheep which has satisfied the breed standard for the relevant fine wool breed. On this basis, there are very few fine wool sheep in China; perhaps less than 1 million. It is commonly believed that starting with local breed ewes, it takes about six generations of crossing with good quality rams which have met the appropriate fine wool breed standard before the progeny are likely to have all the necessary traits to meet a breed standard for fine wool sheep.

Of course, somewhere between the F_1 and F_6 most of the progeny are producing reasonably homogeneous wool which is graded as "fine wool" rather than "improved fine wool". As pointed out in the previous section, the working rule in China is that on average this occurs with the F_4. That is, when fine wool rams which have met the breed standard are mated to local breed ewes, most of the F_1 to F_3 are called "improved fine wool sheep" while most of the F_4, F_5, etc. are classified as "fine wool" sheep.

According to this definition, there are probably about 6 to 8 million fine wool sheep in China. As discussed in Section 5.3.2 and elsewhere in this book, ACIAR Project No. 8811 researchers have not been able to obtain data on the number of "fine" as opposed to "improved fine wool sheep", except in relation to Chifeng City Prefecture in the IMAR and for Hebukesaier County in XUAR.

The total number of fine and improved fine wool sheep in China in 1991 was officially claimed to be 30.8 million or 28% of all sheep in the country. If the estimate of 6 to 8 million fine wool sheep is accepted, then there must be about 23 to 25 million improved fine wool sheep. These improved or crossbred sheep are indeed a very mixed flock.

4.2 Sheep Production Systems and Sheep-Raising Localities

The commercialisation of sheep raising both for mutton and for wool production has been a relatively recent development in China. The production of fine wool in particular has a less than forty year history. In 1950, there was virtually no wool of less than 25µm in average fibre diameter produced in China. The transition from traditional sheep-raising practices to commercial wool production systems is still occurring in many areas. In addition, there is enormous variation in the physical environment in which sheep are raised. Consequently, any description of sheep production systems in China will, of necessity, be a major simplification.

Broadly, there are three different types of sheep production systems: pastoral, agricultural, and mixed pastoral/agricultural. Chinese scientists and officials usually describe sheep-raising localities in terms of the predominant production system. As pointed out in Chapter 3, a high proportion (89%) of the sheep in China are in the 12 pastoral provinces and this book is almost entirely concerned with this part of China. Sheep are raised in the 12 pastoral provinces under all three production systems but in other parts of China, it is relatively rare to find sheep being raised under a pastoral production system.

Since the physical environment of a given locality is the major factor determining how sheep are raised, production systems are best described in relation to specific kinds of locations. Chinese specialists such as Tu (1992) divide the traditional sheep-raising areas in China into three different kinds of locality but also draw attention to the potential of certain new localities in southern China.

4.2.1 Traditional Sheep-Raising Localities

The three traditional groupings of sheep-raising localities in China are the pastoral, the mixed pastoral/agricultural and the agricultural areas. These areas do not correspond neatly with administrative units. For example, as mentioned above, while almost all of the pastoral areas are located in the 12 pastoral provinces, there are also significant areas within this part of China where sheep are raised under mixed and agricultural production systems. However, for the purposes of outlining sheep production systems, it is convenient to organise the discussion according to the three traditional groupings of localities.

- *Pastoral areas*

  Vast expanses of northern and north western China are extremely dry, hot in summer and very cold in winter. Much of this land is semi-desert. As indicated in

Section 3.3, a major portion of the usable native pasture in China is in the 266 pastoral and semi-pastoral counties shown in Plate 2 but there are also sizeable areas of natural grassland in other counties within the 12 pastoral provinces.

Before 1949, the relatively few people who lived on the rangelands of northern and north western China almost all belonged to ethnic groups which are now called minority nationalities. They were usually nomadic herders who moved their livestock substantial distances on an annual cycle. Following the establishment of the People's Republic of China in 1949, there have been enormous increases in both the human populations and the herbivorous animal populations in these parts of the country. Nowadays, most of the sheep-raising farmers live in permanent villages or hamlets but they still move their livestock from one grazing area to another on a seasonal basis. Sometimes the summer pastures are up to 80km from the village, and the farmer and some of his family follow the livestock and live for a few months in tents made of yak hair or coarse wool as in the "old days". However, there are few areas in China today where the grazing system could be described as "nomadic" in the same sense as before 1949.

The seasonal pasture grazing system offers a form of rotational grazing. In winter (November/April), the sheep are grazed near the permanent houses during the day and yarded and/or shedded at night. On extremely cold days, sheep may remain in the sheltered yard or shed all day. In most pastoral localities, this would occur on less than 30 days during the 5- to 6-month winter season.

In May/June and in late-September/October/early-November, the sheep are on the spring/autumn pastures. These pasture areas may be located between the winter and summer pastures but this is not necessarily the case.

Usually but not always, the summer pastures are in areas which are remote and at a high altitude relative to the farmer's place of residence. Sheep are grazed on these areas in July/August/September.

- *Mixed pastoral and agricultural areas*

    In those parts of northern and north western China where irrigation is possible (from underground aquifers or from rivers fed by melting snow) or where rainfall is higher (say above about 300mm per year), sheep are raised as a major sideline activity to cropping. According to Tu (1992), these areas cover about one-tenth of the total area of China, including some parts of the north east of China, North China Plain, the west of Henan, Shanxi, Gansu, Ningxia, Yunnan, Guizhou, and Sichuan.

    Usually the number of sheep raised by each household in these localities is much lower than in the pastoral areas. The general availability of forage materials, village wastes and agricultural by-products means that the sheep are frequently better fed than in the pastoral areas. On the other hand, the quality of husbandry is often lower than in the specialist sheep-raising areas.

    Unlike the situation in pastoral localities where each household (or a small group of households) shepherds its own flock while grazing, the sheep from many households in mixed pastoral/agricultural areas are often herded together as one flock by old men or young boys on a kind of wage basis. The sheep are shepherded around the irrigation areas during all four seasons, eating weeds and crop stubble, etc. as available. In some cases, there may be areas of natural pasture some distance from the village to which the sheep can be driven each day in certain seasons. The sheep belonging to each household are shedded near the house every night of the year and on very harsh days during the winter.

- *Agricultural areas*

Sheep are raised in densely populated agricultural areas in north east China, on the North China Plain, and in Jiangsu, Zhejiang, Shandong, Henan and Anhui Provinces. The climate in these agricultural areas is relatively warm with moderate rainfall, and the availability of roughages and other feedstuffs for the sheep is good. When Chinese officials suggest that the best scope for increasing sheep numbers (and wool production) exists in the agricultural areas, it is to these parts of China they are primarily referring. The agricultural areas in the three pastoral provinces (Liaoning, Jilin and Heilongjiang) of north east China are also considered to have the potential to raise significantly more sheep.

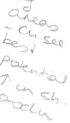

Owing to the pressure on land in these agricultural localities, there is little pasture land available for grazing. As in the mixed agricultural/pastoral areas, sheep are herded by hired labour most days of the year. They act as scavengers, eating weeds, agricultural waste, village refuse, etc. All the sheep are shedded or yarded every night and there may be a substantial component of "cut and carry" feeding involved. Each household usually only raises about three to five sheep. In many places, the quality of sheep husbandry is low and little attention is given to raising the sheep in a way which will enhance wool quality.

The potential for increasing the production of wool in agricultural areas seems to have been already grasped in Shandong Province where the output of wool increased sharply during the 1980s. As noted below in Section 4.3, Shandong is the only non-pastoral province which currently produces a significant quantity of wool (especially fine and improved fine wool).

4.2.2 Potential Sheep-Raising Localities in Southern China

In the early 1980s, it was suggested that there were large areas of underutilised pasture land suitable for sheep raising in southern China. According to a Central government survey, the area of pasture was 997 million mu (i.e. 66.5 million ha or 2.5 times the size of New Zealand). Most of the pasture land (850 million mu) is located on plateaus in mountainous areas. Generally these areas are cool and humid with an average rainfall of 900 to 1,800mm pa. Temperatures range from -8°C in winter to 28°C in summer. The climate in these areas is suitable for perennial grasses and legumes.

The Ministry of Agriculture and the old Ministry of Textile Industries developed the slogan "emigration of northern sheep to southern China" but by 1992 little real progress had been achieved. The project is constrained because: it has not been adopted by the Central government and consequently the financial support of the Central government is lacking; local governments in the project area have only supported the idea on a relatively small scale; and the farmers in the areas concerned are not technologically experienced in relation to sheep since sheep raising is not a traditional activity in these parts of China.

Specifically, the five provinces and the main prefectures or counties which constitute the area under discussion are:

Sichuan Province	(Fuling, Nanchong, Wanxian, Daxian and Liangshan Prefectures);
Hubei Province	(Lichuan, Hefeng, Changyang, Zaoyang, Zhongxiang, Xuanen, and Linfan Counties and Yichang and Louhan Cities);
Hunan Province	(Chengbu County);
Guizhou Province	(Xianming County); and
Yunnan Province	(Yongshan and Xuanlian Counties and Qujing and Zhaotong Prefectures).

Sheep scientists in China do not consider the emigration of northern fine wool sheep in great numbers to the southern pasture areas to be feasible. Instead, it is believed the areas are best suited to semi-fine wool breeds of sheep especially bred for the environment. In particular, imported sheep breeds such as the Romney and Corriedale seem to be well suited. Fine wool rams from the north can also be used to upgrade the local sheep. But, for the most part, these southern pasture areas are not considered suitable for the production of good quality fine wool.

Several substantial experiments have commenced financed by provincial and lower levels of local government. For example, in Hubei Province about ¥5.85 million has been invested in sheep purchases (639 Chinese Merino sheep from Xinjiang in 1988), the establishment of three sheep studs, setting up of AI stations, and the training of sheep technicians. These developments have been concentrated in Zaoyang, Zhongxiang and Xuanen Counties of Hubei Province.

The Australian International Development Assistance Bureau (AIDAB) financed a seven-year pasture development project in Yunnan Province which began in 1983. The sheep sub-program of this project which was located in Qujing Prefecture "expanded and extended the existing Yunnan sheep improvement and upgrading program ... and began development of a ram breeding scheme ... for the production of semi-fine wool sheep by crossing local ewes with imported exotic breeds" (Carrad et al., 1989).

Sichuan Province is another area particularly favoured by the former Ministry of Textiles. There are nine counties which are richly endowed with pasture resources in Fuling, Nanchong, Wanxian, Daxian and Liangshan Prefectures of Sichuan. A major project commenced in 1986 to establish a sheep and wool production base in this part of Sichuan. In the first two years 1986 and 1987, ¥2.3 million was invested and 2,713 stud sheep (males and females) were imported both from elsewhere in China (Xinjiang Fine Wool and Chinese Merinos from Xinjiang) and from Australia and New Zealand (Romney and Corriedale sheep).

Although there is no prospect of the required funds being available, some Chinese spokesmen claim that 150-200 million mu of improved pasture will be planted by the year 2000 and that the southern pasture areas would then be capable of supporting 50-67 million sheep. These unrealistic claims are extended further by suggesting that these sheep could yield 5kg per head of greasy wool. On this basis, the southern pasture areas have the potential to produce between 250,000 and 335,000 tonne of greasy wool per year (i.e. well in excess of the total national wool clip in 1991).

There is no real basis for such optimistic claims. The development of the pastures in southern China for sheep production is progressing very slowly and is likely to concentrate much more on the production of mutton and semi-fine wool than on fine wool. Both for technical and economic reasons, beef cattle may eventually prove to be much better suited to some of the southern pasture areas than sheep.

4.3 Sheep and Wool in the Pastoral Provinces

The 12 pastoral provinces/autonomous regions were discussed in Chapter 3 and the importance of sheep and wool in this part of China was demonstrated in general terms. This section expands the discussion in terms of the three major types of sheep found in China. As in Chapter 3, one of the objectives is to demonstrate the importance of the three case study provinces (IMAR, Gansu and XUAR) considered in Part II of this book.

4.3.1 Sheep Numbers

Table 4.1 presents data for 1981 and 1991 on the sheep population classified by major wool type in the 12 pastoral provinces/autonomous regions and for China as a whole. The full time series 1981 to 1991 corresponding to Table 4.1 for IMAR, Gansu and XUAR are presented and discussed in Chapters 5, 6 and 7 respectively. Unfortunately, as foreshadowed above, it has not been possible to obtain provincial data separating fine/improved fine wool sheep or semi-fine/improved semi-fine wool sheep.

- *Total sheep*

There were 98.6 million sheep in the 12 pastoral provinces in 1991 compared with 94.3 million in 1981. As already mentioned, about 89% of all sheep in China were in these provinces in 1991, while the corresponding figure was 86% in 1981.

Apart from indicating the overwhelming and increasing importance of these 12 provinces in terms of sheep numbers, these data also suggest that the total number of sheep in China has not increased greatly during the 1980s. However, as Fig. 4.2 demonstrates, the 1980s represents an unusual decade in terms of sheep numbers. The Chinese sheep flock grew steadily from 1949 to 1981, declined significantly in the first half of the 1980s and then expanded again to a new peak in 1989. (The gap in the sheep population series between 1965 and 1970 in Fig. 4.2 is a result of the "Cultural Revolution".)

The decline in the sheep population from 1981 to 1985 corresponds with the introduction of the household production responsibility system (HPRS) in the pastoral region. Initially, in some areas households "cashed in" some of their allotted sheep for much the same reasons as in the early 1950s (see Section 3.5.2). No-one knew how long the HPRS would last. However, as the farmers gained confidence in the new policies and wool prices began to rise after 1985, sheep numbers grew rapidly in some areas. In other parts of the pastoral region, as discussed in Section 5.2.5, cashmere prices rose much more than wool prices and there was a big swing to cashmere goats in the second half of the 1980s.

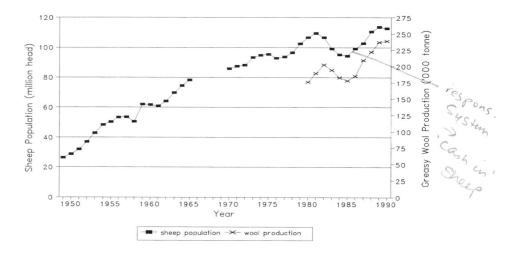

Fig. 4.2 Total Sheep Numbers and Wool Production in China, 1949 to 1991

Table 4.1 Number of Sheep by Wool Type in each of the 12 Pastoral Provinces/Autonomous Regions, 1981 and 1991

Province or autonomous region	Year	Total number of sheep	Fine wool and improved fine wool sheep		Semi-fine wool and improved semi-fine wool sheep		Other sheep	
			No.	Prop. of total no. of sheep in the province	No.	Prop. of total no. of sheep in the province	No.	Prop. of total no. of sheep in the province
		('000)	('000)	(%)	('000)	(%)	('000)	(%)
Hebei	1981	3,749	996	26.6	334	8.9	2,419	64.5
	1991	4,775	1,339	28.0	820	17.2	2,616	54.8
Shanxi	1981	3,569	303	8.5	183	5.1	3,083	86.4
	1991	3,907	576	14.7	269	6.9	3,062	78.4
IMAR	**1981**	**19,588**	**4,520**	**23.1**	**5,095**	**26.0**	**9,973**	**50.9**
	1991	**20,149**	**9,488**	**47.1**	**3,757**	**18.6**	**6,904**	**34.3**
Liaoning	1981	1,742	631	36.2	468	26.9	643	36.9
	1991	1,742	889	51.0	553	31.7	300	17.2
Jilin	1981	1,559	596	38.2	815	52.3	148	9.5
	1991	2,134	1,811	84.9	322	15.1	1	0.0
Heilongjiang	1981	3,044	963	31.6	1,720	56.5	361	11.9
	1991	2,671	928	34.7	1,438	53.8	305	11.4
Sichuan	1981	3,819	569	14.9	91	2.4	3,159	82.7
	1991	3,535	242	6.8	367	10.4	2,926	82.8
Tibet*	1981	13,714	85	0.6	142	1.0	13,487	98.3
	1990	11,107	700	6.3	120	1.1	10,287	92.6
	1991	11,463	--	--	--	--	11,463	100.0
Gansu	**1981**	**8,546**	**708**	**8.3**	**854**	**10.0**	**6,984**	**81.7**
	1991	**8,088**	**1,996**	**24.7**	**569**	**7.0**	**5,523**	**68.3**
Qinghai	1981	14,526	97	0.6	2,727	18.8	11,702	80.6
	1991	14,521	138	1.0	3,169	21.8	11,214	77.2
Ningxia	1981	2,015	137	6.8	31	1.5	1,847	91.7
	1991	1,903	93	4.9	--	--	1,810	95.1
XUAR**	**1981**	**18,408**	**8,637**	**46.9**	--	--	**9,771**	**53.1**
	1990	**23,814**	**10,404**	**43.7**	--	--	**13,410**	**56.3**
	1991	**23,706**	**9,245**	**39.0**	**2,380**	**10.0**	**12,081**	**50.9**
Sub-total	1981	94,279	18,242	19.3	12,460	13.2	63,577	67.4
	1990	100,043	28,570	28.6	11,510	11.5	59,963	59.9
	1991	98,594	26,745	27.1	13,644	13.8	58,205	59.0
All China	1981	109,466	22,687	20.7	15,325	14.0	71,454	65.3
	1990	112,816	32,518	28.8	14,562	12.9	74,736	66.2
	1991	110,855	30,835	27.8	16,629	15.0	63,391	57.2

*No statistical data for fine and semi-fine wool sheep in 1991.
**Number of fine wool and improved fine wool sheep includes that of semi-fine and improved semi-fine wool sheep prior to 1991.
Sources: Statistics of Chinese Animal Husbandry, 1949-1989. Agricultural Yearbook of China, 1992

One of the most important issues of concern in relation to the Chinese sheep and wool industry is whether the long-run upward trend in sheep numbers so clearly evident in Fig. 4.2 can continue. Since almost 90% of the sheep are in the 12 pastoral provinces, the capacity for further expansion in the total flock largely depends on the potential for more sheep in these provinces. The evidence presented later in this book suggests that current rangeland stocking rates, taking into account all types of grazing animals, exceed long-term sustainable rates in almost all pastoral areas. It will be virtually impossible, therefore, for the sheep flock in the pastoral areas of China to continue to grow unless there are compensating reductions in the numbers of other grazing animals. Indeed, as degradation of the pastures reduces the resource base supporting pastoral activities, the reduction in the numbers of animals other than sheep will probably need to be even greater proportionately. That is, the total stocking rate will need to be reduced.

Table 4.2 has been derived from Table 4.1 and shows the proportion of the Chinese sheep flock by wool type in each pastoral province in 1991. In terms of the total number of sheep, IMAR, Tibet, Gansu, Qinghai and XUAR are easily the most significant provinces/autonomous regions. Total sheep numbers, however, actually declined between 1981 and 1990/1991 in three of these five areas (Table 4.1). Only in IMAR and in XUAR did sheep numbers increase during this period.

Table 4.2 Proportion of Chinese Sheep Flock by Wool Type in each of the 12 Pastoral Provinces/Autonomous Regions, 1991

Province or autonomous region	Total sheep	Fine wool and improved fine wool sheep	Semi-fine wool and improved semi-fine wool sheep	Other sheep
	(%)	(%)	(%)	(%)
Hebei	4.3	4.3	4.9	4.1
Shanxi	3.5	1.9	1.6	4.8
IMAR	**18.2**	**30.8**	**22.6**	**10.9**
Liaoning	1.6	2.9	3.3	0.5
Jilin	1.9	5.9	1.9	0.0
Heilongjiang	2.4	3.0	8.6	0.5
Sichuan	3.2	0.8	2.2	4.6
Tibet*	10.3	2.3	0.8	16.7
Gansu	**7.3**	**6.5**	**3.4**	**8.7**
Qinghai	13.1	0.4	19.1	17.7
Ningxia	1.7	0.3	--	2.9
XUAR	**21.4**	**30.0**	**14.3**	**19.1**
Sub-total	88.9	86.7	82.0	91.8
All China	100.0	100.0	100.0	100.0

*No statistical data was available for fine and semi-fine wool sheep in Tibet in 1991. Therefore, the proportion of sheep in each of the three wool types has been assumed to be the same as in 1990.

- *Fine wool and improved fine wool sheep*

Of the 30.8 million sheep in this category in 1991, 87% were in the 12 pastoral provinces/autonomous regions according to Tables 4.1 and 4.2.

The proportion of sheep of each wool type in the total flock for the relevant administrative unit is shown in Table 4.1. For instance, Jilin has the highest proportion of its flock (85%) in the fine and improved fine wool category in 1991. Liaoning is the only other province in which more than half the flock is in this category. For the 12 pastoral provinces/autonomous regions as a whole, 27% of the sheep were in the fine and improved fine wool category in 1991. Unfortunately, prior to 1991 the semi-fine and improved semi-fine wool sheep were included in the statistics for fine and improved fine wool sheep in XUAR. Therefore, the sheep numbers for these two wool types shown in Table 4.1 for 1981 and 1991 are not comparable for XUAR, for the 12 pastoral provinces/autonomous regions, or for all China. Nevertheless, it would seem that the proportion of fine and improved fine wool sheep both in the 12 pastoral provinces/autonomous regions and in China as a whole increased from around 20% in 1981 to about 27% in 1991.

Similarly, although the 1991 data are not strictly comparable with the 1981 data, it seems that the number of fine and improved fine wool sheep in the 12 pastoral provinces/autonomous regions has increased from around 18 million to about 27 million (a 50% increase) between 1981 and 1991. The corresponding estimate for China as a whole is an increase from about 23 million to approximately 31 million.

These estimates confirm that an increasing proportion of the total flock has been mated to fine wool rams and consequently there are an increasing number of sheep in the "improved fine wool" category. Whether they are F_1, F_2, etc. is not recorded in the statistics. Nor is it possible, as already mentioned, to get reliable estimates of how many sheep are actually in the "fine wool" rather than the "improved fine wool" category.

The data in Table 4.2 demonstrate that, of the 12 pastoral provinces/autonomous regions, IMAR and XUAR and to a lesser degree Gansu dominate the fine and improved fine wool sheep sector. However, two other pastoral provinces (Hebei and Jilin) rapidly increased the number of sheep in this category during the 1980s, and both of these areas are now challenging Gansu for third ranking among the pastoral provinces in terms of the number of fine and improved fine wool sheep.

Shandong, as mentioned earlier, is the only province not included in the 12 pastoral provinces/autonomous regions which has a significant number of fine and improved fine wool sheep. In fact, after IMAR, XUAR, and Gansu, Shandong with 1.827 million head in 1990 had the fourth largest provincial flock of this type of sheep in China.

- *Semi-fine wool and improved semi-fine wool sheep*

Of the 16.6 million sheep in China in 1991 in the semi-fine and improved semi-fine wool category, 13.6 million (82%) were in the 12 pastoral provinces. IMAR, Heilongjiang, Qinghai and XUAR dominate this sector of the sheep industry (see Table 4.2) but IMAR and Heilongjiang (and Jilin) recorded large decreases in the numbers of this type of sheep over the 1981 to 1991 period (see Table 4.1). These semi-fine wool sheep may have been replaced with fine wool sheep (perhaps by mating to fine wool rams) or with other grazing animals (especially cashmere goats).

As already mentioned, owing to the change in the XUAR statistics, the 1981 and 1991 figures in Table 4.1 for the 12 provinces/autonomous regions and for all China are only roughly comparable. Nevertheless, it would seem that the total number of semi-fine and improved semi-fine sheep in the pastoral region and in China as a whole increased only a little during the 1980s.

- *"Other" sheep*

This residual category includes a large number of different types of sheep. For the most part, these sheep are raised for their meat (and fat) and hides rather than for wool production. Their wool is extremely coarse but it provides the traditional raw material for the making of carpets and other floor coverings, tents and felt. While most of this wool is utilised locally, the best of it finds a ready market in the commercial carpet-making industry for which parts of China's pastoral region are famous.

In 1991, 57% of the national flock or 63 million sheep were in this very mixed group. Nine out of every ten sheep in this category were in the 12 pastoral provinces/autonomous regions.

4.3.2 Wool Production

Raw wool production in the 12 pastoral provinces/autonomous regions and for all China is shown in Table 4.3 for 1981 and 1991. Shandong is the only province in China producing significant quantities of wool which is not in the list of 12 pastoral provinces/autonomous regions. In fact, wool production (especially fine and improved fine wool) has expanded rapidly in Shandong and in terms of output it is the third biggest producer after IMAR and XUAR. In 1991, Shandong produced a total of 21,574 tonne of raw wool, of which 10,476 tonne was fine and improved fine and 6,660 tonne was semi-fine and improved semi-fine.

- *Total wool*

The graphs in Fig. 4.1 demonstrate that while wool output declined as sheep numbers dropped in the early 1980s, wool production actually rose significantly during the 1980s decade as a whole. By 1991, the total Chinese output of raw or greasy wool was 239,607 tonne compared with 189,062 tonne in 1981, an increase of 27% (Table 4.4). Since total sheep numbers increased by only 1% over the same time period, there was a remarkable improvement in the average wool cut per sheep during the 1980s.

The 12 pastoral provinces/autonomous regions accounted for 84% of the Chinese wool clip in 1991 (and Shandong made up another 9%). Six of the pastoral provinces (IMAR, XUAR, Qinghai, Gansu, Hebei and Heilongjiang) produced in excess of 10,000 tonne of greasy wool in 1991.

- *Fine wool and improved fine wool*

The data in Table 4.4 suggest that fine and improved fine wool output in China rose by 45% between 1981 and 1991. However, the all China total for fine and improved fine wool in 1981 is inflated because it includes the semi-fine and improved semi-fine wool grown in the XUAR in that year. If it is assumed that the proportion of this type of wool wrongly included in the fine and improved fine category in 1981 was the same as in 1991, a conservative assumption, then fine and improved fine wool production increased by over 56% between 1981 and 1991.

The total output of fine and improved fine wool in China in 1991 was 108,613 tonne (greasy weight basis) which was approaching half of all the wool grown in the country in that year. Nevertheless, if it is assumed that this wool has an average clean yield of about 42%, then the total amount of clean scoured wool with an average fibre diameter of $\leq 25\mu m$ produced in China in 1991 was only about 46,000 tonne. Furthermore, for the reasons already discussed, perhaps 70 to 75% of this wool is in the "improved fine" or crossbred category rather than being genuine "fine" wool.

Table 4.3 Wool Production by Wool Type in each of the 12 Pastoral Provinces/Autonomous Regions, 1981 and 1991

Province or autonomous region	Year	Total raw wool prod.	Fine wool and improved fine wool		Semi-fine wool and improved semi-fine wool		Other	
			Amount	Prop. of total raw wool production in the province	Amount	Prop. of total raw wool production in the province	Amount	Prop. of total raw wool production in the province
		(t)	(t)	(%)	(t)	(%)	(t)	(%)
Hebei	1981	5,789	2,994	51.7	638	11.0	2,157	37.3
	1991	11,941	4,881	40.9	2,616	21.9	4,444	37.2
Shanxi	1981	3,532	633	17.9	356	10.9	2,543	72.0
	1991	5,670	2,402	42.4	706	11.5	2,562	45.2
IMAR	**1981**	**47,399**	**16,044**	**33.8**	**14,012**	**29.6**	**17,343**	**36.6**
	1991	**59,701**	**32,672**	**54.7**	**12,077**	**20.2**	**14,952**	**25.0**
Liaoning	1981	5,143	2,265	44.0	1,633	31.8	1,245	24.2
	1991	7,196	3,480	48.4	2,052	28.5	1,664	23.1
Jilin	1981	5,209	2,224	42.7	2,204	42.3	781	15.0
	1991	9,124	7,847	86.0	1,273	14.0	4	0.0
Heilongjiang	1981	10,172	3,117	30.6	7,055	69.4	--	--
	1991	10,766	4,281	39.8	6,389	59.3	96	0.9
Sichuan	1981	2,642	572	21.7	199	7.5	1,871	70.8
	1991	2,798	207	7.4	484	17.3	2,107	75.3
Tibet	1981	8,420	98	1.2	96	1.1	8,226	97.7
	1991	8,064	20	0.2	1,154	14.3	6,890	85.4
Gansu	**1981**	**8,734**	**1,489**	**17.0**	**1,583**	**18.1**	**5,662**	**64.8**
	1991	**15,001**	**4,477**	**29.8**	**1,318**	**8.8**	**9,705**	**62.6**
Qinghai	1981	16,694	400	2.4	4,471	26.8	11,823	70.8
	1991	17,736	323	1.8	6,319	35.6	11,094	62.6
Ningxia	1981	2,545	343	13.5	62	2.4	2,140	84.1
	1991	3,371	302	9.0	--	--	3,069	91.0
XUAR*	1981	34,099	27,961	82.0	--	--	6,138	18.0
	1991	49,578	28,534	57.6	11,740	23.7	9,304	18.8
Sub-total	1981	150,378	58,140	38.7	32,309	21.5	59,929	39.9
	1991	200,946	89,426	44.5	46,128	23.0	65,392	35.5
All China	1981	189,062	74,720	39.5	39,996	21.2	74,346	39.3
	1991	239,607	108,613	45.3	55,839	23.3	75,155	31.4

* Prior to 1991, semi-fine wool was included in the fine wool category.
Sources: Statistics of Chinese Animal Husbandry, 1949-1989
Agricultural Yearbook of China, 1992

The total greasy wool production in Australia in 1990/91 was 989,600 tonne. In that year, 92.3% of all wool sold at auction in Australia had an average fibre diameter of ≤25μm. If the average clean yield of this wool was 65% (which was the average for the whole clip), then on a clean scoured basis Australia produced 593,710 tonne of fine wool (as defined in China) in the 1990/91 season. On this basis, the output of wool of ≤25μm average fibre diameter in Australia in 1990/91 was almost 13 times greater than the output of fine and improved fine wool in China in that year.

Table 4.4, which has been derived from Table 4.3, indicates that most of the fine and improved fine wool is grown in IMAR and XUAR. Other pastoral provinces which produced more than 4,000 tonne of fine and improved fine wool in 1991 were Hebei, Jilin, Heilongjiang and Gansu.

Table 4.4 Proportion of Chinese Wool Production by Wool Type in each of the 12 Pastoral Provinces/Autonomous Regions, 1991

Province or autonomous region	Total wool	Fine wool and improved fine wool	Semi-fine wool and improved semi-fine wool	Other wool
	(%)	(%)	(%)	(%)
Hebei	5.0	4.5	4.7	5.9
Shanxi	2.4	2.2	1.3	3.4
IMAR	**25.0**	**30.1**	**21.6**	**19.9**
Liaoning	3.0	3.2	3.7	2.2
Jilin	3.8	7.2	2.3	0.0
Heilongjiang	4.5	3.9	11.4	0.1
Sichuan	1.2	0.2	0.9	2.8
Tibet	3.4	0.0	2.1	9.2
Gansu	**6.3**	**4.1**	**2.4**	**12.2**
Qinghai	7.4	0.3	11.3	14.8
Ningxia	1.4	0.3	--	4.1
XUAR	**20.7**	**26.3**	**21.0**	**12.4**
Sub-total	83.9	82.3	82.6	87.0
All China	100.0	100.0	100.0	100.0

Over the 1981 to 1991 period, the production of fine and improved fine wool more than doubled in Shanxi, IMAR, Jilin and Gansu. Unfortunately, as with sheep numbers, the 1991 wool production statistics by wool type for XUAR are not comparable with the data for earlier years. In 1991, a significant quantity (11,740 tonne) of semi-fine and improved semi-fine wool was "correctly" listed in the XUAR statistics for the first time. Previously, wool of this type had been included in the fine and improved fine wool category. Nevertheless, after adjusting for this change in the statistics, it would seem that the production of fine and improved fine wool in XUAR did not expand nearly as fast as in IMAR, the other major producer of this type of wool.

- *Semi-fine and improved semi-fine wool*

According to Table 4.3, national production of this type of wool expanded from 39,996 tonne in 1981 to 55,839 tonne in 1991. However, most of this increase was due to the change in the statistical definitions in XUAR in 1991 which have been referred to above several times and which raised semi-fine and improved semi-fine wool output in the XUAR from nil to 11,740 tonne in one year. About 83% of the national production of semi-fine and improved semi-fine wool in 1991 came from the 12 pastoral provinces/autonomous regions. The major growing areas in 1991 were IMAR and XUAR but Heilongjiang and Qinghai were also significant producers.

- *"Other" wool*

As with the residual category for sheep, this residual wool category includes a very large range of wool types. Most of this wool is processed within the household or village where it is grown. The better quality coarse wool, as already noted, is handled by speciality carpet-making enterprises.

The total amount of "other" wool produced in China in 1991 was 75,155 tonne, of which 87% was grown in the 12 pastoral provinces/autonomous regions. The national output of this kind of wool increased only a little during the 1980s.

4.4 Organisation of Production

While precise data are not available, the best available estimate is that around 10% (or 23,500 tonne) of the raw wool grown in China in 1990 was produced by State farms. Virtually all the rest of the national wool clip was grown by private households. While there are some collectively organised production units, they are not major producers of wool.

The private sheep farmer operates under the so-called household production responsibility system (HPRS). This system and the associated property rights reforms were introduced throughout rural China from 1978. These major institutional changes have been well documented in relation to the agricultural areas. However, the background to the reforms and the precise nature of the HPRS in pastoral areas have received almost no attention from Western researchers.

State farms are production units owned and operated by the State. There are many kinds of State farm because they are controlled by different vertical hierarchies and administered by different levels of government. Consequently, they vary significantly in the way they are organised and resourced. The usually better managed and larger State farms are controlled at the provincial or prefectural level. State farms administered at the county level or below are normally the smaller farms. Traditionally, State farms at all levels enjoyed significant State subsidies and preferential access to scarce resources such as machinery, fuel and fertiliser. However, nowadays State farms must support themselves and compete for resources. Consequently, many State farms and in particular those located in pastoral areas have "fallen upon hard times". Indeed, some are bankrupt and many have been forced to radically restructure their activities. One common approach has been to discontinue paying an annual salary to the sheep-herding households on the farm. Instead, the sheep have been contracted out (or leased) to the households under a kind of HPRS. Under these arrangements, the management of the State farm usually retains control over key decisions such as mating and disease control, but the household has responsibility for day-to-day care of the sheep. The wool and surplus sheep are usually sold to the State farm at agreed prices. While the HPRS now in place on most State farms provides considerable independence for the sheep-raising households, they have much less decision-making freedom than private households not living on State farms.

4.4.1 Private Sheep Raising

In late 1978, the Central Committee of the Chinese Communist Party formally approved the implementation of a set of major reforms. One major outcome of these decisions was the introduction of the household production responsibility system (HPRS) in rural areas. Although it was probably not fully anticipated at the time the critical Central Committee decisions were made, the introduction of the HPRS quickly led to the breakup of almost all the communes in rural China and to the privatisation of many aspects of agricultural and pastoral production.

In the pastoral areas, the HPRS was implemented in different ways and at different times in different places. In general, as briefly mentioned in Section 3.5.6, the HPRS was introduced in three steps. First, the livestock owned by the commune were assigned on a kind of rental basis to the households which constituted the commune. Second, the livestock were sold to the households (i.e. they became private property). Third, the land owned by the collectives under the commune was contracted out to the private households (i.e. the village collective/cooperative which replaced the commune administered the leasing of the land to private households).

While the three-step procedure just described captures the essence of the changes in ownership and control created by the introduction of the HPRS in pastoral areas, it is a gross over-simplification. In fact, there were many variations, and some of these modifications together with the implications of these "adjustments to local conditions", as they are called in China, are explored in the provincial and county case studies below.

Selling the animals back to the households reversed the collectivisation of livestock which occurred in 1958/59 (Section 3.5.3). In many pastoral areas, the livestock-raising farmers had embraced the commune concept in the later 1950s as a major step towards the utopian conditions which they were told would eventually come about (Khan, 1984). Indeed, without widespread popular support it is difficult to understand the speed with which the commune system was established. (See, for example, the data for IMAR in Table 3.9.) However, by the early 1980s there must have been widespread dissatisfaction with the commune system because it disappeared almost as quickly as it had been created.

The actual criteria used to disperse the commune flocks to the private households varied somewhat from place to place but were usually similar to those adopted in the IMAR and described in detail in Section 5.5.1. In principle, the households were supposed to purchase their allotted livestock from the commune. However, as pointed out in Section 5.5.1, farmers in some pastoral areas refused to pay for the livestock because they had been forced to "donate" their animals to the commune little more than 20 years earlier. In other areas, book entries were made recording that the households owed their respective village collectives the purchase price of the animals. Whether these loans were actually ever repaid is another question. Despite the obvious problems, privatisation of livestock occurred relatively quickly and smoothly almost everywhere in the pastoral region.

Contracting out the pasture land, on the other hand, has proven to be much more difficult. For many reasons which are explained in the case studies, even in counties where a high proportion of the rangeland has been assigned to households, most of it is still grazed in common.

Despite the obstacles to creating any genuine exclusive rights to the use of designated areas of pasture in many situations, private households now raise sheep under what is called a "double contract" HPRS. They have acquired ownership of the livestock from the collective and in return they are responsible for meeting certain annual obligations to the collective (in the form of welfare taxes etc.). They have also entered into contracts with the collective, which make them responsible for the use of specified areas of pasture. These contracts may or may not involve a "user pays principle"; they may or may not specify the length of time covered by the contract; they may or may not be inheritable; etc. In return for accepting these responsibilities, the households are reasonably free to decide how they will use the resources available to them to achieve their private goals.

Under the "double contract" HPRS, ownership and control over the basic means of production is largely in the hands of private production units. The households are now primarily responsible for the generation of their own income and, ultimately, for their own welfare and survival. The extreme climatic conditions which can occur in most pastoral areas together with present marketing uncertainties mean that a high proportion of the pastoral households face a precarious future, not only in an economic sense but also in a very real physical sense.

4.4.2 State Farms

While State farms produce only about 10% of the total greasy wool grown in China, most of this wool is in the fine and improved fine wool category. Perhaps about one-fifth of the almost 109,000 tonne of fine and improved fine wool produced in 1991 came from State farms. Furthermore, a much higher proportion of the sheep on State farms are genuine fine wools (rather than crossbred sheep being upgraded to genuine fine wools) than would be the case for the privately owned sheep. Therefore, it is probable that over half the genuine fine wool grown in China is produced on State farms. A major proportion of the genuine fine wool produced by the State farm sector is produced by the fine wool sheep studs which are all State farms.

There are wool-growing State farms in all the pastoral provinces except Tibet. The XUAR has easily the largest State farm wool clip, estimated to have been over 16,500 tonne in 1990 or one-third of the total output of wool in XUAR in that year. In 1990, State farms probably produced around 2,700 tonne in the IMAR, 750 tonne in Heilongjiang, 530 tonne in both Qinghai and Jilin, and 350 tonne in both Hebei and Gansu. In other provinces, the wool output from State farms did not exceed 150 tonne.

Commonly, State farms in China are divided into five categories, only two of which include a significant number of wool-growing farms.

- *Reclamation State farms*

In the 1950s, a large number of State farms were established to reclaim wilderness areas and other underutilised land for agriculture, animal husbandry and forestry purposes. Initially, these farms were administered directly by the Ministry of State Farms and Land Reclamation in Beijing. In the 1960s, there was a decentralisation of control over these State farms. Some of the larger farms became the responsibility of provincial or prefectural authorities, but most of the smaller farms were handed over to county-level administrations.

At about the same time, the Ministry of State Farms and Land Reclamation ceased to be a separate ministry and instead became a Department within the Ministry of Agriculture. This Department, which has the same name as the old ministry, still controls the network of reclamation farms although much less directly than in earlier times except in the case of the XUAR (see below).

At provincial and lower levels of administration, the authority responsible for the reclamation farms, while vertically responsible to the Department of State Farms and Land Reclamation in Beijing, is organised differently in various parts of the country. In the IMAR, for example, there is a State Farms and Land Reclamation Bureau within the Agriculture and Animal Husbandry Commission, while in the XUAR and Heilongjiang the reclamation farms are administered by the Production and Construction Corps.

After the establishment of New China in 1949, the People's Liberation Army personnel garrisoned in remote areas such as the XUAR and Heilongjiang Province were ordered to transfer to civilian life where they were located at the time. The troops set about developing wasteland for agriculture, animal husbandry and forestry, initially to feed, clothe and house themselves, but in the longer term to facilitate an influx of new settlers. These civilian undertakings were organised under what has become known as the Production and Construction Corps or PCC.

In the 1950s, the PCC State farms were set up under the control of the Ministry of State Farms and Land Reclamation. As mentioned earlier, this Ministry has become the State Farms and Land Reclamation Department under the Ministry of Agriculture. At present, this Department directly controls the 179 PCC farms in XUAR through the Xinjiang General Bureau of State Farms and Land Reclamation.

Other than in XUAR, the PCC State farms and other reclamation farms administered by the Ministry of Agriculture in Beijing are for the most part under dual direction. Normally, the Central government is concerned with the longer-term development of these farms and assists them to obtain supplies of important inputs such as machinery, fuel and fertiliser. The provincial authorities responsible for the farms, on the other hand, are more concerned with the annual programs on the farms and with day-to-day management.

The fact that the PCC State farms in XUAR are still directly controlled by Beijing is important in relation to the development of the Chinese wool industry. The Xinjiang General Bureau of State Farms, which controls the PCC farms at the provincial level, and its vertical "parent" the State Farms and Land Reclamation Department in Beijing, are in direct competition with the Xinjiang Animal Husbandry Bureau and its "parent" the Animal Husbandry and Veterinary Science Department in Beijing. Both these Departments are within the Central government Ministry of Agriculture. Therefore, within this Ministry there are two vertical hierarchies competing for resources to develop their respective wool-growing State farm networks in XUAR. Ministry officials, no doubt, attempt to share the available resources (including aid projects from Australia and elsewhere) in a more or less equitable fashion. Nonetheless, aid donors need to recognise that assistance to one of these vertical hierarchies is not likely to have much impact on the other. For example, the current AIDAB-funded China-Australia Sheep Research Project centred on the Academy of Animal Sciences under the Xinjiang Animal Husbandry Bureau is unlikely to have any impact on wool production within the PCC State farm network. The two separate vertical bureaucracies in XUAR which control wool growing in that autonomous region are discussed in more detail in the XUAR case study (Chapter 7).

Available data indicate that the network of reclamation farms in China produced about 12,500 tonne of raw wool in 1990. Of this, 8,545 tonne was grown by the PCC farms in the XUAR and a further 2,200 tonne came from reclamation farms in the IMAR, while farms of this type in Heilongjiang contributed 500 tonne.

- *"Conventional" State farms*

Other departments within the Ministry of Agriculture besides the State Farms and Land Reclamation Department are also responsible for State farms. Usually the main activity on these State farms determines to which part of the Ministry the farm in question belongs. For example, farms principally concerned with sheep and wool are under the Animal Husbandry and Veterinary Science Department in Beijing and are administered by the Animal Husbandry Bureaus at lower levels of government. While many agricultural State farms (within the Agricultural Bureau hierarchy) and forestry State farms (in the Forestry Bureau system) also raise sheep, these farms are relatively unimportant producers of wool.

There are a large number of these State farms scattered throughout China. The larger and better resourced are administered at the provincial or prefectural level and many of these farms are concerned with research and development activities of various kinds. Except for the studs within the PCC network, all the important fine wool sheep studs are provincial- or prefectural-level State farms in the Animal Husbandry network. Some county-controlled State farms are also sheep studs but most are referred to as ram-breeding stations. There are also township-administered State farms but these are usually relatively small commercial operations.

According to the limited information available, there are 141 sheep-raising State farms within the Animal Husbandry Bureau network. Many of these State farms are sheep studs or ram-breeding stations but some are commercial wool-growing farms. In total, these farms grew about 10,000 tonne of greasy wool in 1990, of which over 8,000 tonne came from State farms in the XUAR.

- *Other State farms*

The three other categories of State farm are PLA farms, Public Security farms, and farms established by overseas Chinese. The People's Liberation Army (PLA) did not hand over all of its State farms to the State Farm and Land Reclamation Department in the Ministry of Agriculture. It still operates a significant number of farms as business ventures. In recent years, the PLA has been encouraged to train its officers and soldiers in skills and techniques which are applicable in civilian life as well as useful to the PLA. The PLA State farms, therefore, perform a useful role in this regard. While it has been difficult to obtain data, it does not seem that the PLA farms are significant producers of wool.

The Public Security Bureau operates a number of State farms which are correction centres or jails in which the prisoners are required to perform agricultural work. In the past, many State farms in remote areas were used for this purpose even though they were controlled by the Ministry of Agriculture. It is not known precisely to what extent the Public Security Bureau farms contribute to the national wool clip. However, it is not likely to be a significant contribution.

There are a number of State farms, largely in southern China, which have been established by overseas Chinese who have returned to the mainland and set up State farming ventures. These farms, which do not produce wool, are administered through the Office of Overseas Chinese.

4.5 Wool Marketing

Historically, wool has not been a major commercial commodity in China. After the establishment of the People's Republic of China (PRC) in 1949, commodities were classified for purposes of market control. Wool remained a free market commodity until 1956 when it became a category II commodity. Commodities in category II were subject to the Central government plan, and the National Price Bureau was responsible for setting the State prices (Sicular, 1988). Specific production quotas were set for these commodities, and these quotas were procured by the State at State purchase prices. Excess production was normally also sold to the State procurement agency at negotiated prices, in practice usually at the same price as paid for quota production.

This situation existed for wool until 1985 when the Central government suspended the category II classification for wool and announced it would allow provincial governments to set State purchase prices for wool and/or to develop a free market for wool in the areas under their jurisdiction. At the same time, the Central government advised the governments of the four western provinces of IMAR, XUAR, Gansu and Qinghai to develop their wool textile industries by integrating these industries more closely with the wool production sector in their province. This was the so-called "self-produce, self-process, self-sell" policy.

Following the Central government's decision in 1985 to delegate control over wool marketing to the provinces, some provinces such as Gansu and Liaoning declared their wool markets to be free but others such as IMAR and XUAR refused to allow free trade in wool. These decisions led to the so-called "wool war" (Watson and Findlay, 1992). The degree of control over wool marketing varied even within provinces. For example, Gansu, as already mentioned, has supposedly had a free market for wool since 1985. However, the administrative authorities in some autonomous counties in Gansu, such as Sunan, have not allowed free trade in wool. As discussed in detail in Section 5.6.2 below, the IMAR government delegated control over wool marketing to the prefectural-level administrations between 1986 and 1988 (inclusive) but only some prefectures allowed private buyers to operate. In 1989, the IMAR authorities closed the wool market for the whole autonomous region and then in 1992 declared it open again. The period 1985 to 1992 has been a period of great uncertainty and change in relation to wool marketing in China (Longworth, 1993b). Some of these changes and their consequences will be discussed further in the provincial and county case studies in Parts II and III of this book.

4.5.1 Supply and Marketing Cooperatives

When the Central government designated wool as a category II commodity in 1956, it appointed the network of Supply and Marketing Cooperatives (SMCs) as the official State procurement agency. That is, the SMCs were responsible for purchasing and collecting the State production quota on behalf of the State and, in most parts of China, for purchasing all over-quota wool.

Grass roots SMCs are organised at the township (or sumu) level. Besides acting as the state purchasing authority for certain farm outputs such as wool, cashmere and skins, they also operate like a general store selling processed food (powdered milk, biscuits, cakes, etc), alcohol, clothing, household tools and consumer durables (radios, fans, TVs, etc) as well as farm inputs such as fertilisers and pesticides. However, they are not permitted to sell certain farm inputs such as most seeds (there are special seed companies with the State monopoly for the sale of most seeds) or petrol and oils (there are State fuel companies which have this right). The SMCs are also not permitted to purchase grains or meats from farmers since there are special State monopolies called grain companies or food companies established under the Ministry of Commerce to purchase these commodities.

A township-level SMC will have a small branch shop/purchasing depot in most of the villages administered by the township government. Each grass roots/township SMC is a member of the county-level SMC, the county SMCs are members of the prefectural SMC, the prefectural SMCs belong to a provincial SMC, and the provincial-level SMCs form the All China Union of SMCs.

Originally, the SMCs were established as genuine cooperatives in the first half of the 1950s. However, after 1956 and especially during the 1960s, they were absorbed into the State bureaucracy, initially in parallel with the old Ministry of Commerce (now known as the Ministry of Domestic Trade) and eventually becoming part of this Ministry. During the 1980s, there was renewed interest in having an independent, farmer-controlled, agribusiness cooperative movement in China. Some provinces, such as Gansu, have encouraged the SMC to break away from the State bureaucracy and become more independent. Nevertheless, in most provinces and even in Gansu, the SMC and its associated companies such as the Animal By-Products or Native Goods Companies (which trade in wool on behalf of the SMC) remain an arm of the State under the control of the Ministry of Domestic Trade (formerly the Ministry of Commerce and the Ministry of Material Supplies). This relationship became especially important in 1989 when the SMC began to accumulate large stocks of wool. The State has been forced to subsidise the holding of wool stocks by the SMC since 1989. The grass roots SMC was also provided with special loans through the agricultural bank so that it could keep purchasing wool from the farmers in 1989, 1990 and 1991 despite the SMCs having large amounts of capital tied up in their wool stockpiles.

The SMCs have traditionally collected a 27% marketing margin on raw wool. That is, the purchase price paid to the farmers is increased by 27% to determine the price the textile mills must pay. Of this mark-up, 17% is retained by the SMC organisation and shared between the grass roots or township SMC, the county SMC and the prefectural SMC, while 10% is a wool product tax collected by the SMC on behalf of the county governments. As the officially sanctioned marketing margin is specified as a percentage of the State purchase price paid to farmers, the SMCs and the county governments stood to benefit greatly when the price of wool soared in the 1985 to 1988 period, provided of course that the SMCs retained their monopoly over the buying of wool.

The fixed 27% mark-up has created major obstacles to reforming the marketing system for wool. Furthermore, the "wool war" was largely the result of the reactions of the SMCs and the county governments because they feared losing control over this institutionalised marketing margin on wool in 1985 after the wool market was "freed up" in some provinces.

4.5.2 Traditional Wool Pricing Arrangements

Before 1956, wool was traded in a completely free market and prices were not subject to any form of government control. In the pastoral areas, wool was sold by the nomadic herdsmen when they gathered together once or twice each year for trading, celebrations and games. Prices were unstable and unpredictable and inefficient in the sense that they did not reflect any overall interaction of total market supply and total market demand. Communications both over time and over space between the small, isolated local markets were poorly developed.

From 1956 to 1984, in principle, the State completely controlled the marketing of all wool. As a category II commodity, the purchase price for the quota amount (which was usually also the price paid by the SMC for over-quota wool) was established by the State. These administratively-determined prices were set without much reference to costs of production or the welfare of wool growers. The primary objective was to provide the textile mills with the raw material at a price which enabled the end products to be made available to (mainly urban) consumers as cheaply as possible. Consequently, the prices paid for wool remained low and stable between 1956 and the mid-1970s.

The differentials between grades were extremely modest and determined by the quality differentials written into the National Wool Grading Standard. These differentials have remained constant in percentage terms since the Standard was last revised in 1976. Another important feature of the administered prices set for wool was that the prices paid for fine wool and semi-fine wool were the same. Fine wool is wool of 25μm and finer in average fibre diameter and semi-fine wool is wool with an average fibre diameter of 25.1μm to 40μm. Clearly, these two categories of wool would have different end-uses and, under free market conditions, different prices.

In the mid-1970s, there was an upward adjustment to all wool prices followed by a small reduction in 1981. State purchase prices remained constant from 1981 to 1984 at the levels shown in Table 4.5. The fixed quality differentials by which these prices were generated given the price for the reference type and grade (Improved fine and semi-fine wool - grade I) are also shown. The price for coarse wool was established by reference to market conditions rather than a fixed quality differential.

Table 4.5 Official State Purchase Prices for Raw Wool in the Four Years 1981 to 1984

Type of wool	Grade	Price	Quality differential
		(¥/kg greasy)	(%)
Fine and semi-fine wool	- special grade	5.28	124
	- grade I	4.86	114
	- grade II	4.56	107
Improved fine and semi-fine wool	- grade I	4.26	100
	- grade II	3.88	91
Coarse or local wool		3.20	-

It was not until the 1980s that the allocative role of prices gained official attention in China. Previously, prices had been used simply as a means of distributing income in the centrally planned economy. With administratively-determined prices and State planning of production, the role of prices in signalling the preferences of buyers with respect to quality characteristics was completely neglected. Consequently, there was no incentive for sellers to allocate either resources or effort to improving the quality of their product. However, with the emergence of free markets in China, the allocative role of prices is becoming more important.

With free markets, prices signal how resources should be allocated (i.e. prices serve as a means of communication between buyers and sellers) as well as providing the major determinant of the incomes for the owners of the resources. When market-determined prices generate politically unacceptable distributions of income between groups of producers (i.e. groups of resources owners) or between producers and consumers, governments of all political persuasions come under great pressure to modify the market outcomes. One policy approach is to interfere with the free market so that more politically appropriate prices are generated. Another policy response is to leave the free market alone and to address the unsatisfactory income distribution outcomes directly. After more than half a century of focusing on the former approach, the governments of Western developed nations now seem to be shifting the emphasis to the second approach.

In general terms, the rural pricing reforms of the mid-1980s demonstrate that the Chinese government wants to take advantage of the efficiency gains which free markets and prices can offer in relation to the allocation of resources. However, the potentially unacceptable implications for the distribution of income impose major constraints on how far and how fast the Chinese government is prepared to move in the direction of free markets.

The situation in relation to wool prices is especially difficult. While premiums for finer, better quality wool would in the longer run encourage farmers to produce more of the "right" kind of wool from the viewpoint of the textile mills (and ultimately the final consumers), in the immediate future premiums could only be achieved by substantially discounting the prices paid for semi-fine and coarse wool. At present, most of the finer, better quality wool is produced by State farms and better-off private farmers with larger flocks. Furthermore, as the provincial case studies in Chapters 5, 6 and 7 make abundantly clear, the production of this better wool is concentrated in relatively few of the pastoral and semi-pastoral counties. Consequently, the overwhelming majority of farmers raising sheep, the majority of SMCs which purchase wool, and the majority of county governments (which depend on the 10% wool product tax for fiscal revenue) would be strongly opposed to a wool pricing system which created more appropriate premiums and discounts for quality.

This is a major political obstacle to the development of a more appropriate pricing mechanism for wool of different types and grades in China. In fact, it is probably the underlying reason why the free wool market experiment which began in 1985 was so vigorously opposed and why there is so much resistance to the further development of wool auctions (Longworth, 1993b and Zhang, 1993). A properly functioning free market system would automatically generate the appropriate premiums and discounts for the various types of wool.

PART II

PROVINCIAL CASE STUDIES

Chapter Five

Inner Mongolia Autonomous Region

On 1 May 1947, more than two years prior to the founding of New China, the Communist Party of China set up The People's Government of the (Inner) Mongolia Autonomous Region (IMAR). It was the first autonomous region government established. At that time, perhaps 40% of the population of about six million people belonged to minority nationalities, especially the Mongolian Nationality. Nowadays about one in seven residents are Mongolian while the other minority groups in total represent little more than 3% of the population.

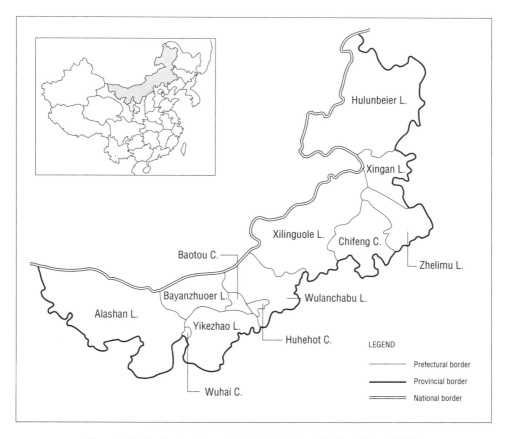

Map 5.1 Prefectural-Level Administrative Units of the IMAR

The IMAR is one of the two major wool growing provinces/autonomous regions in China. Roughly one-quarter of all the rangeland in the whole country is in the IMAR and it is the most economically important provincial-level administrative unit in the pastoral region. In recent years, IMAR has produced more wool, cashmere and camel hair than any other province/autonomous region.

5.1 Identifying the Pastoral Areas of IMAR

There are 12 prefectural-level administrative units in the IMAR. Eight of these are called leagues and four are cities (Map 5.1 and Table 5.1). As explained earlier, one of these cities (Chifeng) is not a relatively small, densely populated, geographic area as might be suggested by the name. Indeed, Chifeng City Prefecture, as it is usually called, is an administrative area larger than several of the leagues.

Under the prefectural level, there are 100 administrative units at the county level (Table 5.1). The Central government and the IMAR government use different criteria and different base periods to define pastoral and semi-pastoral counties (or banners). The IMAR government recognises 24 of these counties/banners as pastoral while a further 19 are classified as semi-pastoral. Below the county level, there are 1,534 township level administrative units of which 468 are in the pastoral areas and are referred to as sumus.

Table 5.1 Number of County/Banner Level Administrative Units in the IMAR by League/City Recognised by Governments as being Pastoral or Semi-Pastoral

Leagues or cities (prefecture)	Administrative units at the county level				
	Total	IMAR government recognised		Central government recognised	
		Pastoral banners	Semi-pastoral banners	Pastoral banners	Semi-pastoral banners
Huhehot C.	6	-	-	-	-
Baotou C.	8	-	-	-	-
Wuhai C.	3	-	-	-	-
Chifeng C.	12	-	6	5	2
Hulunbeier L.	13	4	-	4	3
Xingan L.	5	-	3	1	3
Zhelimu L.	8	-	5	3	4
Xilinguole L.	12	9	1	9	1
Wulanchabu L.	15	2	2	2	2
Yikezhao L.	8	4	1	4	4
Bayanzhuoer L.	7	2	1	2	2
Alashan L.	3	3	-	3	-
Total	100	24	19	33	21

Note: The ACIAR Project No. 8811 surveys included five counties/banners in IMAR. Four of these are located in Chifeng City Prefecture and one is in Yikezhao League.
Source: IMAR Statistics, 1989

According to the Central government, there are 33 pastoral counties and 21 semi-pastoral counties in IMAR which are included in the 266 counties defined as the pastoral region of China (Table 5.1). These 54 counties are listed in Table 5.2 together with data for each county on a number of key characteristics.

The five counties in the IMAR surveyed and presented as case studies later in this book, Balinyou, Wongniute, Aohan, Alukeerqin and Wushen, are highlighted in Table 5.2. As indicated in the table, all are considered to be poor counties by the Central government. With the exception of Aohan and to a lesser extent Wongniute, these counties have a relatively high proportion of minority nationality people in their populations. Furthermore, all five are major wool producing areas with Wushen and Aohan having respectively the third and fourth largest wool clips of all counties in the IMAR in 1990.

Over time, the administrative areas which are regarded as being in the IMAR have changed. Some prefectures have moved both into and out of the IMAR. Chifeng City Prefecture, for example, was not part of IMAR in the late 1940s and early 1950s. After belonging to the IMAR for nearly 20 years, it was reallocated to Liaoning Province in 1970. Similarly, some of the southern areas adjacent to Ningxia Province were allocated to that province in 1970. In the late 1970s, both of these areas were returned to the IMAR. Because of these administrative changes, time series data for the IMAR need to be examined closely.

5.2 Physical Details

5.2.1 Geography

The IMAR sits on the northern frontier of China and shares most of the national border with Mongolia (Plate 1). It extends 2,400km from east to west and has a total area of 1.183 million km² which is 12.3% of China and more than twice the size of France. Most of the IMAR is located more than 1,000m above sea level. Large areas of the region are windswept and sandy. The rainfall varies from a high of 450mm per year in the north east to 150mm and even lower in the west. A significant proportion of the IMAR is desert or semi-desert.

5.2.2 Population

Most of the 6 million people living in IMAR when it was established in 1947 would have been dependent on pastoral and, to a lesser extent, agricultural pursuits. The total population of IMAR just over four decades later, that is in 1989, was almost 21 million of which about 14.5 million (or 75%) were in agricultural areas and 1.9 million (9%) lived in pastoral areas. There are 44 different minority nationalities in the IMAR with by far the largest ethnic group being of the Mongolian Nationality. In 1989, there were 3.07 million Mongolians (which represented 14.7% of the total population) while all of the other minorities totalled only 600,000 (3.1% of the population).

The distribution of the population by prefectural-level administrative unit in 1988 is shown in Table 5.3. Chifeng City Prefecture, in which four of the five counties surveyed are located, had the largest population. However, it ranked only fifth in total land area, suggesting that the population density in Chifeng City Prefecture is considerably greater than in some of the other leagues. This is borne out by the population densities for the pastoral counties shown in Table 5.2.

Table 5.2 Some Characteristics of the Central Government Recognised Pastoral and Semi-Pastoral Counties/Banners in IMAR, 1990

Prefecture/city and banner		Poverty status[1]	Population		Total area	Population density	Number of sheep[2]	Wool production	Total arable land	Usable pasture area
			Number of people	Prop. of minority nationality						
			('000)	(%)	(km²)	(person/km²)	('000)	(tonne)	('000 mu)	('000 mu)
Pastoral banners/counties										
Chifeng City	- **Alukeerqin***	nsp	**278**	**33.6**	**14,277**	**19.47**	**402**	**1,440**	**981**	**15,138**
	- Balinzuo	nsp	335	24.0	6,644	50.42	153	445	1,107	4,500
	- **Balinyou***	nsp	**160**	**43.0**	**10,256**	**15.60**	**279**	**803**	**376**	**11,287**
	- Keshiketen	nsp	242	8.7	19,790	12.22	502	1,606	929	22,720
	- **Wongniute***	nsp	**446**	**12.1**	**11,807**	**37.77**	**283**	**1,437**	**1,655**	**8,830**
Hulunbeier	- Ewenke	-	129	36.6	18,630	6.92	118	160	100	12,000
	- Xinbaerhuyou	-	32	-	25,194	1.27	387	649	9	30,168
	- Xinbaerhuzuo	-	39	-	22,200	1.75	281	434	33	24,580
	- Chenbaerhu	-	54	46.8	21,192	2.54	172	300	401	21,779
Xingan	- Keerqinyouyizhong	nsp	222	81.8	15,613	14.21	94	158	982	9,000
Zhelimu	- Keerqinzuoyizhong	psp	487	-	9,646	50.48	153	516	2,196	7,978
	- Keerqinzuoyihou	psp	372	68.9	11,535	32.24	153	540	1,775	10,500
	- Zhalute	-	259	38.8	17,254	15.01	298	694	1,192	10,820
Xilinguole	- Xilinhaote	psp	123	26.1	18,750	6.56	538	1,211	234	19,998
	- Abaga	-	40	53.4	27,495	1.45	529	751	3	37,220
	- Sunitezuo	-	30	14.2	33,469	0.89	377	785	1	47,178
	- Suniteyou	psp	68	32.5	26,700	2.54	389	935	76	35,285
	- Dongwuzhumuqin	-	58	66.7	47,328	1.22	957	1,200	201	64,461
	- Xiwuzhumuqin	-	68	64.8	22,960	2.96	770	1,200	34	30,529
	- Xianghuang	-	29	63	4,960	5.84	293	921	15	6,600
	- Zhengxiangbai	psp	72	27.7	6,229	11.55	416	1,204	216	7,621
	- Zhenglan	psp	76	35.4	10,182	7.46	578	2,152	232	12,499
Wulanchabu	- Damao	psp	108	-	18,177	5.94	652	1,556	1,173	21,700
	- Siziwang	nsp	207	6.95	25,513	8.11	681	2,336	2,083	29,810

Table 5.2 continued

Prefecture/city and banner		Poverty status[1]	Population		Total area	Population density	Number of sheep[2]	Wool production	Total arable land	Usable pasture area
			Number of people	Prop. of minority nationality						
			('000)	(%)	(km²)	(person/km²)	('000)	(tonne)	('000 mu)	('000 mu)
Yikezhao	- Etuokeqian	nsp	63	32.9	12,504	5.03	245	793	111	9,765
	- Etuoke	-	103	22.5	21,476	4.79	326	1,120	100	20,747
	- Hangjin	nsp	132	17.5	18,914	6.97	239	894	383	14,695
	- **Wushen***	**nsp**	**90**	**32**	**11,640**	**7.73**	**480**	**2,047**	**166**	**7,840**
Bayanzhuoer	- Wulatezhong	-	128	18.8	22,744	5.62	437	930	542	29,877
	- Wulatehou	-	48	25	25,000	1.92	780	249	30	30,110
Alashan	- Alashanzuo	-	123	-	80,412	1.52	371	615	118	69,000
	- Alashanyou	-	23	-	73,443	0.31	570	80	18	72,635
	- Ejina	-	15	31.7	114,606	0.13	30	30	32	8,910
Sub-total above 33 pastoral banners			4,659	n/a	826,538	5.63	12,933	30,191	17,504	765,780
Semi-pastoral banners/counties										
Chifeng City	- Linxi	psp	231	4.0	3,950	58.48	142	614	1,036	1,800
	- **Aohan***	**nsp**	**538**	**4.3**	**8,294**	**64.86**	**278**	**1,702**	**1,875**	**4,450**
Hulunbeier	- Zhalantun	-	418	8.5	16,926	24.69	141	139	1,607	1,870
	- Arong	-	315	-	12,063	26.11	88	91	1,676	3,545
	- Mulidawa	-	268	10.0	10,600	25.28	22	29	1,919	3,212
Xingan	- Keerqinyouyiqian	-	408	38.2	27,752	14.70	514	625	1,711	15,000
	- Zhalaite	-	382	40.3	11,155	34.24	190	283	1,790	8,330
	- Tuquan	-	302	16.0	4,890	61.75	220	272	1,327	2,770
Zhelimu	- Tongliao	-	696	30.4	3,518	197.83	105	356	1,801	1,305
	- Kailu	-	353	10.7	4,488	78.65	183	496	1,246	2,779
	- Kulun	psp	164	55.0	4,650	35.26	91	300	1,421	2,780
	- Naiman	nsp	397	32.2	8,159	48.65	355	1,610	1,405	5,181
Xilinguole	- Taipusi	-	215	4.1	3,415	62.95	229	903	1,217	2,880

Table 5.2 continued

Prefecture/city and banner		Poverty status[1]	Population		Total area	Population density	Number of sheep[2]	Wool production	Total arable land	Usable pasture area
			Number of people	Prop. of minority nationality						
			('000)	(%)	(km²)	(person/km²)	('000)	(tonne)	('000 mu)	('000 mu)
Wulanchabu	- Chayouzhong	psp	231	1.5	4,190	55.13	270	1,000	1,573	2,740
	- Chayouhou	nsp	216	4.9	3,910	55.24	271	924	1,161	3,230
Yikezhao	- Dongsheng	psp	142	7.4	2,200	64.54	97	246	321	1,877
	- Dalate	psp	305	3.4	8,200	37.19	246	583	1,098	5,000
	- Zhungeer	nsp	237	7.7	7,692	30.81	215	554	862	7,646
	- Yijinghuoluo	psp	131	6.2	6,000	21.83	382	1,110	354	4,777
Bayanzhuoer	- Dengkou	-	101	-	4,166	24.24	125	369	224	3,279
	- Wulateqian	-	312	4.3	7,476	41.73	386	1,081	1,087	6,359
Sub-total above 21 counties			6,362	n/a	163,694	38.86	4,550	13,287	26,711	90,810
Grand total above 54 pastoral and semi-pastoral banners/counties			11,021	n/a	990,232	11.12	17,483	43,478	44,215	856,590
Total for all of IMAR			21,490	19.53	1,144,650	18.78	20,749	57,000	73,000	1,180,000

*These five banners were surveyed as part of ACIAR Project No. 8811.

[1] Poverty status "nsp" means recognised as a poor county under the Central government poverty alleviation scheme. These banners receive special assistance from both Central and IMAR governments. In addition to these banners, the IMAR government identifies certain other counties for special poverty relief. Such counties are shown with a "psp" poverty status. (See Section 3.4.3 for more details.)

[2] End of year data.

Table 5.3 Population and Land Use in the IMAR by Prefectural Administrative Unit, 1988

League or city (prefecture)	Population	Land use			
		Total land area	Cropland	Total pasture land*	Usable pasture land
	('000)	('000 mu)	('000 mu)	('000 mu)	('000 mu)
Huhehot C.	1,333	9,118	2,340	2,686 (29%)	2,240
Baotou C.	1,695	14,785	3,225	7,508 (51%)	6,366
Wuhai C.	278	2,705	45	2,344 (87%)	1,877
Chifeng C.	3,979	130,279	12,075	82,402 (63%)	69,620
Hulunbeier L.	2,465	372,389	9,060	169,470 (46%)	149,706
Xingan L.	1,498	89,333	5,955	45,511 (51%)	39,183
Zhelimu L.	2,700	89,841	10,680	68,542 (76%)	55,704
Xilinguole L.	861	303,870	2,925	292,794 (96%)	264,914
Wulanchabu L.	3,221	125,070	18,765	84,679 (68%)	76,266
Yikezhao L.	1,173	131,142	3,240	82,981 (63%)	71,838
Bayanzhuoer L.	1,472	98,327	4,650	80,127 (81%)	69,366
Alashan L.	155	384,471	135	263,023 (68%)	146,786
Total	20,830	1,751,332	73,065	1,182,067 (67%)	953,866

*The percentage in brackets is the proportion that total pasture land represents of the total land area.
Source: Inner Mongolia Statistics, 1989

5.2.3 Land Use

Almost 70% of the IMAR is pasture land (i.e. 1,182 million mu) but only four-fifths of the total pasture land is regarded as usable (Table 5.3). Pasture scientists claim that there are over 900 different herbage grasses and other plants which grow in the pasture areas of IMAR but of these only about 210 species are eaten by livestock. Given the enormous areas and the range of ecological environments, it is not particularly helpful to record that the average annual dry matter production per mu is considered to be around 50kg per year. Nevertheless, this average figure, which corresponds to 750kg of dry matter per ha per year, provides some idea of the maximum potential carrying capacity.

The best pasture land is harvested for hay and is referred to as "cutting land". There is no aggregate estimate of the area of "cutting land" but around 4 million tonne of pasture hay is made each year in IMAR.

The total crop land in the region is about 73 million mu. Woodland forest areas are estimated to be about 240 million mu.

The prefectural distribution of cropland, pasture land and usable pasture land within IMAR is illustrated by the data in Table 5.3.

5.2.4 Pasture Land Degradation

Degradation of rangelands is a major problem in the IMAR. In the past, two large areas of IMAR native grassland were famous throughout China, namely the Keerqin or Eastern Grasslands (mainly in Xingan and Zhelimu Leagues and northern Chifeng City Prefecture) and the Eerduosi Grasslands (in south eastern Yikezhao League). These parts of IMAR are still major sheep-raising districts but large areas of the once lush native pastures have become deserts and the remaining grassland is badly degraded. Three of the case study counties (Balinyou, Wongniute and Alukeerqin) are located in the Eastern Grasslands and one (Wushen) is in the Eerduosi Grasslands. Another major pasture area which has been turned to desert by overgrazing lies along the Kelulun River which flows out of Mongolia through Xinbaerhuyou Banner in western Hulunbeier League and into Lake Hulun.

Apart from these three especially serious examples, there are many areas of IMAR in which the pastures have been badly degraded. The official estimates of the extent of pasture degradation in 1988 are presented in Table 5.4. These estimates are likely to be conservative. Nonetheless, they indicate that 36% of the total pasture land in IMAR was regarded as degraded in 1988 and over half of the degraded pasture is said to be exhibiting medium to heavy degradation.

Table 5.4 Extent of Pasture Degradation in the IMAR by Prefectural Administrative Unit, 1988

League or city (prefecture)	Areas of pasture degraded	Percentage of total pasture area which is degraded	Proportion of pasture area where degeneration is		
			Light	Medium	Heavy
	('000 mu)	(%)	(%)	(%)	(%)
Huhehot C.	1,406.7	52	29	32	39
Baotou C.	-	-	-	-	-
Wuhai C.	1,200.0	51	60	40	0
Chifeng C.	69,089.2	84	37	34	29
Hulunbeier L.	23,340.9	14	59	34	7
Xingan L.	11,381.2	25	40	26	34
Zhelimu L.	46,082.4	67	32	27	40
Xilinguole L.	143,632.6	49	48	42	10
Wulanchabu L.	22,235.2	26	64	24	12
Yikezhao L.	61,280.8	74	39	45	19
Bayanzhuoer L.	25,649.1	32	70	29	1
Alashan C.	23,418.4	9	53	45	2
Total	428,714.7	36	46	37	17

Source: IMAR Animal Husbandry Bureau

In the context of the counties surveyed in the IMAR, it is interesting to note that Chifeng City Prefecture (with 84% of pasture land degraded) is easily the worst affected of all the prefectural-level units and Yikezhao League (with 74%) is the second most seriously degraded area. Four of the five case study counties were in Chifeng City Prefecture and the fifth was in Yikezhao League. Therefore, the comments about the extent of degradation in the county case studies in this book may overstate the current general situation. On the other hand, the extent of degradation is increasing almost everywhere and action is desperately needed to prevent the rest of IMAR "catching up" to Chifeng City Prefecture in regard to the seriousness of the pasture degradation problem. For a detailed discussion of pasture degradation in Chifeng City Prefecture see Lin (1990).

The usefulness of the estimates of the extent of pasture degeneration shown in Table 5.4 depends upon the definitions used. Although the precise definitions of light, medium and heavy degradation change from place to place (and over time) in China, the data in Table 5.4 has been based on three broad criteria:

(a) Types of herbage grass:
   light – small changes in the types of grasses present;
   medium – more significant changes with the appearance of a few weed species;
   heavy – large change in the types of herbage grass, with most edible grasses being replaced by weeds.

(b) Height and the yield of the grass:
   light – the annual yield of usable dry matter declines by less than 30%;
   medium – annual yield of usable dry matter declines by between 30% and 80%; and
   heavy – annual yield of usable dry matter declines by more than 80%.

(c) Grazing potential:
   light – can be grazed seasonally, but not the whole year round;
   medium – usable only in winter and spring; and
   heavy – no longer usable.

Chinese pasture scientists have conducted extensive research into the nature and the extent of degradation and the volume of literature on the topic (in Chinese) is considerable. For example, precisely what light, medium and heavy degradation means in Chifeng City Prefecture can be gauged from the information in Tables 5.5 and 5.6. Obviously, the ecology of pastures will vary from location to location and different researchers are likely to establish different criteria in different places and at different times. For these and other reasons it is not easy to obtain reliable and comparable time series data to indicate how the extent of pasture degradation has changed both over time and in different parts of the IMAR (or in China generally). However, there have been three major surveys undertaken in IMAR: in 1958; in 1961 to 1965; and the most recent set of studies which took eight years to complete between 1981 and 1988. The data in Table 5.4 summarises the findings of the most recent investigation.

Policy makers in the IMAR have become increasingly aware of the pasture degradation problem. Overstocking is recognised as the major direct cause of the problem although droughts, rodent infestations and past cultivation mistakes exacerbate the situation.

In a legalistic attack on the problem, the IMAR government adopted a Rangeland Management Regulation in June 1984. This Regulation was an important precursor to the National Rangeland Law promulgated by the Central government in 1985 (Standing Committee of the Sixth National People's Congress, 1985). The National Law lacks specific details about how the rangelands should be managed. It was expected that the various provincial, prefectural and county level administrations would develop increasingly precise interpretations and guidelines for action. The IMAR government revised the Rangeland Management Regulation in August 1991 and a translation of the revised law is presented in Appendix 5A.

Underpinning the Regulation is a set of scientific guidelines or standards which establish the procedure for calculating the proper stocking rate for natural pasture land (see Appendix 5B). Establishing the legal framework for managing the use of native pastures and policing the implementation of these policies are two different things. Consequently, as several of the county case studies make abundantly clear, there are major deficiencies at the local level in terms of devising incentives and setting and imposing penalties to ensure that the private households comply with the general regulations.

Table 5.5 Production Characteristics of the *Gramineae* Group of Grasslands* in Various Stages of Degradation in Chifeng City Prefecture, 1981-1985

Production characteristics	Undegraded	Lightly degraded	Medium degraded	Heavily degraded
Number of perennial forage grass genera	19	15	16	8
Number of mono or bi-annual grass genera	1	4	5	7
Average grass height (cm)	28	18	11	4.9
Grass coverage (% of total area)	35	30	26	18
Yield (kg/mu)	93.4	68.5	48.7	52.7

*The *Gramineae* group of grasslands include the following grass genera: *Achnatherum beauv.; Agrostis L.; Agropyron gaertn; Aeluropus trin; Alopecurus L.; Aneurolepidium nevoli; Aristida L.; Bromus L.; Elymus L.; Festuca L.; Melica L.; Phalaris L.; Stipa L.* and *Tristerum pers.*
Source: Hu, M.G. (ed.) (1990). *Chifeng Pastoral Resources*, Agricultural Publishing House, Beijing [in Chinese].

Table 5.6 Yields of Different Grades of Forage Grasses for the *Gramineae* Group of Grasslands* in Various Stages of Degradation in Chifeng City Prefecture, 1981-1985

Grade of forage grass*	Undegraded	Lightly degraded	Medium degraded	Heavily degraded
	(kg DM/mu)	(kg DM/mu)	(kg DM/mu)	(kg DM/mu)
Grade 1	19.8	16.8	6.7	-
Grade 2	22.7	19.6	17.2	20.9
Grade 3	33.3	18.6	16.9	0.2
Grade 4	17.6	10.5	3.1	1.7
Inferior	-	3.1	4.8	29.9

*Graded according to digestible levels of energy and protein. Grade 1 is the highest and inferior grade is the lowest.
Source: Hu, M.G. (ed.) (1990). *loc. cit.*

5.2.5 Livestock Numbers

The increases in the herbivorous livestock populations of IMAR since 1949 have exceeded the changes which have occurred in the national totals (Table 5.7). In particular, the IMAR sheep flock has grown much faster than the national flock. At the end of June 1989, there were 47.57 million large and small animals in the IMAR. The corresponding figure at the end of June 1970 was 35 million. On this basis, the total number of livestock in IMAR increased by over 12 million (or 36%) in the twenty years ended 1990.

It should be noted that Chinese livestock statistics are collected at "the end of June" and at "the end of the year". It is very important to distinguish between these data sets. For example, the end of June 1989 total livestock figure for IMAR of 47.57 million head referred to above contrasts with an end of year 1989 figure of 37.28 million head. The reason for the large difference between the June and December livestock numbers is that all the natural increase for the year is on hand at the end of June while large numbers of animals are slaughtered for meat in the later months of the year before the onset of winter. The data in Table 5.7 are end of year livestock numbers.

Table 5.7 Long-Term Changes in Herbivorous Livestock Numbers in the IMAR and All China, 1949 and 1990

Livestock type	IMAR			All China		
	1949	1990	Increase*	1949	1990	Increase*
	('000)	('000)	(%)	('000)	('000)	(%)
Large livestock	2,999	7,075	136	60,023	130,213	117
Sheep	3,749	20,749	453	26,221	112,816	330
Goats	2,069	9,490	359	16,126	97,205	503

*Calculate as $[\frac{1990-1949}{1949}] \times 100$

Source: Statistical Yearbook of China, 1991
             Collection of Statistical Data by Province (1949-1989)

According to officials of the IMAR Animal Husbandry Bureau, the major reasons for the dramatic increase in the number of herbivorous animals in the last two decades, have been the increased attention paid to animal husbandry activities by the government of IMAR which has recognised that animal husbandry is an economic cornerstone of the IMAR economy; the introduction of the household production responsibility system; and the investment in animal shedding and pasture improvement made by both the State and the private households.

The IMAR has the largest goat herd of any province/autonomous region. In 1990, there were 9.5 million goats in IMAR, which was more than twice as many as in XUAR, the province/autonomous region with the next largest herd.

Similarly, there were more large livestock (cattle, horses, mules, donkeys, camels and yaks) in the IMAR than in any other province/autonomous region.

In connection with sheep, the IMAR is slightly behind the XUAR in terms of total sheep numbers, but there were more fine and semi-fine wool sheep in IMAR in 1991. As discussed in Section 4.3.1 in relation to Tables 4.1 and 4.2, until 1991 the statistics have always indicated that XUAR had the most fine and improved fine wool sheep. However, in 1991 a total of 2.38 million XUAR sheep previously in the fine and improved fine category were re-classified as semi-fine and improved semi-fine wool sheep.

Table 5.8 expands the data in Table 4.1 of Chapter 4 to provide more detail on the trends in sheep numbers in IMAR during the 1980s. While the total flock shrank by almost 10% between 1981 and 1984, numbers recovered steadily to a new peak in 1989 and seem to have stabilised at around 20 million. On the other hand, fine and improved fine wool sheep increased steadily over the eleven-year period, with the 1991 number being a little more than double the 1981 figure. As a result, sheep of this type now represent almost half the sheep in IMAR. No reliable data could be obtained to indicate how many of the 9.5 million fine and improved fine wool sheep in IMAR were genuine fine wools rather than cross-breds undergoing upgrading to the fine wool category. Evidence from Chifeng City Prefecture (discussed in Section 5.3.2 below) and informed opinion both indicate that less than 10% of the 9.5 million sheep in the fine and improved fine wool category in IMAR are genuine fine wools and that this proportion is not increasing rapidly.

Table 5.8 Number and Proportion of Total Sheep Represented by Each Major Category of Sheep in the IMAR, 1981 to 1991*

Year	Total sheep		Fine and improved fine wool sheep		Semi-fine and improved semi-fine wool sheep		Other sheep	
	('000)	(%)	('000)	(%)	('000)	(%)	('000)	(%)
1981	19,588	100	4,520	23.1	5,095	26.0	9,973	50.9
1982								
1983								
1984	17,717	100	5,619	31.7	3,576	20.2	8,522	48.1
1985	18,148	100	5,880	32.4	4,220	23.3	8,048	44.3
1986	18,323	100	6,521	35.6	3,860	21.1	7,942	43.3
1987	18,464	100	6,378	34.5	4,265	23.1	7,821	42.4
1988	20,597	100	8,617	41.8	3,561	17.3	8,419	40.9
1989	20,845	100	8,782	42.1	3,521	16.9	8,542	41.0
1990	20,749	100	9,218	44.4	3,802	18.3	7,729	37.2
1991	20,149	100	9,488	47.1	3,757	18.6	6,904	34.3

*Year end data.
Source: *China Agriculture Yearbooks*. Agriculture Publishing House, Beijing.

The distribution of sheep, goats, large livestock and pigs in IMAR by prefectural government administrative unit is presented in Table 5.9. Chifeng City is clearly a major pastoral prefecture. In 1988, of the twelve prefectural level units in IMAR, Chifeng City had the greatest number of large livestock, the second largest goat herd, and the third biggest sheep flock.

Since overstocking is acknowledged as the principle direct cause of pasture degradation and since Chifeng City is the prefecture with the greatest proportion of degraded pasture, changes in stocking rates over time in this prefecture are of particular interest. Figure 5.1 presents long-term statistics on sheep numbers and total sheep equivalents in each of the four counties surveyed in Chifeng City Prefecture.

In the two more pastoral counties, Balinyou and Alukeerqin, total stock numbers as measured by sheep equivalents rose sharply until 1963/64, plunged dramatically in the late 1960s due to the excesses of the Cultural Revolution, and have exhibited a modest but erratic upward trend since 1970. Sheep numbers on the other hand did not peak until 1983. Since the full introduction of the household production responsibility system, there has been a downward trend in sheep numbers in both Balinyou and Alukeerqin. The large decline in end of year sheep equivalents in Balinyou in 1989 was due to the severe drought in that year. While precise data are not available, livestock numbers increased sharply in Balinyou in 1991 and 1992, which were years of above average rainfall.

Table 5.9 Livestock Numbers in the IMAR by Prefectural Administrative Unit, 1988*

Leagues or cities (prefectures)	Sheep	Goats	Large livestock	Pigs
	('000)	('000)	('000)	('000)
Huhehot C.	318	84	99	95
Baotou C.	577	286	94	99
Wuhai C.	14	21	2	6
Chifeng C.	2,684	1,337	1,518	1,193
Hulunbeier L.	1,150	205	765	359
Xingan L.	947	196	741	458
Zhelimu L.	1,249	515	1,359	1,248
Xilinguole L.	5,038	1,145	1,350	57
Wulanchabu L.	3,791	583	673	388
Yikezhao L.	2,294	2,012	277	278
Bayanzhuoer L.	2,075	1,284	280	476
Alashan L.	462	661	188	7
Total	20,597	8,331	7,346	4,664

*End of year data.
Source: Inner Mongolia Statistics, 1989

In Wongniute, which is a mixed pastoral-agricultural banner, and in Aohan which is more agricultural, total sheep equivalents have trended downwards since the mid-1960s, with the decline being more marked in Aohan. Sheep numbers have followed essentially the same pattern as in Balinyou and Alukeerqin, perhaps with a more pronounced decline since about 1983.

The graphs in Fig. 5.1 demonstrate that if the current level of total sheep equivalents is considered excessive, overstocking is not a recent phenomenon in these four counties. Of course, as more and more of the pasture becomes degraded, a given number of sheep equivalents will exert increasing pressure on the remaining pasture resources. Therefore, the grazing pressure, at least in Balinyou and Alukeerqin, has almost certainly increased since the early 1960s.

Another feature of the graphs is the apparent decline in sheep numbers since the introduction of the household production responsibility system (HPRS) in the early 1980s. This trend reflects the economically rational response of herdsmen to the sharp decline in the relative profitability of wool (and especially fine wool) production as compared with cashmere production. Between 1984 and 1988, wool prices doubled but cashmere prices increased five-fold. Farmers also consider goats easier and less costly to keep than sheep (especially fine wool sheep). Sheep and goats are almost perfect substitutes in a production sense. They are herded together; they eat essentially the same forage (although goats can survive harsh grazing conditions better); they are shedded together; and external parasite control measures (dipping) are the same. Therefore, the proportion of sheep and goats kept by households is sensitive to the relative prices paid for wool and cashmere.

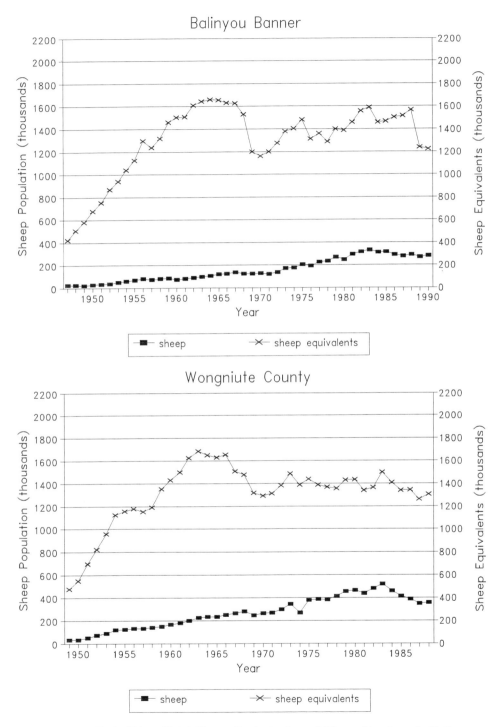

Fig. 5.1(a) End of Year Total Sheep Equivalents and Sheep Numbers in Balinyou and Wongniute Counties of Chifeng City Prefecture, IMAR

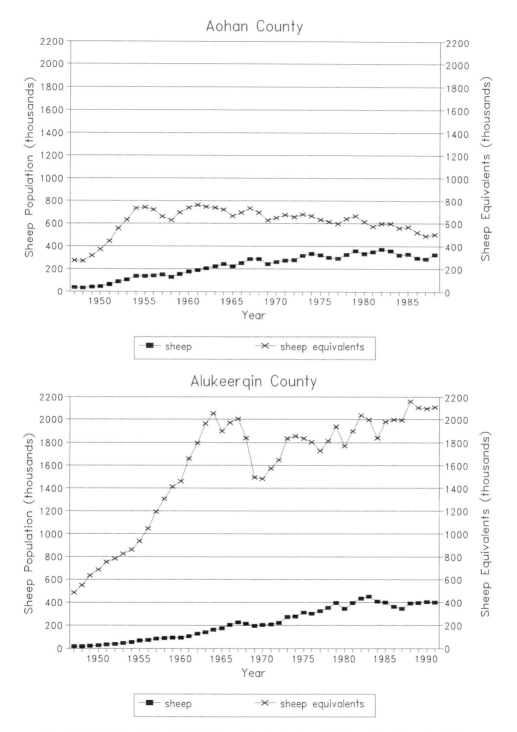

Fig. 5.1(b) End of Year Total Sheep Equivalents and Sheep Numbers in Aohan and Alukeerqin Counties of Chifeng City Prefecture, IMAR

As the pastures are further degraded and consequently the grazing conditions become harsher, the preference for goats and hardier types of sheep (i.e. coarser woolled sheep) will intensify. A substantial improvement in the price of fine and improved fine wool relative to both coarser wool and cashmere will be needed if the number of fine wool sheep being raised in the more degraded pastoral areas of the IMAR is to continue to increase.

5.2.6 Location of Pastoral Production

The distribution of livestock numbers by prefectural-level administrative unit presented in Table 5.9 provides a broad indication of the location of pastoral activities within the IMAR. However, the production data in Table 5.10 indicate in more detail which parts of the IMAR constitute the main production areas in relation to the major pastoral products.

Xilinguole League (see Map 5.1) is the top producer of beef, mutton and goat meat, total wool and fine and improved fine wool. As mentioned previously, Chifeng City and Yikezhao League are the prefectures in which the five IMAR case study counties discussed in Part III of this book are located. Chifeng City is the third and fifth largest producer of beef and mutton and goat meat respectively, and the third largest wool growing area. Perhaps more importantly, Chifeng City ranks second in terms of fine and improved fine wool output. Yikezhao League is a significantly larger producer of mutton and goat meat than Chifeng City but it produces a little less wool in total and less than half as much fine and improved fine wool. Chifeng City and Yikezhao League together account for one-third of the fine and improved fine wool grown in the IMAR.

Table 5.10 Output of Major Pastoral Products in the IMAR by Prefectural Administrative Unit, 1990

League or city (prefecture)	Beef	Mutton and goat meat	Raw wool			
			Total raw wool	Fine wool and improved fine wool	Semi-fine wool and improved semi-fine wool	Coarse/ local wool
	-- (tonne) ------------------------------					
Huhehot C.	460	1,707	722	160	309	253
Baotou C.	676	3,136	1,153	86	552	515
Wuhai C.	3	173	18	0	0	18
Chifeng C.	15,564	13,671	9,521	7,328	826	1,367
Hulunbeier L.	14,194	7,323	2,517	970	368	1,179
Xingan L.	5,564	4,183	2,308	1,494	51	763
Zhelimu L.	18,572	5,773	4,982	3,646	0	1,336
Xilinguole L.	24,849	34,897	12,253	8,282	241	3,730
Wulanchabu L.	4,149	18,241	10,950	4,499	4,874	1,577
Yikezhao L.	820	18,094	7,731	3,308	2,736	1,687
Bayanzhuoer L.	516	17,009	6,323	2,613	2,296	1,414
Alashan L.	516	3,570	725	1	0	724
Total	85,883	127,777	59,203	32,387	12,253	14,563

Source: *Statistical Yearbook of IMAR, 1991*, Statistical Press of China, Beijing

5.3 Livestock Improvement Programs

In 1976, the IMAR government decided to concentrate on improving a limited number of animal species and breeds in certain specific areas. Not only in IMAR but in China generally, these specific areas are referred to as "bases". The leagues and cities can be grouped roughly according to the main animal husbandry activities to be developed in those areas (refer to Map 5.1). To a large extent these groupings reflect the location of current production as discussed in the previous section in relation to the major pastoral products.

The areas mainly for the production of milk: Hulunbeier League;
Xingan League;
Huhehot City;
Baotou City; and
the eastern areas along the railway lines also produce milk.

The areas mainly for meat and milk cattle (dual purpose cattle):
Xilinguole League;
Chifeng City Prefecture; and
Zhelimu League.

The coarse wool production areas are: Wulanchabu League (northern part);
Zhelimu League; and
Bayanzhuoer League (northern part).

Fine wool and semi-fine wool production is mainly in:
Wulanchabu League (southern part);
Xilinguole League;
Chifeng City Prefecture;
Zhelimu League;
Xingan League;
Bayazhuoer League (southern part); and
Yikezhao League (south/eastern part).

The central part of Bayanzhuoer League is mountainous country where farmers raise goats for cashmere production. Yikezhao League has fine wool production in the southeast and goats in the north.

The major types of livestock in Alashan League are goats and two-humped camels. Fifty percent of all camels in China are in the IMAR. The wool or camel cashmere from these two-humped camels in Alashan League is especially famous and valuable. It is used as a substitute for feathers and is very light and fluffy. The camels are also used for meat, milk and transportation.

5.3.1 Cattle

Dairy cattle in IMAR are almost all friesians. The average milk yield per cow is 4,000kg per year.

There are about 400,000 dual-purpose cattle in Inner Mongolia. Animals belonging to the local improved dual-purpose breed are called Three River cattle. The local breed was crossed with Shorthorns to establish a new breed which was recognised in 1986. The average milk yield from these dual-purpose animals is 1,000 to 2,000kg per year, and is produced only on a seasonal basis. There is also a crossing program with Simmentals which aims to establish a new breed in the near future.

Around 300,000 to 320,000 cows are artificially inseminated each year with frozen semen.

5.3.2 Fine and Semi-fine Wool Sheep

Before 1949, there were no fine wool or semi-fine wool sheep in IMAR. All of the sheep at that time were local breeds (overwhelmingly of the Mongolian breed type) kept for their mutton and coarse wool. The development of fine wool and semi-fine wool breeds especially suited to the IMAR environment began in 1958 (Longworth and Williamson, 1993).

- There are now two dual-purpose fine wool breeds of sheep:
  - *Inner Mongolian Fine Wool Sheep* which was developed in the Xilinguole League. This breed was registered in 1976.
  - *Aohan Fine Wool Sheep* which was developed at the Aohan Stud in Aohan County of Chifeng City Prefecture. This breed was first recognised officially in 1978.

Although both the Inner Mongolian Fine Wool Sheep and the Aohan Fine Wool Sheep are said to be dual-purpose sheep, they are not ideal mutton type animals. Nevertheless, they have been bred, at least in part, for their meat production.

new breeds - not such good meat

- There are two specialist fine wool breeds:
  - *Erdos Fine Wool Sheep* which is still being developed in Wushen County of Yikezhao League. This breed was recognised as a local breed in 1985.
  - *Keerqin Fine Wool Sheep* which is a type of Chinese Merino developed at Gadasu Stud in Zhalute Banner of Zhelimu League.

Of the four fine wool breeds listed above and already recognised by the IMAR government, only one, the Keerqin type of Chinese Merino, is also fully recognised by the Central government.

- In addition to these four fine wool breeds, there are two other types of fine wool sheep still being developed as local breeds in Inner Mongolia:
  - *Wulanchabu Fine Wool Sheep* which is mainly based in the Wulanchabu League. It was expected that this breed would be recognised in IMAR in 1992. In 1990, there were around 250,000 sheep in various stages of upgrading, being F_1, F_2, F_3 etc. animals. To be recognised as a separate breed by the IMAR government, there must be at least 30,000 F_3 (or better) ewes which meet the standard established for the breed.
  - *Xingan Fine Wool Sheep* which is being developed in Xingan League and which was also to be recognised in 1992. There were about 180,000 sheep which had reached the standard for this breed in 1990.
- The IMAR government is also trying to develop a number of other breeds, especially the *Inner Mongolian Semi-fine Wool Sheep*.

The IMAR imported about 400 merino rams from Australia in the 1985 to 1989 period, mainly from Western Australia. Many of these imported sheep did not do well in the harsh environment in northern China. For example, of the 206 rams imported in 1989, about 50 had become sick by mid-1990.

The advantages of the Australian merino are said to be that the wool is much longer on the belly and the density of the fleece is much better than Chinese fine wool breeds, yielding a much more closed staple which reduces the amount of sand and other rubbish entering the fleece.

It was claimed that in the IMAR where artificial insemination (AI) has been used to introduce superior fine wool breeds of sheep (Chinese Merino and Australian merino), after two generations the new blood has increased the staple length by 1.5cm; increased the wool yield by 6-10% (i.e. increased the clean wool yield per sheep by 0.5 to 0.8kg for adult ewes); increased the density of the staple; and improved the colour of the fleece (the older-type sheep tended to have yellow to brown grease while the new ones have a more creamy colour). Longworth and Williamson (1993) present some specific data demonstrating the impact of Australian merino blood.

Each good quality ram can be crossed with 2,000 to 3,000 ewes per year using AI. The greatest number of ewes mated to one ram in 1989 was 9,000. There are about 10,400 AI stations in the whole of IMAR, and these stations use about 15,000 rams. It was said that in 1989, 3 million ewes were mated using AI and a further 3.5 to 4 million ewes mated in the natural fashion.

Since the introduction of the household production responsibility system in the early 1980s, the proportion of ewes mated by AI has declined in many parts of IMAR (and in the pastoral region in general). The official explanation for this development is that there has been a growing shortage of labour at mating time and the farmers do not believe the equipment is good enough to give high conception rates. In reality, it reflects the increased freedom of the farmers to breed from rams they consider better suited to their particular situations than the pure-bred fine wool rams available through the AI stations. While each of the county case studies includes a section on the use of AI for sheep improvement, the extent of the decline in the use of AI in the early 1980s is considered in detail in relation to Balinyou County (Chapter 8).

The development of new fine wool breeds especially suited to the local conditions, the infusion of imported superior genetic material, and the widespread availability of AI services are three major steps towards upgrading the local sheep to genuine fine wools. On the other hand, if the private households do not regard fine wool sheep as desirable (for economic, culinary, social or whatever reasons), then the upgrading program may stall.

It is extremely difficult to get any reliable information which makes the distinction between genuine fine wool sheep and improved fine wool sheep. Table 5.11 presents the only longish time-series data of this kind obtained during the whole four years of research in China. Similar figures for just three years (1967, 1977 and 1987) were obtained for Hebukesaier County in XUAR (see Section 17.2.3). The reliability and hence usefulness of the information in Table 5.11 is uncertain. For instance, none of the sheep are classified as being in the semi-fine category although, according to Table 5.10, Chifeng City Prefecture produced 826 tonne of semi-fine and improved semi-fine wool in 1990. In addition, there are several extraordinarily large year-to-year changes in the number of fine wool sheep. Perhaps the precise definition of what was counted as a "fine wool" sheep may not have been the same in all years. However, it is understood that allowing for an element of judgement about the quality of the sheep, animals which were progeny of four or more matings to pure-bred fine wool rams were included in the fine wool sheep category.

With the above reservations about the figures in Table 5.11 in mind, the following points are worth noting. First, the data suggest that for the prefecture as a whole, sheep numbers tended to decline between 1980 and 1990, although the mid-year figures for 1990 were distorted by the unusually heavy sell-off which occurred in late 1989, due to the severe drought in that year. Secondly, local sheep numbers declined markedly during the 1980s but the number of improved sheep did not increase despite the official emphasis on the use of fine wool sires for upgrading the local sheep to improved status. Instead, as mentioned in Section 5.2.5 above and discussed in the case studies (especially Alukeerqin), herdsmen have replaced sheep with goats. The third, and perhaps most important observation, is that fine wool sheep both as a proportion of the total flock and in absolute numbers seem to have declined since the household production responsibility system was introduced in the early 1980s.

The last point is emphasised by the figures in Table 5.11 for Balinyou and Alukeerqin. As discussed in the case studies for these two banners (Chapters 8 and 11), it would appear that given the freedom to choose, private sheep farmers in relatively remote pastoral banners like Balinyou and Alukeerqin, do not consider it worthwhile to continue upgrading their flocks to genuine fine wool status. In Aohan County, as pointed out in the case study for that county (Chapter 10), it is compulsory to use good quality fine wool rams via the AI stations. Therefore, a major share of the genuine fine wool sheep in Chifeng City Prefecture are probably in Aohan County and perhaps the proportion of fine wool sheep in the prefecture would have dropped even more had the farmers in Aohan enjoyed more freedom of choice in relation to sheep breeding.

As already mentioned, Table 5.11 provides the only long-term time series data available on the issue of how successful the fine wool sheep improvement program has been in China. However, anecdotal evidence suggests that the situation in Chifeng is representative of many genuinely pastoral localities. (See also, for example, Section 17.2.3). Given the freedom to choose their own rams, a high proportion of the private sheep farmers in the more remote pastoral localities apparently prefer not to use pure-bred fine wool rams. In these localities the introduction of the household production responsibility system (HPRS) seems to have severely retarded official plans to upgrade large numbers of sheep to genuine fine wools.

At the same time, as Table 5.8 demonstrates for the IMAR and Table 4.2 suggests for the whole pastoral region, both the total number and the proportion of fine and improved fine wool sheep continued to rise after the HPRS was introduced. It would seem, therefore, that the fine wool sheep improvement program must have been particularly successful in at least some semi-pastoral and agricultural localities to compensate for the lack of enthusiasm for the program in many pastoral localities.

5.3.3 Meat Sheep

The local Mongolian type sheep which have traditionally been raised for mutton and coarse wool, are extremely hardy and well adapted to the harsh arid pastoral conditions of IMAR. However, these sheep and the remote pastoral banners in which they are primarily raised are not well suited to commercial mutton production.

In the 1980s, as the market demand for mutton expanded, the possibility of developing specialist meat sheep breeds for the agricultural areas of China was suggested. The first national project to develop improved types of meat sheep began in 1985 and was based at the IMAR Academy of Animal Husbandry Sciences in Huhehot. At that time, there were five banners regarded as specialising in the raising of meat sheep. By 1989, more than 20 banners in agricultural areas had significant numbers of improved meat sheep and over 20,000 ewes were mated to meat breed rams by AI.

A number of different imported breeds have been used in the national meat sheep improvement program including Suffolks (imported from Australia and New Zealand in 1986), Dorset Downs (from New Zealand), Dorsets (from Australia) and Charolais (imported from France in 1987). Of these, the heavy-milking, large (rams up to 120kg) and fecund (80% twinning rate) Charolais breed has attracted a lot of attention.

Table 5.11 Number of Sheep by Wool Type in Chifeng City Prefecture as a Whole and in Balinyou and Alukeerqin Banners, 1980 to 1991*

Year	Total sheep	Fine wool sheep		Improved fine wool sheep		Local wool sheep	
		Number	Proportion of total	Number	Proportion of total	Number	Proportion of total
	('000)	('000)	(%)	('000)	(%)	('000)	(%)

Chifeng City Prefecture

Year	Total sheep	Number	Proportion	Number	Proportion	Number	Proportion
1980	3,620.5	259.0	7.2	2,224.9	61.4	1,136.6	31.4
1981	3,723.9	190.5	5.1	2,315.5	62.2	1,217.9	32.7
1982	3,971.6	194.1	4.9	2,546.4	64.1	1,231.1	31.0
1983	4,007.9	477.7	11.9	2,515.8	62.8	1,014.4	25.3
1984	3,620.0	492.8	13.6	2,278.7	62.9	848.5	23.5
1985	3,395.1	433.5	12.8	2,338.0	68.9	623.6	18.3
1986	3,159.7	359.0	11.4	2,348.4	74.3	452.3	14.3
1987	3,065.0						
1988	3,197.0	n/a		n/a		n/a	
1989	n/a						
1990	2,451.8	273.2	11.1	1,912.3	78.0	266.3	10.9
1991	n/a	n/a		n/a		n/a	

Balinyou Banner

Year	Total sheep	Number	Proportion	Number	Proportion	Number	Proportion
1980	299.1	6.5	2.2	214.1	71.6	78.5	26.2
1981	332.0	8.5	2.6	220.7	66.4	102.8	31.0
1982	375.1	10.1	2.6	259.8	67.5	105.2	29.9
1983	408.0	9.1	2.2	320.9	78.7	78.0	19.1
1984	394.0	5.6	1.4	317.2	80.5	71.2	18.1
1985	401.0	3.7	1.0	341.4	85.1	55.9	13.9
1986	376.0	5.8	1.5	329.9	87.8	40.3	10.7
1987	368.0						
1988	398.0						
1989	490.0	n/a		n/a		n/a	
1990	390.1						
1991	438.9						

Alukeerqin Banner

Year	Total sheep	Number	Proportion	Number	Proportion	Number	Proportion
1980	434.6	15.2	3.5	305.8	70.4	113.6	26.1
1981	440.0	11.9	2.7	295.6	67.2	132.5	30.1
1982	496.7	2.6	0.5	355.1	71.5	139.0	28.0
1983	519.4	11.9	2.3	375.4	72.3	132.1	25.4
1984	487.1	12.4	2.5	342.8	70.4	131.9	27.1
1985	487.7	14.7	3.0	382.3	78.4	90.7	18.6
1986	436.3	10.6	2.4	352.8	80.9	72.9	16.7
1987	420.0						
1988	455.0						
1989	517.9	n/a		n/a		n/a	
1990	500.1						
1991	538.9						

*Mid-year data.
Source: IMAR Animal Husbandry Bureau

5.3.4 Goats

At the end of June 1989, there were 11 million goats being kept for cashmere production in IMAR. In the last few years, there has been an effort to reduce coloured goats which previously constituted 40% of the herd and now most goats are white. The Liaoning white cashmere goat has been introduced to the eastern parts of IMAR. This breed of goat is selected in the Gaixian County in Liaoning Province for its production of cashmere, which is normally 16-17μm in fibre diameter.

The Mongolian white cashmere goat is used in the western part of IMAR, and has been selected from the traditional Mongolian cashmere goat herds. It has a finer cashmere (13-15μm) than the Liaoning white cashmere goat but its yield is significantly lower. When female goats are selected on the basis of their cashmere production, the normal expectation is that in eastern IMAR, the female will yield around 400g of cashmere per year, while 350g per head is acceptable in western IMAR. Inner Mongolia produces about 1,900 tonne of cashmere per year.

5.4 Animal Husbandry Research and Extension Organisations

Under the Agricultural and Animal Husbandry Commission of the IMAR government, there are two bureaus concerned with administering the animal husbandry affairs of IMAR, namely, the Animal Husbandry Bureau and the Economic Management Bureau.

5.4.1 Animal Husbandry Bureau

The Animal Husbandry Bureau of IMAR (AHB/IMAR) is vertically connected to the Animal Husbandry and Veterinary Science Department of the Ministry of Agriculture in Beijing. The AHB/IMAR has responsibility for the Pasture Management Station; the Animal Husbandry Station; the Veterinary Station; the Livestock Improvement Station; the Academy of Animal Husbandry Sciences of IMAR; and an Animal Sciences College in Huhehot (the staff of which conduct some research). The AHB/IMAR also works closely with the Grassland Research Institute, which is a national research institute within the Chinese Academy of Agricultural Sciences located in Huhehot (Henzell *et al.*, 1987), and with research units in several universities.

In most provinces in China, there is usually a pasture management station, an animal husbandry station, a veterinary station, and an economic management station at the provincial, prefectural and county level. However, at the township level, all four are commonly combined into one station. The first three types of station (pasture management, animal husbandry and veterinary stations) are primarily responsible for the development and dissemination of production technology information. The economic management station is responsible for a number of things, including the collection of statistics and the provision of marketing information. The major source of marketing information for farmers, however, is likely to be the Supply and Marketing Cooperatives (SMCs) and the Ministry of Domestic Trade food companies with which the farmer has traditionally been forced to trade most of his outputs, and, in the case of the SMCs, from which he must buy most of his inputs.

The roles of the provincial level stations under the AHB/IMAR are mainly concerned with administration and training. For example, the Pasture Management Station in Huhehot, which has approximately 70 staff, is mainly concerned with administering various programs relating to pasture improvement; management of pastures; utilisation of pastures; and small-scale experiments (used normally only as demonstration plots). The Pasture Management Station is also responsible for training county level technicians, some more senior level people, and perhaps even some household heads. It does this by running demonstrations, field days and short-term training programs. The primary way in which this station disseminates information is to provide it to the staff of the prefectural and county level pasture management stations, who in turn are expected to carry out the extension role at lower levels.

The Academy of Animal Husbandry Sciences of IMAR was established in 1954, and represents the research arm of the AHB/IMAR. Within the Academy there are four divisions or institutes:
- Animal Husbandry Research Division
- Animal Husbandry Economics Division
- Grasslands Research Division
- Veterinary Research Division.

In 1990 the Academy had almost 300 staff of whom about 10% were senior research workers. The Academy (and the Animal Husbandry Research Division in particular) was one of the major collaborating research institutes involved in another ACIAR project concerned with mineral nutrition problems in sheep (Lehane, 1993).

5.4.2 Economic Management Station/Bureau

The IMAR Economic Management Station for Agriculture, Animal Husbandry and Fisheries is unusual in that it is on the same level as the AHB/IMAR. That is, unlike the situation in most provinces/autonomous regions where the economic management station is one station under the AHB, in IMAR it has equal status.

The roles of this organisation are to collect economic statistics; to calculate the cost of producing animal products; to assess technical and economic efficiency; to manage land contracts; to manage collective capital; and the dissemination of economic information to farmers. It has 52 staff at the provincial level, 18 stations at the prefectural level, 114 stations at the county level, and 570 stations at the sumu level.

Every village in IMAR has an accountant who is responsible for collecting the economic statistics from all households in that village on behalf of the Economic Management Station. There are about 10,000 village-level accountants located throughout the IMAR. The county/banner government is responsible for training these accountants. There are also another 3,500 accountants working at the township level, and usually the prefectural-level economic management station is responsible for training these people. The provincial economic management station is primarily responsible for training the prefectural and county level staff. The Ministry of Agriculture in Beijing provides training for the provincial level staff, mainly at an economic management college. Most of the provincial-level economic management station staff are university graduates.

5.4.3 Vertical Bureaucratic Connections

Both the AHB/IMAR and the IMAR Economic Management Station are provincial-level government organisations which are vertically connected with the corresponding organisation in the Ministry of Agriculture in Beijing. It was claimed that there is no direct vertical control exercised by the corresponding Beijing group. The higher level only provides "advice and training". There are, however, special project funds which are passed down the vertical channels which help cement the vertical linkages between the instrumentalities. The Director of the Economic Management Station for IMAR indicated that cooperation at the prefecture and county level could be enhanced by judicious use of these project funds. Nevertheless, the staff working for the stations at the various levels are paid and promoted by the governments at those respective levels. Their primary allegiance, therefore, would be to the governments at the level at which they are operating, rather than to their vertically superior organisation.

5.5 Introduction of the Household Production Responsibility System

The so-called "double track" or double contract household production responsibility system which now exists in the pastoral areas was introduced in the IMAR in a series of steps. The first step was taken in 1977 when experiments were conducted in some communes to investigate the possibility of reintroducing the cooperative systems which existed in the mid 1950s before collectivisation (see Section 3.5.3).

5.5.1 Privatising the Livestock

The second stage was introduced first in Zhelimu Prefecture in 1981 when the animals were leased to the farm households. The farm households paid a six-monthly lease fee to the collective for the use of the animals and also paid a tax into the collective accumulation fund, and another tax into the collective public welfare fund. After these payments to the collective, the remaining profit could be kept by the households. At this time, the animals were still owned by the collective.

The third stage began in 1983 when, throughout the IMAR, the animals were sold to the farm households by the collectives. In most cases, the farmers were not charged any interest and had five to seven years in which to pay the collectives for their animals. The official view in 1990 was that most farmers had repaid their collectives. To the extent that this is the case, it would be interesting to know what the village collectives did with the money.

In some cases, the collectives had other commercial enterprises which made them relatively wealthy and they did not enforce the repayment of the livestock loans. In other areas, the farmers simply refused to repay their loans on the grounds that they had contributed their animals free to the collectives in the 1956 collectivisation program, and they did not appreciate being forced to buy them back. This attitude appeared to receive an official blessing when the IMAR government approved special arrangements for two banners. In these banners, the collectives were allowed to distribute the livestock free of charge to the households. The official reason for the change of policy was that pastoral incomes were very low in these banners. However, the two banners involved were Xianghuang Banner and Abaga Banner, both in Xilinguole League. These banners are not classified as being poor banners by either the Central or the Autonomous Region government (Table 5.2). In fact, they are considered to be representative pastoral banners in Xilinguole League, which is one of the two best pastoral leagues in the IMAR. The real reason for permitting the animals to be returned to the farmers free of cost may be related to the high proportion of people of minority nationality in these banners (Table 5.2) and the political sensitivity of the issue.

When the animals were sold to the farmers in 1983, the values were set for each individual animal, and in general they were a little higher than the value of the animals at the time of collectivisation in 1956, but somewhat lower than the market prices which existed in 1983. The animals were valued and distributed to the households taking account of the household size and the number of labourers in the household, as well as the various species and types of animals to be distributed. The distribution of the collective livestock to the various households which were members of the collective must have been a remarkable exercise.

5.5.2 Contracting Out the Pasture Land

By 1990, all of the cutting land and about 90% of the remaining usable pasture in the IMAR had been "contracted out" to households. That is, while ownership of the pasture land remained with the village collectives, the right of use had been assigned, under contract, to the private households. The rangeland not yet allocated in this way was usually in remote areas and had no water or was badly infested by weeds. In some areas it had proven difficult to identify and allocate separate pieces of each type of pasture to each household. In other areas, although the land had been nominally allocated, it was impractical to manage the individual areas separately. Therefore, much of the rangeland was still grazed in common. This was especially true in the semi-pastoral and agricultural areas where the best land was used for crops and the available pastures were interspersed between the crops or scattered throughout the territory controlled by the village collective in question.

The allocation of particular pasture areas to particular households was determined on the basis of the history of the family using the pasture, the willingness of the farmers to accept the responsibility for taking charge of the pasture, the number of people in the household, the number of livestock they had to take care of, the number of workers, and the species of livestock involved. Roughly speaking, the land was allocated to the households on the basis of a 70% weighting for the number of people in the family and a 30% weighting for the number of animals owned by the family. In some areas, the number of labourers in the household was taken into account either instead of or as well as the number of people in the household.

The original pasture contracts were issued in the early 1980s, and the system has continued to evolve. Most households have operated under a "double contract" system since about 1985. That is, the households which bought the animals from their collective and which at that time entered into a contract to pay certain taxes and levies to the collective in return for private ownership of the animals, also signed a second contract for the right to use the pastures owned by the collective. The original contracts were simple agreements giving the household access to certain pasture areas for a specific time period, usually 15 years. In recent times, some counties have introduced new pasture use contracts (or leases) which extend the "right to use" period up to 30 years (with inheritance rights). The county governments enjoy a considerable degree of autonomy over the administration of rangeland contracts. As illustrated by the case studies, the differences in the approaches adopted to these contracts by different counties, even in the same prefecture, are remarkable. For a comprehensive review and critique of the rangeland contracting system, see Williamson and Longworth (1993).

Initially the use-rights were assigned to the households free of charge. However, by 1987, experiments were underway with what is now called the "user pays" system. The idea is that households with pasture use contracts should pay an annual fee to the collective. This fee is variously described as a pasture management fee, a pasture use fee, a grazing levy, etc. but it is really an annual rental charge. The principal motivation for introducing the fee was to collect funds to help maintain and develop the pastures. However, depending upon how it is structured, it may also provide an incentive for farmers to reduce overgrazing and encourage farmers to increase the quality (as distinct from the quantity) of their livestock.

By 1990, five counties in the IMAR had adopted a "user pays" system still more or less on an experimental basis. These counties were Alukeerqin and Balinyou (two of the counties surveyed in Chifeng City Prefecture), Xianghuang and eastern Xiwuzhumuqin (both in Xilinguole League), and Wulatezhong (in Bayanzhuoer League). The pasture use fee was sometimes set on a per mu basis (see, for example, the Appendix to Chapter 8) and sometimes on a per sheep equivalent basis (see Appendix to Chapter 11). In principle, it would seem that a per sheep equivalent basis would provide an incentive to graze better quality, more productive livestock. It would also be simpler to administer because there would be no need to identify and classify all the pastures (so that differential fees could be charged for different quality pastures).

The pasture use fees have been introduced along with new pasture land contracts. As the Appendices to Chapters 8 and 11 demonstrate, these new contracts are remarkably detailed and sophisticated legal documents. Not only do they define the "user pays" system but they also establish maximum allowable stocking rates for each type of pasture and define the penalties for exceeding these rates. On the other hand, the new arrangements confirm the rights of the household under the original contracts, extend the term of the "lease", and (in some cases) permit limited sub-contracting out of the pastures to a third party.

The experiments with the "user pays" principle must have been considered successful in the IMAR because, according to a newspaper report (*China Daily - Business Weekly*, April 18-24, 1993, p.7), the policy of charging an annual fee for the use of pasture had been extended to apply to over half the herdsmen in the Region by the end of 1992. However, as discussed in the Balinyou County case study (Chapter 8), having a policy of "user pays" as outlined above and actually collecting the fees, are two different things. The application of the "user pays" principle to pastures, at least in some areas, seems to be a case of "policy mirage". Officially the policy is in place and, as far as the higher levels of government are concerned, everything is under control. At the grass roots on the other hand, while local cadres pay lip service to the official policy, they do not implement it in any meaningful way because it is unacceptable to their constituents.

5.6 Trends in Incomes and Wool Pricing

Of the 54 Central government recognised pastoral and semi-pastoral banners in IMAR, 17 pastoral and 10 semi-pastoral banners are designated as poor counties (Table 5.2). That is, exactly half of the banners/counties in IMAR which are included in the Central government's list of 266 pastoral and semi-pastoral counties are considered to be poverty stricken.

Wool sales are a significant determinant of many household incomes in those poor banners. Wool pricing policies in IMAR, therefore, have a major impact on incomes.

5.6.1 Trends in Rural Incomes

Time series data on average net income per capita for Balinyou, Wongniute, Aohan and Alukeerqin Banners are presented in the case studies (Chapters 8 to 11). Some of these data have been plotted in Fig. 5.2.

It is interesting to note the variation in the graphs even though these four banners are all located close together in one prefecture. Incomes in Alukeerqin and Balinyou, the two more remote pastoral banners, are higher than in Wongniute which is a semi-pastoral banner. Aohan, the most agricultural of the four banners, has the lowest incomes. This ranking of income levels is common throughout the pastoral region. Incomes tend to be higher in the sparsely populated pastoral counties and lower in the more densely populated areas where there is more agricultural activity. Whether the real standard of living is highly correlated with monetary income estimates is another question. Households living in remote, sparsely settled pastoral areas must purchase a high proportion of their personal food grain requirements and have much less access to public facilities such as education and health services.

Nominal and real incomes increased sharply in all four banners up until about 1983. Since then nominal and real incomes have diverged, with real incomes plateauing and then tending to fall in the latter part of the 1980s. Although for obvious reasons Chinese officials emphasise the rising trends in nominal incomes, real incomes determine the economic welfare of the people. In this regard, the graphs in Fig. 5.2 suggest that real incomes in the pastoral region of IMAR have not improved since about 1983. On the other hand, real incomes in many other parts of China have increased significantly since the early 1980s.

It is important to appreciate that pastoral households (i.e. households in which more than 50% of the income comes from the sale of wool and other pastoral commodities) constitute a relatively small fraction of total households in many parts of the pastoral region. For example, in Balinyou Banner only about one-third of the households are classified as pastoral; in Alukeerqin the proportion is about one-quarter; it is less than a fifth in Wongniute; and is only a tiny 2% in Aohan. The average rural incomes plotted in Fig. 5.2, therefore, may not reflect simply the changing fortunes of pastoral households *per se*. At the same time, sheep-raising is a major sideline activity of many rural households which are not classified as pastoral. Therefore, wool and mutton sales are a significant determinant of household income for a much higher proportion of rural households than represented by the percentage of pastoral households.

The potential for wool and other pastoral commodities to contribute further to the improvement in rural incomes in the pastoral banners of IMAR is severely limited. The current widespread overstocking problem and the continuing degradation of the native pastures impose a major physical constraint on future production increases. While there is little if any capacity to increase the total output of pastoral commodities, a switch to better quality and hence higher value outputs could improve incomes to some extent. Nevertheless, in general, the further development of the pastoral resources in many poor banners of IMAR does not represent a realistic approach to raising average rural incomes in these banners.

5.6.2 Wool Pricing Since 1984

The Provincial Price Bureau under the general guidance of the National Price Bureau and in consultation with the prefectural-level textile mills, the IMAR Bureau of Textiles, the IMAR Bureau of Animal Husbandry and other interested groups, has traditionally had the responsibility for setting the official State wool purchasing prices for the IMAR. Prior to 1985, wool prices in the IMAR were usually those established on a national basis by the National Price Bureau. These prices for 1981 to 1984 were discussed in Section 4.5.2 and presented in Table 4.5.

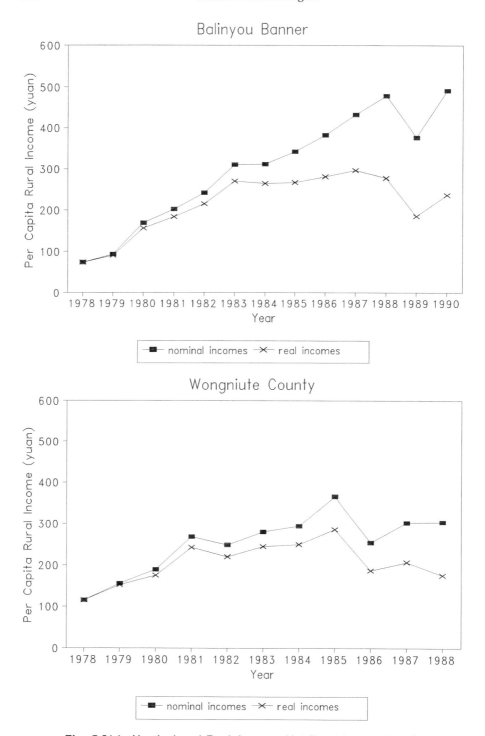

Fig. 5.2(a) Nominal and Real Average Net Rural Income Per Capita for Balinyou and Wongniute Counties in IMAR

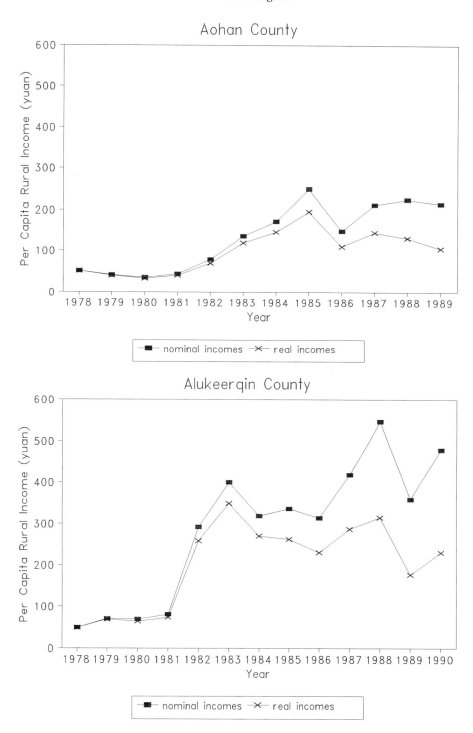

Fig. 5.2(b) Nominal and Real Average Net Rural Income Per Capita for Aohan and Alukeerqin Counties in IMAR

Despite the 1985 Central government decision to allow provinces to open their wool markets if they wished, the government of the IMAR did not allow prefectural governments to operate free wool markets until 1986. Consequently, at least in theory, the official State prices were the purchase prices paid to all farmers by the SMCs in the IMAR in 1985. These official 1985 prices were set by the Provincial Price Bureau as being the same as those shown in Table 4.5 for 1984, plus 20%. In 1986, 1987 and 1988, the prefectural administrations were given permission by the provincial government to permit free trading in raw wool. At the same time, the Provincial Price Bureau continued to set official State prices which were supposed to apply by default in those areas where the prefectural administrations and/or county governments did not permit a free market to operate. For 1986, these official purchasing prices were increased to levels which were 25% above 1984 prices. In 1987, the Price Bureau established the official prices 35% above the 1984 level.

In 1988, the IMAR government aimed at stabilising wool prices at their 1987 level. But because the so-called "wool war" was raging and there was a large number of buyers active in the market (Watson and Findlay, 1992), the prices paid for wool rose to more than double the 1984 prices in some areas.

During the 1985 to 1988 period despite the fixed prices set by the IMAR Price Bureau, purchase prices paid to farmers by the SMC in those areas where the market remained officially closed, varied widely from prefecture to prefecture (and even between counties within the same prefecture). There were two main reasons for this situation. First, some prefectures (and counties) wanted to encourage wool production and were not concerned about the cost of the raw material to the textile mills. The prefectural (and county) Price Bureaus in these areas tried to set the official purchase prices as high as possible. On the other hand, local governments (at the prefectural and/or county level) which owned and operated textile factories were keen to keep raw wool prices down to increase the profitability of their factories. The second reason for the growing regional variation in official raw wool prices during the 1985 to 1988 period was the emergence of the "black market" for wool in these officially closed areas, with many government units and individuals competing with the SMC for the available wool. This "private" illegal competition was especially important in counties (and prefectures) adjoining provinces such as Liaoning where a vigorous free market in wool had developed from 1985 onwards.

In those prefectures which opened their wool markets, the SMCs faced even more vigorous competition for the available wool from other government units and private dealers. The rush to buy and deal in raw wool pushed prices up sharply in 1987 and 1988. For example, in June and July 1987 the average price per kg of raw (greasy) wool in the IMAR was ¥9.80 to ¥10.00 but this rose to ¥14 to ¥16 per kg in June/July of 1988. The market became very disorderly with farmers and wool dealers adulterating wool because buyers were often not experienced and would buy almost any kind of wool.

The wool-buying season in China commences in June and is most active in June and July because most sheep are shorn at that time. Traditionally, official State prices for wool are announced just before the buying season commences, that is, normally in May or early June.

By mid-1989, the market conditions for raw wool were much less buoyant than at the beginning of the 1988 season, which had been the peak of the 1980s wool boom in China. The strong deflationary monetary and fiscal policies implemented by the Central government from September 1988 had severely dampened the demand for finished textile goods, for textile fabrics and for woollen yarns and consequently for textile raw materials such as raw wool. Political events in China in May 1989 which culminated in the June 4 incident in Beijing, also affected the demand for raw wool. In fact, the "private" wool buyers and dealers who had been so active in 1987 and 1988 had virtually disappeared by the time the 1989 wool-buying season opened in June of that year.

Consequently, the provincial government of the IMAR decided to close the wool market in all prefectures for the 1989 and subsequent seasons. Once again, the grass roots SMCs became the only organization permitted to buy wool from farmers in the IMAR.

In recognition of the depressed market, the IMAR Price Bureau, on behalf of the provincial government, changed the way the official purchasing prices for the IMAR would be established for 1989. They announced the set of maximum and minimum (reserve) prices shown in Table 5.12 rather than a single set of fixed prices. There continued to be no differentiation in prices for fine and semi-fine wool. Furthermore, both the maximum and minimum prices were set relative to the reference type and grade (Improved fine and semi-fine wool - grade I) according to the quality differential written into the National Wool Grading Standard. (See Section 4.5.2.)

Table 5.12 Maximum and Minimum Prices for the Various Types of Wool Established by the IMAR Price Bureau for the 1989 Wool Purchasing Season

Type of wool	Grade	Maximum price	Minimum price (reserve)	Quality differential
		¥/kg (greasy)	¥/kg (greasy)	(%)
Fine and semi-fine wool	- special grade	10.88	6.86	124
	- grade I	10.00	6.32	114
	- grade II	9.40	5.92	107
Improved fine and semi-fine wool	- grade I	8.78	5.54	100
	- grade II	7.98	5.04	91
Coarse wool		8.18	3.20	-

Source: No. 3 Woollen Mill, Huhehot, 1990

Although officially the prices paid to farmers by the grass roots SMCs are supposed to be the same throughout the whole wool year, in the post-1985 era many grass roots SMCs had begun to vary their prices during the season so that they could compete with "private" buyers. In the 1985 to 1988 seasons, most farmers who stored their wool after shearing were able to obtain higher prices later in the season. However, because the market became progressively more depressed as the 1989 season advanced, the reverse was the case in 1989. In fact, due to the build-up in stocks at the end of 1989 and into 1990, the SMC stopped buying raw wool and many farmers who had initially held their clips off the market in the second half of 1989 could not sell their wool.

Despite the deteriorating market situation, the IMAR Price Bureau set maximum official prices in 1990 which were the same as in 1989 (and shown in Table 5.12). Furthermore, it actually raised the minimum (reserve) prices for all types by 27% relative to 1989. This kind of official pricing policy effectively priced domestic raw wool out of the textile fibre market in China in 1989 and 1990.

Appendix 5A

Rangeland Management Regulation of the Mongolia Autonomous Region

Adopted at the Second Session of the Sixth Mongolia Autonomous Region People's Congress (June 7, 1984). Revised According to the 22nd session of Standing Committee of the Seventh Mongolia Autonomous Region People's Congress (August 31, 1991).

CONTENTS
The First Chapter	General Principles
The Second Chapter	Rangeland Ownership and Right of Use
The Third Chapter	Protection, Management, Construction and Use of Rangeland
The Fourth Chapter	Organisation of Rangeland Management
The Fifth Chapter	Rewards and Punishments
The Sixth Chapter	Supplementary Articles

FIRST CHAPTER   GENERAL PRINCIPLES
Article 1
The regulation is formulated in accordance with the Constitution of the People's Republic of China, the Law of the National Autonomous Regions of the People's Republic of China and Rangeland Law of the People's Republic of China in order to further the protection, management, construction, and reasonable use of rangeland, improve the ecology, raise rangeland productivity and develop animal husbandry.

Article 2
This regulation is applicable to all rangeland, including pasture area, semi-pasture area, agricultural area, suburban district and forest zone. Included also are wild animals and plants found in these areas, within The Inner Mongolia Autonomous Region.

Article 3
It is an important duty of the different levels of government to further the protection, management and use of rangeland.

All the grassland under the jurisdiction of the relevant levels of government must be surveyed, protected, used and constructed in accordance with a general program in order to preserve the ecological balance and everlasting use of grassland.

Article 4
Protecting grassland is the duty of all units and citizens, which have right of supervision, accusation and complaint when discovering actions related to defacing of the grassland.

SECOND CHAPTER   RANGELAND OWNERSHIP AND RIGHT OF USE
Article 5
All rangeland within the Inner Mongolia Autonomous Region is state and collective property.
1 Rangeland which has been approved for enterprise and institutional use by county governments, or which is not developed and used or which does not belong to collectives is state property.
2 Rangeland that is used by pastoral area and rural collective organisations is collective property.
3 Small tracts of forest, brushes, reed, medicinal herbs (wild plants on rangeland) are the property of rangeland ownership units, except the rangeland which is allocated to state enterprise and institutions by the above county government.

Article 6
After right to use the rangeland is evidenced, state-owned and collective-owned rangeland must be registered. Collective-ownership of rangeland is provided by a License of Rangeland Ownership. State-ownership of rangeland is provided by a License of Using Rangeland. The two licenses are printed uniformly by the Autonomous Region government and provided by the county government.

Article 7
The units which have the ownership rights to the rangeland are able to contract rangeland to a management unit and an individual for long-term use, which fit in with the animal contract system.

The contractor has the management rights and use rights to rangeland contracted, and also has the duty of rangeland protection and construction. Contractors who degrade the rangeland and damage the rangeland vegetation because of careless protection and management along with less than adequate levels of improvement and restoration shall have their contracting rights rescinded.

Article 8
The management rights to, and the right to use, a rangeland are protected by law. No unit or individual shall violate these rights.

Article 9
No one shall buy or sell the rangeland by any way. No one shall illegally transfer or occupy the rangeland.

Article 10
A dispute over the ownership, or the right to use, a rangeland shall be settled through negotiations between the parties concerned, with good understanding and compromise and an approach conducive to harmony and unity. If the negotiations proved unsuccessful, the county level people's governments concerned shall initiate the negotiations between the two parties.

1. A dispute over the ownership of, or the right to use, a rangeland, arising between collective organisations, between collective organisation and the unity of the county, shall be arbitrated by the people's government of the county level.
2. A dispute over the right to use a rangeland between counties and between the county and the unit of the prefecture or city shall be arbitrated by the people's government of the prefecture and city level.
3. A dispute over the right to use a rangeland between prefectures or city and between the prefecture or city and autonomous region shall be arbitrated by the people's government of the autonomous region level.
4. A dispute over the right to use a rangeland between the autonomous and neighbouring province, or between the autonomous region and the department of the State Council, shall be arbitrated by the people's government of the autonomous region or the State Council.

Article 11
If a disputing party refuses to accept the government's ruling, the party, within fifty days from the date on receiving the government's ruling, may apply for reviewing a case to a higher government. If refusing to accept the higher government ruling, the party within fifty days from the date on receiving the government's ruling may appeal to the local people's court for hearing. If the party neither apply for reviewing a case nor appeal for hearing, or accept the government's ruling as well, the ruling will be carried out by order of the local people's court.

Article 12
A dispute over the boundary is arbitrated, according to the following principle:

1. The boundary between the autonomous region and neighbouring province is delineated in accordance with the boundary prior to the Liberation.
2. The boundary between the counties and between collective production units is delineated in accordance with the boundaries devised during the time of Agricultural Cooperation.

Article 13
No one is allowed to damage a disputed rangeland/or its facilities, pending a proper settlement of the dispute over the ownership of, or the right to use, the rangeland.

THIRD CHAPTER   PROTECTION, MANAGEMENT, CONSTRUCTION AND USE OF RANGELAND

Article 14
The unit which has been allocated the ownership right and the right to use the rangeland shall survey the rangeland and organise a program for the construction and use of the rangeland, in order to develop animal husbandry production, industry and sideline production in accordance with the status and characteristics of the rangeland resources.

Article 15

The people's government at the county level shall determine a suitable grazing rate to be followed by those units having jurisdiction over the rangeland.

The unit using the rangeland shall survey the rangeland, then set the end of year animal population in accordance with the observed level of grass output in order to maintain a balance between range production and animal population.

If a rangeland turns sandy and degrades, the user shall be ordered to spell and fence the pasture, or otherwise regrow grass in order to restore the vegetation.

Article 16

The contractors are asked to uniformly devise a management program to improve and construct the rangeland contracted in a planned and organised manner.

The unit undertaking rangeland construction owns the right to that construction. No one unit or individual shall violate this right. When the right to use is transferred, the unit or individual who receives the right shall pay compensation to the last contractor for any rangeland construction.

Article 17

In order to requisition rangeland, the relevant unit shall apply to the rangeland management office, in order to examine and verify the rights to the land. Finally, the unit must obtain approval from an administrative level higher than the people's government at the county level.

Any requisition for less than ten mu of rangeland can be approved by the people's government at the county level.

Requisitions for more than ten mu but less one hundred mu, must be approved by the people's government at the prefectural or city levels.

Requisitions for more than one hundred mu but less two thousand mu, must be approved by the people's government at the autonomous region.

The unit which requisitions rangeland shall pay compensation to those units which have interests in the ownership and use of the rangeland.

Article 18

In a special situation such as a natural calamity when it is necessary for an unaffected rangeland between the counties and between the prefectures or cities, or between the collective production unit to temporarily accommodate the herds of the afflicted area, the two parties concerned, shall negotiate the terms, sign a contract and define limits and time limit on the basis of volition and mutual benefit, which shall be approved by the rangeland management office of the higher level.

Article 19

No unit or individual is allowed to illegally farm the rangeland and damage vegetation. Permission to farm the rangeland shall only be given by the people's government at the autonomous region.

Article 20

No one is allowed to farm rangeland with the following characteristics:
1 Areas which are prone to desertification following removal of protective vegetation.
2 Areas with slopes greater than 20 degrees.
3 Non-irrigation areas where annual rainfall is less than 335 mm.
4 Areas where ownership is disputed.
5 Areas which constitute animal paths and watering points.

Article 21

With respect to rangeland which has been farmed in circumstances outlines in 1 and 2 below, the local government shall order the rangeland to be closed within a definite time period and the vegetation restored:
1 Cultivation has caused the rangeland soil to deteriorate or to erode through water run-off.
2 The rangeland has been farmed illegally.

Article 22

Hunting of wildlife on rangeland shall be controlled rigorously according to the law of protecting wild animals of the state and autonomous region.

Article 23

Anyone wishing to collect wild plants from a rangeland shall obtain the permission of the unit of the rangeland before applying to the rangeland management office of county level.

Article 24
The autonomous region is able to delineate designated areas for the protection of rare and precious species of animals and plants. No one shall cut, collect and hunt in the areas without first obtaining the permission of the people's government at the autonomous region.

Article 25
The government undertakes to protect the ecology of the rangeland, and to combat pollution where it occurs. Units which have polluted the rangeland shall compensate for any losses in accordance with the related laws and regulations.

Article 26
Main roads shall be delineated and indicated by traffic signs issued by the traffic office. Roads which have been built without permission shall be closed.

Article 27
The commerce department or SMC shall drive animals along fixed ways appointed by the traffic office.

Article 28
Units or individuals which have permission to prospect for the purpose of mining, dig wells, pick stones, dig medicinal plants and so forth on the rangeland, shall fill in any holes arising from these works in order that the vegetation be restored.

Article 29
Safety measures shall be adopted and strictly adhered to, more particularly, during specified seasons in which range fires are likely to occur. Should a fire break out, the local authorities shall organise the largest number of people to extinguish it quickly.

Article 30
Units or individuals which damage the rangeland vegetation through grazing without permission, prospecting for mines, picking stones, digging wells, and driving animals are required to pay any fees necessary for the maintenance of the rangeland.

The fees for rangeland maintenance will be charged by the rangeland management office at the county level and are to be used for rangeland construction.

FOURTH CHAPTER   ORGANISATION OF RANGELAND MANAGEMENT

Article 31
The department of animal husbandry at the different levels of government is responsible for work arising out of rangeland management, as well as rangeland management organisation.

Article 32
The main duties involved in rangeland management organisation are as follows:

1. To publicise the regulations and policies contained within Rangeland Law of the People's Republic of China, and supervise their enforcement.
2. To take responsibility for the examination, registration and management of the rights to the ownership and use of the rangeland and to provide Certificates of Rangeland Ownership and Rangeland Use.
3. To regulate rangeland requisitions and the temporary allocation of rangeland to the provisional department of land management.
4. To handle applications for the opening up of the rangeland and to provide certificates for specified work on the rangeland.
5. To handle disputes surrounding the ownership rights or the use rights to the rangeland and, where necessary, to adjust to the rights to use rangeland.
6. To check and ratify the grazing capacity and supervise protection, management, construction and use of the rangeland.
7. To investigate and take appropriate action where the rangeland law has been violated and the rangeland damaged.
8. To charge, manage and supervise the use of fees for rangeland maintenance and management.

Article 33
The people responsible for rangeland supervision must wear uniform clothing and badges.

Article 34
Duties for forest police assigned to rangeland areas:
1 To publicise the Rangeland Law of the People's Republic of China and this regulation.
2 To take concerted action with the organisation responsible for rangeland management to supervise and stop anyone damaging the rangeland.
3 To prevent and extinguish fires in cooperation with the organisation of the fire-prevention department.
4 To protect wild animals as specified in the rangeland law.

FIFTH CHAPTER   REWARDS AND PUNISHMENTS
Article 35
The production unit or individual which satisfies one of the following conditions shall be rewarded by the relevant level of the people's government and the organisation for rangeland management.
1 A distinctive contribution towards the protection, management, and construction of a rangeland, and to the development of animal husbandry.
2 A distinctive contribution towards fighting actions which damage the rangeland.
3 A distinctive contribution towards extinguishing rangeland fires.
4 A distinctive contribution towards rangeland research regarding surveys of rangeland resources, application of new technology and dissemination of scientific information about the rangeland, and so on.

Article 36
The following actions are regarded as criminal and punishable by the law.
1 To illegally occupy, buy or sell the rangeland and transfer the possession of the rangeland.
2 To illegally farm the rangeland or refuse to close the rangeland that has been farmed.
3 Driving motor vehicles which do not follow fixed roadways and subsequently lead to a disruption of the rangeland surface.
4 Damage to the rangeland due to overstocking and arbitrary grazing.
5 Violation of fire-prevention rules leading to rangeland fires.

Article 37
Administrative punishment is decided by the rangeland management organisation of the people's government at an administrative level higher than the county level.

Article 38
A person who obstructs staff employed as a member of the organisation of rangeland management and in the execution of public affairs will be punished. The person breaking the law, shall be found responsible for a crime.

Article 39
Staff employed in the organisation of rangeland management who abuse their power, work for their own ends while undertaking public affairs, corruptly accept bribes, violate a citizens legal right, shall be punished by their servicing organisation or a higher organisation. Any person found to be violating the law, will be found responsible for a crime.

SIXTH CHAPTER   SUPPLEMENTARY ARTICLES
Article 40
The people's government of the autonomous region is able to make further detailed rules and regulations in accordance with this regulation.

Appendix 5B

The Local Standard of the Inner-Mongolia Autonomous Region

The Standard of Calculation For a Proper Stocking
Rate for Natural Pasture Land

Issued by: Standard & Measurement Bureau of Inner-Mongolia Autonomous Region
Date of Issue: 17.12.1990
Date of Enforcement: 1.4.1991

1 The contents of the subject and its suitability

The standard is suitable for the calculation of a proper stocking rate on natural pastures.

2 Calculating indexes

(1) The utilized rate of pastures

The utilized rate of pastures represents the grazing intensity for a proper stocking rate. Pastures appear neither to be overgrazed, nor to be degraded. Animal's normal growth and production can be maintained.

The indexes of utilized rate for every type of pasture are as follows:

70-75% for plain pastures in low land areas;
65-70% for plain pastures in mountain areas;
60-65% for plain pastures of grassland;
55-60% for typical pastures of grassland;
50-55% for desert pastures of grassland;
45-50% for desert pastures;
40-50% for sandy pastures.

(2) Conserved rate of pastures

The conserved rate of pastures refers to the percentage or the conserved pasture production during non-growing seasons divided by the highest production during the peak production month.

The indexes of the conserved rate of major pasture types are as follows:

55-60% for shrub and semi-shrub pastures;
60-65% for root and stem grass pastures and cyperaceae pastures;
50-55% for thick growth grass pastures
45-50% for weed pastures.

(3) The usable production of pastures

The usable production of pastures refers to the pasture production which could be produced under the conditions of pastures properly grazed and used. The usable pasture production can be broken down to warm seasons (growing period) and cold seasons (withered seasons) according to the utilized period.

The unit in use is kg/ha.

The conversion formula of usable pasture production is as follows:

- usable pasture production in warm seasons equals the highest pasture production in the peak production month multiplied by the utilized rate (%).
- Usable pasture production in cold seasons equals the highest pasture production in the peak production month (December?) multiplied by the conserved rate. The annual changing rate of pasture production is mainly determined by an annual precipitation.
- The average annual precipitation of several years is regarded as the precipitation in a normal year. 25% increased relative to the precipitation in a normal year is regarded as a good year. 25% decreased relative to the precipitation in a normal year is regarded as a bad year. Natural pasture production is mainly constrained by the imbalanced distribution of the precipitation and heat factors in different seasons and different places. It changes from year to year and from region to region.
- The index of the annual changing rate is shown in the following table:

The annual changing rate of pasture production in different areas

Pasture Types - the annual changing rate (%)	Good Year	Normal Year	Bad Year
Plain grassland	110	100	90
Typical grassland	120	100	80
Desert grassland	130	100	65
Desert pastures of grassland	135	100	60
Desert pasture	130	100	70

(4) The division between warm seasons and cold seasons and the grazing period

The division between warm seasons and cold seasons is determined by local actual conditions and the beginning and ending dates on which the average daily temperature is stabilized at or more than 5 degree centigrade ($\geq 5°C$) respectively.

Grazing period and days for warm and cold seasons in different areas are shown in the following table.

Areas	Season	Beginning period	Ending period	Grazing days
Hulunbeier Plateau	warm	15 May–8 June	5 Oct – 21 Oct	135–160
	cold	6 Oct–22 Oct	14 May–7 June	230–205
The Plain of Western Xiliaohe	warm	25 April–15 May	21 Oct–31 Oct	160–190
	cold	22 Oct–1 Nov	24 April–14 May	205–175
Plain and Plateau of Xilinguole	warm	10 May–20 May	16 Oct–26 Oct	150–170
	cold	17 Oct–27 Oct	9 May–19 May	215–195
Plain and Plateau of Wulanchabu	warm	6 May–15 May	21 Oct–31 Oct	160–180
	cold	22 Oct–1 Nov	5 May–14 May	205–185
Plain and Plateau of Eerduosi	warm	20 Apr–30 Apr	26 Oct–5 Nov	180–200
	cold	26 Oct–1 Nov	24 Apr–8 May	195–175
Plain and Plateau of Alashan	warm	20 Apr–30 Apr	26 Oct–5 Nov	180–200
	cold	27 Oct–6 Nov	19 Apr–29 Apr	185–165

(5) The unit in use

Sheep equivalent refers to one adult sheep

Standard sheep equivalent conversion:

one adult sheep	=	one sheep equivalent
one goat	=	0.9 sheep equivalents
one horse	=	6 sheep equivalents
one donkey	=	3 sheep equivalents
one cattle	=	5 sheep equivalents
one mule	=	5 sheep equivalents
one camel	=	7 sheep equivalents

3 young animals less than one year old can be converted into one adult animal;

3 large animals less than two years old but more than one year old can be converted into two adult animals.

(6) Daily fodder intake

Daily fodder intake refers to the quantity of fodder required daily per animal to maintain normal development and a certain level of production performance. The standard is 2kg of hay.

3 The Calculating Method for a Proper Stocking Capacity of Grassland

A proper stocking capacity refers to the number of animals that can be carried on a certain area of pasture land for a certain period of time under the circumstances of full use of pasture resources, no pasture degradation, and the maintenance of normal development and production of animals.

(1) The calculation of pasture area required by one sheep equivalent
   (a) The unit in use: 3mu/sheep equivalent
   (b) Calculation formula:
   . Pasture area required by one sheep equivalent in warm seasons
     = grazing days in warm seasons x daily fodder intake/available pasture production per mu in warm seasons
   . Pasture area required by one sheep equivalent in cold seasons
     = grazing days in cold seasons x daily fodder intake/available pasture production per mu in cold seasons
   . Pasture area required by one sheep equivalent in a year
     = pasture area required by one sheep equivalent in warm seasons plus pasture area required by one sheep equivalent in cold seasons
(2) The calculation of pasture stocking capacity
   (a) Unit in use is sheep equivalent.
   (b) Calculation formula:
   . Stocking capacity in warm seasons
     = usable pasture area in warm seasons/pasture area required by one sheep equivalent in warm seasons
   . Stocking capacity in cold seasons
     = usable pasture area in cold seasons/pasture area required by one sheep equivalent in cold seasons
   . Stocking capacity
     = total usable pasture area/pasture area required by one sheep equivalent in a year

Appendix 5B(i)

The Measurement of Natural Pasture Productivity

1 Measuring time

The measurement of natural pasture productivity is conducted in July and August in which pasture productivity is higher during the year. If the measurement is conducted before or after that, the productivity measured should be converted into the highest in July and August according to the monthly dynamic coefficient of local pasture productivity.

2 Measuring Method

(1) Selection and measurement of samples.

A typical piece of land should be selected as a measurement sample within the measurement areas.

The measurement samples which can represent the pasture productivity of the sample areas should be selected within the sample areas.

(2) The measurement of pasture productivity

(a) Measurement method for low grasses and low semi-shrub.

The measurement area is one times one square meter, repeated four times. Grasses are cut at the surface of the earth, and then weighed by economic types. After drying, they are weighed again and again until the weights are constant.

(b) Measurement method for shrub and *Achnatherum splendens* (Trin.) Nevski

100 square meters is measured for the shrubs and *Achnatherum splendens* (Trin.) Nevski with a clear thicket. The shrubs and *Achnatherum* (Trin.) Nevski in sample areas can be divided into big, medium and small classes according to the scope and height of each kind of grass. Then count them and average their scopes and select three thickets from each class, cut down the branches which are new ones in the current year.

If the thicket of the shrubs and *Achnatherum splendens* (Trin.) Nevski are relatively small and the differences are not clear, 4 times 4 square meters are measured and cut down all the branches which are new ones in the current year and weigh them respectively.

3 The calculation of pasture production

(1) The calculation of pasture production for low grasses and small semi-shrubs

(a) pasture production per square metre = $\dfrac{\text{sum up the pasture productions of all samples}}{\text{total sampled area (m}^2\text{)}}$

(b) pasture production per mu = pasture production per square meter × 666.7

(2) The calculation of pasture production and occupied area for shrubs and *Achnatherum splendens* (Trin.) Nevski

(a) The calculation of pasture production for shrubs and *Achnatherum splendens* (Trin.) Nevski.

The calculation of average production for each of big, medium and small thickets.

$$A = \dfrac{E}{G}$$

A - the average production of the standard pasture thicket
E - the total weight of a standard pasture thicket production
G - the number of the standard pasture thickets

The calculation of the number of pasture thickets for each class within one mu of pasture land

$$M = \dfrac{F \times 666.7}{N}$$

M - pasture thicket numbers of certain class in one mu of pasture land
F - pasture thicket numbers of the measurement area
N - measurement area

The calculation of the production of shrubs and *Achnatherum splendens* (Trin.) Nevski for one mu of pasture land (W).

$$W = A_{big} \times M_{big} + A_{medium} \times M_{medium} + A_{small} \times M_{small}$$

(b) The occupied area of shrubs and *Achnatherum splendens* (Trin.) Nevski.

The standard thicket area of a certain class of pastures

$$S = \pi \cdot R^2$$

S - the standard thicket area for a certain class
π - ratio of the circumference of a circle to its diameter (3.1416)
R - radius of the standard thicket

The area of shrubs and *Achnatherum splendens* (Trin.) Nevski in one mu of pasture land (S_m)

$$S_m = S_b \times G_b + S_m \times G_m + S_s \times G_s$$

$_b$ - big; $_m$ - medium; $_s$ - small

pasture production of small grasses in one mu of pasture land (P_s)

$$P_s = P_m \times (666.7 - S_m)$$

P_m - pasture production of small grasses per square meter.

The calculation of the total pasture production for shrubs and *Achnatherum splendens* (Trin.) Nevski.

$$T_p = S_p + S_g$$

T_p - total pasture production
S_p - shrub production
S_g - small grass production

$$T_p = A_p + S_g$$

T_p - total pasture production
A_p - the production of *Achnatherum splendens* (Trin.) Nevski
S_g - small grass production

Appendix 5B(ii)

The standard of a proper stocking capacity for different pastures in different areas

Distribution	Pasture Category	Proper stocking capacity mu/sheep equivalent		
		Warm Seasons	Cold Seasons	Whole Year
Hulunbeier Plateau	plain grassland	2 – 4	6 – 9	8 – 13
	steppe	5 – 7	14 – 17	19 – 24
	sandy pastures	7 – 9	16 – 19	23 – 28
	plain pastures	1.5 – 4	4 – 7	5.5 – 11
The Plain of Xiliaohe	plain grassland	3 – 6	7 – 10	10 – 16
	steppe	7 – 10	13 – 17	20 – 27
	desert pastures	7 – 10	10 – 14	17 – 24
	plain pastures	3 – 6	5 – 8	8 – 14
Xilinguole Plateau and Plain	plain grassland	3 – 6	8 – 10	11 – 16
	steppe	8 – 11	16 – 20	24 – 31
	desert grassland	24 – 28	56 – 62	80 – 90
	sandy pastures	8 – 11	16 – 20	24 – 31
	grassland oriented desert pastures	22 – 26	44 – 50	66 – 76
	plain pastures	5 – 8	8 – 12	13 – 20
Wulanchabu Plateau and Plain	plain grassland	4 – 7	8 – 12	12 – 19
	steppe	8 – 12	18 – 22	26 – 34
	desert grassland	12 – 16	28 – 34	40 – 50
	grassland oriented desert pastures	15 – 20	30 – 34	45 – 54
	desert pastures	30 – 35	46 – 52	76 – 87
	plain pastures	4 – 7	9 – 12	13 – 19
Eerduosi Plateau and Plain	steppe and desert grassland	14 – 18	25 – 29	39 – 47
	sandy pastures	10 – 14	17 – 21	27 – 35
	grassland oriented desert and desert pastures	12 – 17	18 – 24	30 – 41
	plain pastures	5 – 8	6 – 10	11 – 18
Alashan Plateau and Plain	desert grassland	12 – 16	15 – 21	27 – 37
	grassland oriented desert pastures	22 – 26	28 – 34	50 – 60
	desert pastures	41 – 47	54 – 60	95 – 107
	plain pastures	5 – 8	6 – 9	11 – 17

Chapter Six

Gansu Province

Gansu is a poor province located in the near north west of China. Indeed, on the basis of six different indicators of development or wealth, Walker (1989, p.466) ranked Gansu as the least developed province in the country.

Total output of greasy wool in Gansu is only about one quarter that of the IMAR. Nevertheless, after IMAR, XUAR and Qinghai, Gansu had the fourth largest wool clip of the 12 pastoral provinces/autonomous regions in 1991. More particularly, Gansu has become an important producer of fine and improved fine wool. As mentioned in Chapter 4, the output of this type of wool in Gansu trebled between 1981 and 1991.

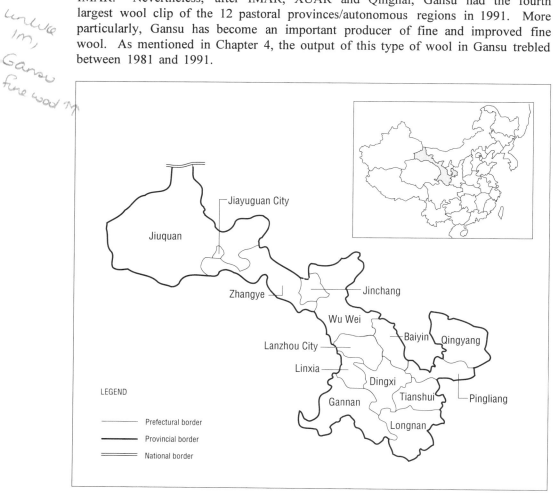

Map 6.1 Prefectural-Level Administrative Units in Gansu Province

Just over 8% of the population of Gansu belong to the 41 recognised ethnic minorities living in the province. The most numerous minorities are Hui, Tibetan and Donxiang.

6.1 Identifying the Pastoral Areas of Gansu

There are 14 prefectural level administrative units in Gansu as shown in Map 6.1 and listed in Table 6.1. Five of the prefectures (Gannan, Jiuquan, Zhangye, Wu Wei and Qingyang) have large areas of grassland and about 60% of the sheep and goats in Gansu are raised in these prefectures.

Table 6.1 Number of Sheep and Goats in Prefectures and Cities of Gansu in 1988

Prefecture/city	Total number of sheep and goats	Proportion of total
	('000)	(%)
Lanzhou	607.3	5.54
Tianshui	254.0	2.32
Baiyin	736.3	6.72
Jinchang	295.8	2.70
Jiayuguan	29.4	0.27
Qingyang	1,186.3	10.82
Pingliang	350.2	3.19
Longnan	408.2	3.72
Dingxi	897.6	8.19
Wu Wei	1,260.0	11.50
Zhangye	1,456.2	13.29
Jiuquan	1,065.0	9.72
Linxia	731.9	6.68
Gannan	1,572.5	14.35
State Farms	109.1	0.99
Total	10,959.8	100.00

Source: Gansu Statistics Yearbook, 1988

The Gansu provincial government recognises the seven pastoral counties and two semi-pastoral counties listed in Table 6.2. Most of the fine wool and improved fine wool sheep in Gansu are raised in these nine counties. However, these counties account for less than one-third of all sheep in Gansu.

The Central government includes the seven pastoral counties in Table 6.2 and a total of 12 semi-pastoral counties in Gansu in its list of 266 pastoral and semi-pastoral counties. These 19 counties are shown in Table 6.3 by prefecture. Two of the seven pastoral counties (Sunan and Tianzhu) were surveyed. A third county (Dunhuang in Jiuquan Prefecture) was also surveyed although it is regarded as an agricultural county and, therefore, is not included in the official Central government list of 266 pastoral and semi-pastoral counties.

Sunan and Tianzhu are easily the two most important wool-producing counties (Table 6.3). They also have two of the three largest sheep flocks of all counties in Gansu (Table 6.2). Sunan is one of the most sparsely populated counties while Tianzhu (with a population density of over 29 people per km^2) is the most heavily populated of the seven pastoral counties in Gansu (Table 6.3). Sunan is a Yugur

Minority Autonomous County in which just over half the population are of minority nationality, mostly Yugur and Tibetans. Tianzhu is a Tibetan Autonomous County in which about one-third of the people belong to minority groups, of which the Tibetans are easily the largest.

Table 6.2 Number of Sheep in the Seven Pastoral and Two Semi-Pastoral Counties Recognised at the Provincial Level in Gansu, 1987

Prefecture	County		Number of sheep
	Pastoral	Semi-pastoral	
			('000)
Wu Wei	Tianzhu		430.7
Zhangye	Sunan		540.5
Jiuquan	Subei		227.0
	Akesai		138.0
Gannan	Xiahe		613.4
	Liuqu		262.7
	Maqu		296.8
		Zhuoni	117.5
		Diebu	71.7
Total for the above nine counties			2,698.3

Source: Gansu Statistics Yearbook, 1988

Dunhuang County is representative of the counties in the desert areas not only of Gansu but also of China generally, where oases or localised pockets of irrigated agriculture enable significant populations to live and prosper. There are many such areas in the Hexi Corridor of Gansu, in Qinghai and in XUAR relying on melting snow for water to irrigate ancient fertile lake beds. One popular viewpoint in China is that these irrigated oases could become the foci for an expansion in the sheep (and goat) population. The small ruminants are considered to be complementary to the existing irrigation-based agriculture in these areas. The Dunhuang case study enabled the feasibility of this proposal to be investigated.

6.2 Physical Details

6.2.1 Geography

Gansu sits astride the old silk road which began in Xi'an, passed through Lanzhou (the capital of Gansu) and then proceeded up through the Hexi Corridor to XUAR, Central Asia and Europe. In land area, Gansu is 0.456 million km² which represents 4.7% of China and an area almost exactly twice the size of the Australian state of Victoria. Most of the land is barren low hills but there are considerable areas of high mountainous alpine country and sandy or stony desert.

Table 6.3 Some Characteristics of the Central Government Recognised Pastoral and Semi-Pastoral Counties in Gansu, 1990

Prefecture/city and county		Poverty status[1]	Population		Total area	Population density	Wool production	Total arable land	Usable pastoral area
			Number of people	Prop. of minority nationality					
			('000)	(%)	(km²)	(person/km²)	(tonne)	('000 mu)	('000 mu)
Pastoral counties									
Jiuquan	- Subei	-	10	41.3	66,748	0.15	318	10	41,877
	- Akesai	-	7	14.6	33,374	0.21	208	4	12,577
Zhangye	- **Sunan***	-	**36**	**50.1**	**23,887**	**1.51**	**1,497**	**39**	**21,030**
Wu Wei	- **Tianzhu***	-	**209**	**31.8**	**7,150**	**29.23**	**830**	**368**	**5,871**
Gannan	- Maqu	-	30	-	-	-	165	-	12,880
	- Luqu	-	27	81.5	5,298	5.10	154	41	5,902
	- Xiahe	-	138	61.4	8,697	15.87	413	313	9,952
Semi-pastoral counties									
Lanzhou	- Yongdeng	nsp	457	1.5	6,090	75.04	466	1,386	-
Jinchang	- Yongchang	-	227	-	7,439	30.52	392	573	-
Baiyin	- Jingyuan	nsp	399	-	5,809	68.69	186	1,386	-
Jiuquan	- Anxi	-	72	0.7	24,130	2.98	92	224	-
Zhangye	- Shandan	-	185	0.7	5,402	34.25	288	590	-
Wu Wei	- Minqin	-	263	-	15,909	16.53	194	924	-
Tianshui	- Zhangxian	nsp	163	0.1	2,164	75.32	58	479	-
Dingxi	- Minxian	nsp	395	-	3,500	112.86	111	639	-
Qing Yang	- Huanxian	nsp	286	-	9,236	30.97	407	1,359	-
	- Huachi	nsp	110	0.6	3,776	29.13	119	401	-
Gannan	- Zhuoni	nsp	88	58.6	5,420	16.24	76	175	4,832
	- Diebu	-	47	70.3	4,826	9.74	15	84	2,401
Total for above 19 counties			3,149	n/a	238,853	13.18	5,989	8,995	117,322
Total for all of Gansu			22,400	8.25	456,460	49.07	15,542	521,500	n/a

*These 2 pastoral counties were surveyed as part of ACIAR Project No. 8811.
[1] Poverty status "nsp" means recognised as a poor county under the Central government poverty alleviation scheme. These counties receive special assistance from both the Central and provincial governments. In addition to these counties, the Gansu provincial government identifies certain other counties for special poverty relief. None of these provincially recognised poor counties in Gansu are classified as being pastoral or semi-pastoral. (See Section 3.4.3 for more details.)

The climate varies a great deal. For example, the mean annual precipitation varies from 30mm to 860mm and the mean annual growing period ranges from 35 to 285 days. Nevertheless, most of Gansu is very dry and the environment is extremely harsh, with mean monthly temperatures ranging from as low as -10°C in January to 26°C in July.

6.2.2 Population

The total population of Gansu Province in 1989 was 21.17 million living in 4.68 million households. The agricultural population was about 18 million consisting of 3.72 million households. The total number of households in the provincially recognised seven pastoral and two semi-pastoral counties is about 90,000 but not all of these would be heavily dependent on pastoral activities for their livelihood. Therefore, only a very small proportion of the households in Gansu are genuine pastoral households.

As mentioned earlier, there are 41 minority nationalities living in Gansu and together they represent 8.3% of the provincial population. Only ten of these nationalities had a population greater than 1,000 in 1990 (Table 6.4). Most of these people live in the pastoral areas in autonomous prefectures (Gannan – Tibetan, Linxia – Hui) or autonomous counties (Subei – Mongolian, Akesai – Kazak, Sunan – Yugur, and Tianzhu – Tibetan). People of the Dongxiang and Yugur Nationality are mostly located in Gansu Province. There are very few people of these nationalities living elsewhere in China.

Table 6.4 Minority Nationalities with Population Greater than One Thousand in Gansu, 1990

Nationality	Population
	(no.)
Mongolian	8,354
Hui	1,094,354
Tibetan	366,718
Manchu	16,723
Kazak	3,148
Dongxiang	311,457
Tu	13,000
Sala	6,739
Bao'an	11,069
Yugur	11,809

6.2.3 Land Use

The total land area of 0.456 million km^2 is equivalent to about 681 million mu. Of this, about 71 million mu is State-owned woodlands and 52 million mu is cropland (including vegetable-growing land). All the cropland is contracted to households. There is about 0.2 million mu of cutting land (i.e. good pasture land used for haymaking), all of which has also been contracted out. Over 5 million tonne of pasture hay is made each year.

Most of the remaining land area is either native grasslands or unusable (deserts, high mountains, devoid of water, etc.). Precise estimates of the total area of usable grassland are not available but it is over 200 million mu. There are many different kinds of native grassland and these pastures vary a great deal in terms of dry matter production.

6.2.4 Types of Grassland

A major survey of grasslands was conducted in Gansu between 1978 and 1985. The details provided in this section are based on a verbal discussion of the results of this survey since written reports on the grassland survey were not available. There had been two previous less comprehensive studies of the grasslands in Gansu, conducted in 1958 and 1964, but it was not possible to obtain any data from these earlier studies. Although not available to outsiders, it seems that a wealth of information exists about the grasslands in Gansu, dating from the late 1950s and there can be little doubt that the relevant Gansu authorities are aware of the massive changes (degradation and desertification) which have occurred in the grasslands of Gansu since the 1950s.

Grasslands of economic importance to animal husbandry in Gansu Province can be divided into four major forms.

- *Desert grassland*

This grassland is distributed along the southern side of the Hexi Corridor and covers an area of approximately 100 million mu. Goats, camels and sheep (fine wool and improved fine wool, and coarse wool breeds) are the dominant animals of economic importance which graze these pastures. Approximately 50% of the desert grassland is usable by livestock. The pasture "usability" is low due to:
- the very low rainfall associated with this particular grassland; [The annual rainfall is said to vary between 30 and 100mm per annum.]
- the low average vegetation coverage; [On average the vegetation is said to cover only around 15% of the ground.] and
- low availability of water for livestock.

Within the desert grassland form, the major categories are as follows:
(a) Low and flat grassland – this grassland is found in areas with relatively high water tables. Dunhuang County, one of the case study counties, has a number of grassland areas which could be described as low and flat grassland. These areas are known as the Xihu and Beihu Lake areas (see the Dunhuang County case study in Chapter 14). The total area of low and flat grassland in Gansu Province is around 5 million mu. The average carrying capacity of this grassland is said to be 16 to 17mu per sheep unit (calculated on an annual basis) with a range from 3mu per sheep unit to 30mu per sheep unit.

The carrying capacity of a particular grassland is determined by Chinese scientists as follows: average annual grassland productivity in terms of dry matter divided by the average annual dry matter requirements of a sheep unit, adjusted for the dry matter utilisation efficiency rate (usually 50% for Gansu Province). The carrying capacity determined in this manner is a theoretical long-term sustainable carrying capacity and tends to underestimate the actual carrying capacity of the pasture land. It does not take into account any short-term fluctuations due to drought or high rainfall. The productivity of pasture varies sharply between years and variations of between 50% and 60% were said to be relatively common. It is a recommended rate and as such is not enforceable throughout the Province. It was said that in almost every area in the Province, the theoretical sustainable carrying capacity for each category of grassland was being substantially exceeded.

Of all the grasslands in the Desert Grassland form, the low and flat grasslands are said to have the highest productivity. Pasture production from these grasslands is said to average 200kg of herbage fresh weight per mu per annum, with the air-dried dry matter content of the herbage being around 36% of the fresh weight. The major grass species found on the low and flat grasslands are *Achnatherum splendens* and *Phragmites adans*. Palatability of grasses and shrubs found on these grasslands is around 50%. The high lignin content of the slow growing vegetation is the major reason for this particularly low palatability percentage. Interestingly, it was claimed that once the vegetation is cut for hay, the palatability percentage increases to around 60%. The proportion of weeds and toxic vegetation in this category of pasture land is low and is usually less than 10% of total vegetation. The amount of crude protein contained in the major native grass species described above is said to average around 12%. This percentage varies seasonally and reaches a peak in spring.

(b) Alpine cold desert grassland - these grasslands, which have a total area of 2.4 million mu, are distributed along a line between the western end of the Qilian Mountains and the Rajin Mountains and are predominantly located in the counties of Akesai, Subei and Tianzhu. The carrying capacity of these grasslands is very low and averages around 40mu per sheep unit. Sheep, goats and yaks are the animals of major economic importance found on this type of grassland. The extreme cold associated with the high altitudes of this grassland limits the grazing time for sheep and goats to approximately one month per year. Yaks, however, are able to graze the pasture land the whole year round. Traditionally, the dominant variety of sheep on this pasture land was the Kazak sheep. This sheep is a traditional coarse wool sheep located mainly in Akesai County. Nowadays, about 60% of the sheep in this county are crossbreeds between the indigenous Kazak sheep and fine wool breeds from XUAR. The remaining 40% of the sheep are Xinjiang Fine Wools.

(c) Temperate desert grassland - this particular grassland is the largest in area of all the categories of grassland in the Desert Grassland form. The total area of the grassland is around 70 million mu and the carrying capacity is around 80mu per sheep unit. The major sub-types of grassland within this grassland are:
  (i) semi-desert grassland (said to be the dominant and economically most important of the sub-types of grassland);
  (ii) sand desert grassland; and
  (iii) stony (gibber) desert grassland.

(d) Steppe and semi-desert grassland - the total area of this grassland is 860,000mu and it has a carrying capacity of 40mu per sheep unit.

• *Form II grassland*

This particular form of grassland is scattered throughout the Province and is predominantly located in agricultural areas. The Form II grassland is mainly located east of the Yellow River with some notable exceptions, such as the desert grasslands located in Sunan County, which is one of the counties surveyed and which is located well west of the Yellow River. In total, the Form II grassland has

an area of approximately 84 million mu. The major categories of grasslands within this form are:
(a) meadow pasture land - occupying an area of 15 million mu;
(b) desert grassland - occupying an area of 15 million mu;
(c) steppe pasture land - occupying an area of 40 million mu; and
(d) alpine and cold pasture land - occupying an area of 14 million mu.

The first three of the above four categories of grassland are located primarily east of the Yellow River. Of these categories, the meadow and steppe categories of grassland are the most important in terms of area and productivity. The major part of these grasslands is said to be planted to improved pastures such as lucerne and *San folia*. Meadow, desert and steppe grasslands are all said to be found in the winter pasture lands in Sunan County. The alpine and cold pasture lands are located in the southern edge of the Hexi Corridor, west of the Yellow River. These grasslands, as the name suggests, are located in high altitude areas (above 3,000m) with cold and dry climates. Some of the summer pasture lands in Sunan County are said to contain alpine and cold grasslands. The dominant native species on the alpine and cold grasslands are *Steppa purpurea*, and *Agropyron cristatum* (L.) *gaertn.* (crested wheat grass). The pasture production from the alpine and cold grassland is less than 100kg of fresh matter per annum. Carrying capacity is around 30mu per sheep unit in summer. The dominant animals utilising the grassland in summer are sheep and goats, with yaks dominating in winter.

- *Real meadow grassland*

This form of grassland is found mainly in Gannan Tibetan Autonomous Prefecture and along the Qilian Mountains. The real meadow pastures in Sunan County located in the Qilian Mountains are the major summer pastures in that county. The real meadow grassland is the largest single connected grassland area in Gansu. It is a grassland located primarily in areas characterised by high altitudes (3,000 to 4,000m), high rainfall (400 to 700mm per annum), and low temperatures (0°C to 3°C average annual temperature). Tibetan sheep and yaks are the principal animals of economic importance found on real meadow grasslands. Fine wool sheep such as those in Sunan County also graze the real meadow grasslands in summer.

The total area of the real meadow grasslands is 59 million mu. The grassland yield is around 340kg fresh weight per mu per annum, and it has a carrying capacity of between 7 to 8mu per sheep unit. The principal native species of grasses found on this grassland are *Kobresia* genus (various species) and *Cyperaceae* genus (various species). Both genera are said to be of good quality and are suitable for sheep production.

- *Bush and grass grassland*

This form of grassland is mainly located in the mountainous region of Longnan Prefecture, and covers a total area of 12 million mu. The productivity of the grassland is around 450kg per mu fresh weight per annum, and it has a carrying capacity of between 5 to 6mu per sheep unit.

6.2.5 Degradation of Grasslands

Officials of the Animal Husbandry Bureau in Gansu stated that almost all pastures in Gansu were overstocked in 1990. As the total number of herbivorous animals increased only a little during the 1980s, the pastures have been under excessive grazing pressure for a long time.

It was estimated that in 1990 around one-third (or 70 million mu) of the available pasture land was degraded. Of this, 30 million mu was considered badly degraded, 20 million mu was medium and 20 million mu was lightly degraded.

In the grassland areas, each county has a grassland management station, and each station is responsible for a fixed area of pasture which it is expected to monitor each year to keep track of the degradation problem. In some counties, monitoring began as early as 1954, with certain fixed or semi-fixed areas being examined each year. Usually these areas are 30-50mu in size. Each grassland management station might have from two to six such monitoring areas.

Despite the substantial amount of monitoring which is supposed to occur, hard data on the changes in the extent of the degradation problem were not available. One specific example of the county monitoring program is briefly discussed in the Tianzhu County case study (Chapter 15).

Very few studies have been undertaken in Gansu Province to investigate pasture land recovery following heavy degradation. Only one example of such a study was cited. This was a five-year study conducted in Yongchang County, a semi-desert area in the central part of the Hexi Corridor. The experiment examined the impact of progressive reductions in the stocking rate on the degraded pasture land. The pasture land used in the experiment was not replanted with useful species of grasses and consequently weed species tended to dominate the regenerated vegetation.

6.2.6 Livestock Numbers

Over the 1949 to 1990 period, the number of large livestock in Gansu increased a little faster than the national total but sheep and particularly goat numbers increased much less rapidly than in the rest of China (Table 6.5).

Table 6.5 Long-Term Changes in Herbivorous Livestock Numbers in Gansu and All China, 1949 and 1990

Livestock type	Gansu			All China		
	1949	1990	Increase*	1949	1990	Increase*
	('000)	('000)	(%)	('000)	('000)	(%)
Large livestock	2,498	5,840	134	60,023	130,213	117
Sheep	3,018	8,789	191	26,221	112,816	330
Goats	1,301	2,309	77	16,126	97,205	503

*Calculated as $[\frac{1990 - 1949}{1949}] \times 100$

Source: Statistical Yearbook of China, 1991
Collection of Statistical Data by Province (1949-1989)

While sheep are easily the most numerous of all herbivorous animals in Gansu (Fig. 6.1), they represent only about one-quarter of the total sheep equivalents. From 1981 to 1984, sheep numbers and total sheep equivalents both declined, probably because private households "cashed in" unusually large numbers of livestock in the early phases of the introduction of the household production responsibility system when farmers were very unsure about the permanence of the new system. Beginning in 1985, meat and wool prices increased sharply and this encouraged farmers to increase their flocks and herds in the second half of the 1980s. Sheep numbers and total sheep equivalents rose to record levels in 1989 but the sheep population declined sharply in 1990 and 1991 due to the collapse of the wool price boom from 1989 onwards (Fig. 6.1).

While the total number of sheep in Gansu in 1991 was below the corresponding number in 1980, there was a major change in the composition of the flock during the 1980s. According to the data in Table 6.6, fine wool and improved fine wool sheep numbers more than trebled between 1981 and 1989. While such a change is consistent with the threefold increase in the output of fine and improved fine wool noted earlier, there are some year to year increases in the number of fine and improved fine wool sheep shown in Table 6.6 which seem extraordinarily large in percentage terms.

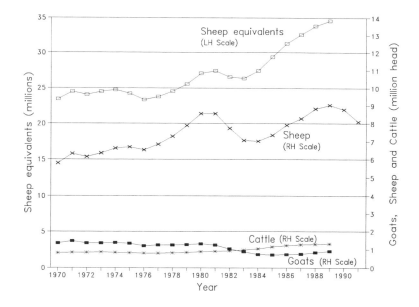

Fig. 6.1 Livestock Numbers in Gansu Province, 1970 to 1991

Despite some reservations about the data, there appears to have been a substantial swing toward the use of fine wool rams in Gansu during the 1980s. Whether this trend will continue is an issue examined in the county case studies. The upgrading of the sheep currently in the improved fine wool class to good quality fine wools is likely to continue at a slower pace than in the past. At the same time, it is unlikely that many farmers who have not yet commenced upgrading by using pure-bred fine wool rams, will decide to go down this path in the near future. That is, while the number of genuine fine wool sheep will continue to increase slowly, the total number of fine wool and improved fine wool sheep may actually decline.

Table 6.6 Number and Proportion of Total Sheep Represented by Each Major Category of Sheep in Gansu, 1980 to 1991

Year	Total sheep		Fine and improved fine wool sheep		Semi-fine and improved semi-fine wool sheep		Other sheep	
	('000)	(%)	('000)	(%)	('000)	(%)	('000)	(%)
1980	8,586							
1981	8,546	100	708	8.3	854	10.0	6,984	81.7
1982	7,729							
1983	7,062							
1984	7,017	100	935	13.3	599	8.5	5,483	78.2
1985	7,349	100	1,091	14.8	664	9.0	5,594	76.2
1986	7,923	100	1,199	15.2	802	10.1	5,922	74.7
1987	8,285	100	1,534	18.5	840	10.1	5,911	71.4
1988	8,823	100	1,781	20.2	831	9.4	6,211	70.4
1989	9,021	100	2,205	24.4	728	8.1	6,088	67.5
1990	8,789	100	2,078	23.6	703	8.0	6,008	68.4
1991	8,088	100	1,996	24.7	569	7.0	5,523	68.3

Note: Year end data
Source: China Agriculture Yearbooks

The official policy in Gansu is to stabilise sheep numbers at a little below 9 million head and goats at below 4 million head. At the same time, it is intended that the quality of these animals should be continuously upgraded by using good quality sires.

6.2.7 Wool Production

Total sheep numbers increased by only about 5% between 1980 and the peak in 1989 (Table 6.6), but total greasy wool production expanded by over 94% before declining slightly in 1990 (Table 6.7).

As one would expect, given the significant change in the composition of the Gansu sheep flock during the 1980s in favour of fine and improved fine wool sheep, wool of this type also became a much larger proportion of the Gansu clip. As a result of the threefold increase in output referred to earlier, fine and improved fine wool represented about one-third of all wool produced in Gansu in 1990. The production of semi-fine and improved semi-fine wool did not change much during the 1980s while coarse wool output increased modestly.

6.3 Sheep Breeds and Improvement Programs

While there is considerable geographical overlap, the three broad types of sheep (fine and improved fine wools, semi-fine and improved semi-fine wools, and the coarse wool/local sheep) are generally found in different parts of Gansu. During the last decade, there has been a definite strategy in place to increase the proportion of the Gansu sheep flock which are fine or improved fine wool sheep.

Table 6.7 Production of Greasy Wool, Goat Hair and Cashmere in Gansu, 1980 to 1990

Year	Wool	Goat hair	Cashmere
	(tonne)	(tonne)	(tonne)
1980	8,066.4	873.4	184.2
1981	8,066.6	914.9	166.7
1982	8,967.1	870.6	169.1
1983	8,452.6	755.1	153.9
1984	8,971.7	688.9	136.1
1985	10,050.8	720.4	128.8
1986	12,071.2	910.0	170.2
1987	13,472.7	1,013.8	191.0
1988	15,193.2	1,058.5	278.4
1989	15,660.0	1,069.0	232.0
1990	15,542.0	1,132.0	259.0

Source: China Agriculture Yearbooks

6.3.1 Breeds and their Location

Most of the good quality fine wool rams used in Gansu are now Gansu Alpine Fine Wool sheep. However, there are still significant numbers of Xinjiang Fine Wool rams imported each year and a few Australian and New Zealand merino rams become available from time to time.

The typical mature pure-bred Gansu Alpine Fine Wool ewe is expected to produce about 3.5kg of greasy wool per year. The yield of clean wool averages about 40% of the greasy wool weight. In terms of spinning count the wool is 60s to 64s, but the average fibre diameter may be finer than the spinning count would suggest, perhaps in the 20 to 22µm range.

As discussed in Section 4.1.2, it usually takes at least four generations or "doses" of pure-bred rams to produce a sheep which can be called a genuine fine wool sheep. Sheep which have not yet been upgraded to this extent are called improved fine wools. In 1990, seven out of eight of the 2 million (approx.) fine and improved fine wool sheep in Gansu remained in the improved fine wool category. Of the 250,000 fine and improved wool sheep in Gansu which are genuine fine wools, most are raised in the two counties surveyed (Sunan and Tianzhu) or on the three nuclear studs (Huang Cheng, Song Shan and Yong Chang). For more information about both the Gansu Alpine Fine Wool breed of sheep and the three nuclear studs, see Longworth and Williamson (1993).

Fine wool and improved fine wool sheep are concentrated in the foothills of the Qilian Mountains on the southern side of the Hexi Corridor which extends north west from Lanzhou. A high proportion of the sheep in Subei, Akesai, Sunan, Tianzhu, Yongchang, Shandan, and Minqin counties are fine wools or improved fine wools (see Table 6.3). For example, in the two counties surveyed (Sunan and Tianzhu), it was claimed that 90% and 87% of the sheep respectively were of this type.

The 5 million coarse/local wool sheep in Gansu consist of many different breeds, perhaps reflecting the central location of Gansu in relation to the traditional pastoral region of China. Tibetan sheep breeds (such as Ganjia and Oula) are kept mainly in the cold alpine areas of Gannan Tibetan Autonomous Prefecture (especially Xiahe, Liuqu, Maqu, and Zhuoni counties) and in Lintao, Minxian and Tianzhu counties. Mongolian sheep breeds (and their derivatives such as the Tan, famous for lamb's fur pelts, and the Lanzhou Big-tail) are widely distributed but are especially common in the agricultural areas of the Hexi Corridor.

6.3.2 The "One Million" Fine Wool Sheep Base

As just mentioned, only about 250,000 of the 2 million (approx.) fine wool and improved fine wool sheep in Gansu in 1990 could be called genuine fine wool sheep. Nevertheless, the potential is there to continue upgrading the 1.75 million improved fine wool sheep.

The Gansu provincial government has established a project to develop a base of one million pure-bred Gansu Alpine Fine Wool sheep. It takes at least six generations using pure-bred rams to obtain sheep which meet the standard for the Gansu Alpine Fine Wool breed. The project is centred on the three Gansu Alpine Fine Wool nuclear studs (Huang Cheng, Song Shan and Yong Chang) and embraces three prefectures in the Hexi Corridor (Wu Wei, Zhangye and Jiuquan). There are also four eastern and southern prefectures involved in the project (Pingliang, Dingxi, Linxia, and Qingyang). These eastern and southern prefectures are all agricultural areas, some of them on the loess plateau, and none of them have any fine wool studs. The Gansu government has financed the purchase of good quality old fine wool ewes from households and State farms such as Song Shan in the Hexi Corridor, for relocation to Pingliang and Qingyang Prefectures. Households in these eastern prefectures raise one lamb from the old ewes and then sell them for mutton. Farmers in these poor areas usually only keep from one to four ewes since their primary activities are agricultural.

During the second half of the 1980s, the major push to increase wool production by crossing local and semi-fine wool ewes with Gansu Alpine Fine Wool rams led to a shortage of good quality rams. In fact, there was a general shortage of fine wool rams throughout China in the 1986 to 1988 period when wool prices were at record levels. While the resulting higher prices and increased ram sales boosted the incomes of sheep studs, it also led to a deterioration in the average quality of rams being sold.

When Chinese wool prices dropped sharply in 1989, the demand for rams also declined. Even the best studs such as Huang Cheng and Song Shan found it hard to sell rams in 1990 and 1991. Indeed, Huang Cheng sold over 1,000 young lambs for slaughter before winter in 1991. It would seem that the shortage of good quality rams in the second half of the 1980s was a short-lived phenomenon.

During the early stages of the development of the Gansu Alpine Fine Wool sheep, a breeding committee was responsible for distributing the rams from the three nuclear studs to the various counties (about 12) in the area running from Yongdeng westward up the Hexi corridor. Before 1980, the breeding committee provided rams from the nuclear studs free of charge to the various county ram breeding stations and to certain collectives that were also breeding rams. These "free" rams were financed by the Gansu Animal Husbandry Department.

The breeding committee was replaced in late 1980 by the General Extension Station for Livestock Technology, which operates under the Gansu Animal

Husbandry Department. The change from the breeding committee to the extension service was regarded as being a natural transition once the breed had become firmly established. Extending the breed to farmers was considered a duty of the extension service. The General Extension Station has a number of extension stations in the counties. This organisation is still supposedly responsible for the distribution of rams from the nuclear studs but its role in this regard is now greatly diminished. Nowadays, as discussed by Longworth and Williamson (1993), the State farms which operate the nuclear studs are expected to pay their own way. Consequently, they sell their rams to "all comers" in a more or less free market.

6.3.3 Other Sheep and Goat Breeding Programs

In addition to the major project aimed at upgrading 1 million sheep to pure-bred Gansu Alpine Fine Wools, there are also breed improvement programs for semi-fine wool sheep, for coarse wool (local) sheep breeds, and for cashmere goats.

There are two major semi-fine wool studs with a total of over 1,000 breeding ewes and two major coarse wool studs with over 2,000 breeding ewes. The better known of the latter studs is the Minxian Sheep Breeding Farm in Minxian County in the south of Dingxi Prefecture (see Map 6.1). This stud is using imported Border Leicester rams to improve the local breeds.

Gansu is a significant producer of cashmere. A cashmere goat base is being built in Akesai, Subei and Huanxian counties in Jiuquan Prefecture (Map 6.1). At the end of 1988, there were 370 thousand head of goats in this base and cashmere production had reached 83 tonne, which was about one-third of the total cashmere production of Gansu in that year.

6.4 Animal Husbandry Research and Extension Organisations

The provincial government of Gansu has a number of commissions, one of which is the Agriculture Commission. The Animal Husbandry Department/Bureau is a department under the Agriculture Commission. As in IMAR, the Animal Husbandry Bureau of Gansu is divided into a number of stations including the Pasture Management Station, the Livestock Improvement Station, the Veterinary Station, the Economic Management Station, etc. These stations also exist at prefectural, county and township levels. The staff of the higher level stations are concerned mainly with administration and the training of lower level staff. It is the county and especially the township level technicians who are primarily responsible for extending new technologies etc. to the village collectives and to individual farm households.

In addition to the Gansu Animal Husbandry Bureau, there are three other important organisations concerned with research, extension and training in relation to pastures, sheep, and wool growing with their headquarters in Lanzhou. These are:
- Gansu Grassland Ecological Research Institute;
- Lanzhou Research Institute of Animal Science; and
- Department of Animal Science at the Gansu Agricultural University.

The Gansu Grassland Ecological Research Institute (GGERI) was a joint initiative of the Central government Ministry of Agriculture and the Gansu provincial government. It was established in 1981 to undertake research in grassland ecology in various parts of China; to train postgraduate students in grassland sciences; and to provide extension, consulting and other services in grassland sciences and technologies. The Institute has a number of branches in other parts of China. It was the principal Chinese institution engaged in the AIDAB Grassland Agricultural Systems Research and Development Project which was implemented from 1986 to 1992. This project was managed by the Australian agricultural consultants Hassall and Associates and was centred mainly on the Loess Plateau Research Station which is a GGERI station located near the centre of Qingyang Prefecture east of Lanzhou (Chandra, 1993).

In the 1950s and 1960s, the North West Institute of Animal Science in Lanzhou was part of the Chinese Academy of Agricultural Science (CAAS) under the Central government Ministry of Agriculture. In 1971, the Institute of Animal Science was transferred to the Gansu Animal Husbandry Bureau but it was reassigned back to CAAS in 1978. The Lanzhou Research Institute of Animal Science, as it has been called since 1978, is now vertically responsible through CAAS to the Ministry of Agriculture in Beijing. Although there are lingering informal ties with the Gansu Animal Husbandry Bureau, there are no formal vertical linkages. ACIAR Project No. 8456 concerned with sheep breeding research was a collaborative project with the Lanzhou Research Institute of Animal Sciences (Lehane, 1993). Much of the field experimental work associated with this Project was conducted on Huang Cheng State Farm and Sheep Stud which is directly under the control of the Gansu Animal Husbandry Bureau. It is not known to what extent the lack of vertical control over the management of the State Farm hindered the staff of the Institute of Animal Science in their collaboration with Australian scientists working on ACIAR Project No. 8456.

The Lanzhou Research Institute of Animal Sciences has published the professional journal *Sheep in China* on a quarterly basis since 1981. This journal is the most important professional publication in China in regard to sheep and wool production.

The Gansu Agricultural University has a Department of Animal Husbandry, some staff of which conduct research on sheep husbandry and related issues.

6.5 Introduction of the Household Production Responsibility System

In Gansu, the household production responsibility system (HPRS) was first tried in the early 1960s. The system began to take shape in Linxia Hui Autonomous Prefecture but in 1965 it was criticised and stopped. In 1979 when government policy was relaxed, the people in this prefecture quickly began to apply the HPRS once again. Indeed, the Linxia Hui Autonomous Prefecture, an agricultural area, was one of the first areas in all China to adopt the HPRS in 1979. The system was introduced in pastoral areas later than in agricultural areas.

In 1983, the first major province-wide steps were taken to introduce the HPRS in pastoral areas of Gansu. At the beginning, the system in Gansu was very similar to the early arrangements in the IMAR. Initially, it was based on contracts setting fixed production goals and fixed rewards in terms of work points.

In 1984, however, more radical reforms were permitted and almost all animals belonging to the collectives were sold to households. The livestock became the private property of the households and could be inherited from one generation to the next. In 1985, most of the better quality pasture land was contracted to households, groups of households (production teams) or joint households (2 to 5 households). In particular, about 30% of the pasture land in Gannan Tibetan Autonomous Prefecture was contracted to joint households (2 to 5 households).

By 1990, about half the total pasture land in Gansu had been contracted to farmers. Almost all of the pasture not contracted was summer and autumn grazing land located at considerable distances from the villages. It is not easy to manage these areas on an individual basis.

The original pasture contracts used in Gansu Province did not involve any "user pays" concept, but the farmers did have to make annual contributions to the collective accumulation and welfare funds. By 1990, as discussed in the Sunan County case study (Chapter 13), steps had been taken to clarify the rights and obligations of both the collectives and the households under the pasture land contracts. In Sunan this was achieved by issuing "pasture utilisation certificates" (see the Appendix to Chapter 13 for an example).

6.6 Economic Conditions in Pastoral Areas of Gansu

As can be seen from Table 6.3, none of the seven counties in Gansu classified as pastoral by both the Gansu government and the Central government are included in the 46 poor counties in Gansu. However, seven of the 12 Central government recognised semi-pastoral counties are classified by both the Central and the provincial government as in need of poverty alleviation assistance.

While the pastoral counties in Gansu do not qualify as poor counties in terms of average net per capita incomes, they are not areas in which the people enjoy high living standards. The remoteness and the consequent high cost of transport and other services, together with the need for many herdsmen households to buy food grains since they often cannot grow their own needs, mean that living costs are much higher than in most other areas.

6.6.1 Growth in Incomes since 1978

Nominal net income per capita is the usual criterion used in China to assess changes in income. Available data on rural net incomes in two of the three case study counties have been plotted in Fig. 6.2. The raw figures and additional information are presented in the case studies (Chapters 13 and 14).

One of the striking features of these graphs is the difference deflating the income data to allow for inflation makes to the overall picture which emerges. The price index used in this case was the "Overall Retail Price Index" published by the State Statistical Bureau of China. While this index is unlikely to be an entirely appropriate deflator, it was considered by Chinese economists to be the most suitable index available. The base year was taken as 1978.

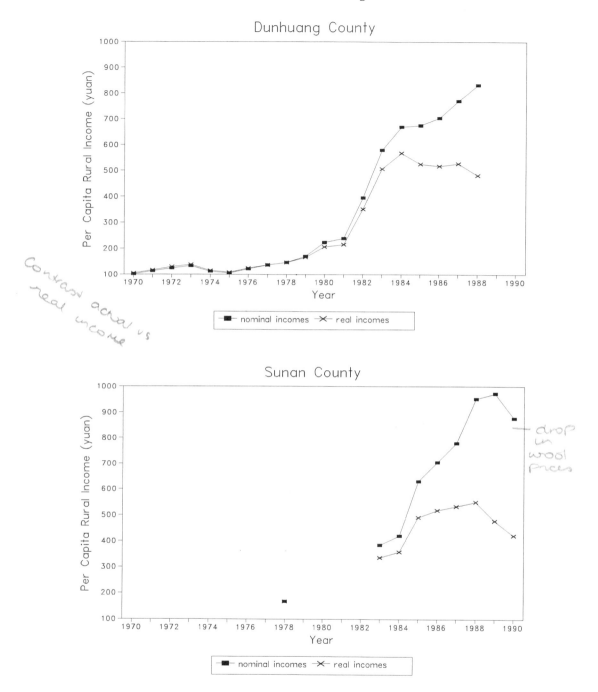

Fig. 6.2 Nominal and Real Average Net Rural Income Per Capita for Two Case Study Counties in Gansu

Real incomes in the agricultural county (Dunhuang) rose rapidly between 1981 and 1984 but have declined slowly since that year. In Sunan, the purely pastoral county, there is an observation for 1978 but the data series really begins in 1983. It would appear that real incomes in Sunan continued to rise until 1988 (as wool and meat prices rose) and then declined rather sharply (as wool prices dropped). The nominal income graphs suggest that incomes in both counties continued to rise until 1990 when a sharp fall was recorded in Sunan.

The nominal rural net income per capita in 1989 for the whole of Gansu was ¥366 and for all China ¥602. On this basis, average rural incomes in both Dunhuang and Sunan were well above the national average in 1989 and approaching three times the provincial average.

6.6.2 Poverty Alleviation Policies

Although the seven pastoral counties in Gansu are not poor counties, at least not when the criterion is net per capita income, more than half the semi-pastoral counties are regarded by the Central government as poverty-stricken. In fact, of the 12 counties in Gansu which are designated as "poor" under the Central government poverty alleviation scheme, seven are semi-pastoral counties (Table 6.3).

In addition to the 12 poor counties recognised by the Central government, the Gansu provincial government also defines another 34 counties as "poor". In 1990, the arrangements to assist the total of 46 poor counties in Gansu were based on a seven-point program.
- Financial assistance to poverty-stricken counties. Each year, a total of about ¥110 million is invested in the poverty-stricken counties both by the Central government and by the Gansu government to assist economic development in these areas.
- The poverty-stricken counties are given some advantages in terms of paying government taxes. For example, new township enterprises in these counties do not pay tax at the beginning for at least three to five years (they are given tax holidays).
- A certain amount of grain at State list prices is allocated to the poverty-stricken counties. The total amount available is fixed for several years at a time.
- The poverty-stricken counties have advantages in terms of getting access to building materials at State list prices.
- Some large factories, especially for mining coal and other minerals (such as iron ore), are encouraged to provide technicians and technology to the poor county areas to help them establish their own mining enterprises.
- Many technicians have been sent to the poverty-stricken counties to assist these areas to develop their science and technology.
- The poverty-stricken counties are targets for the anti-hunger projects of the nation. The aim is to use new technology to help farmers produce more food grains so that these areas will no longer be threatened by famine.

Most of the poverty-stricken counties are located in very arid areas, in high altitude, humid areas where minority groups are concentrated, or in old revolutionary bases. Two especially poverty-stricken areas are the Long Dong (East Mountains) and the Long Nan (South Mountains) regions.

6.6.3 Development of Pastoral Product Processing

The analysis presented by Walker (1989) provides a number of explanations for why Gansu is such a poor province. In the context of this book, one of the most important reasons is that secondary and tertiary industries are not well developed in rural Gansu. The fiscal reforms in China between 1980 and 1983 created incentives for local governments (township and county) to invest in industrial and other commercial undertakings.

For remote pastoral counties such as Sunan, wool scouring is one of the few feasible forms of industrial development. However, Brown and Longworth (1992b) argue that the construction of county wool scouring plants in China has so far proved to be a major backward step for the wool growing industry and for economic development in the pastoral region generally. The experiences with such plants in Sunan and Tianzhu, which are described in some detail in the county case studies, support this conclusion.

Chapter Seven

Xinjiang Uygur Autonomous Region

The Xinjiang Uygur Autonomous Region (XUAR) is located in the extreme north west of China. It is geographically the largest province/autonomous region in China, occupying just over one-sixth of the country (Map 7.1). By Chinese standards, it is extremely sparsely populated. The southern border of XUAR is in the Kunlun Mountains, the northern border lies in the Altay Mountains, and through the middle runs the Tianshan Range. All of these mountain ranges extend more or less east–west and reach altitudes of 4,000m and above in many places.

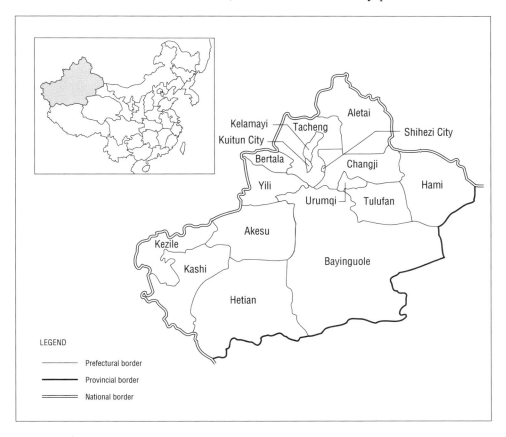

Map 7.1 Prefectural-Level Administrative Units of the XUAR

The Junggar Basin and the Gurbantunggut Desert are located between the Altay and Tianshan Ranges. Between the Tianshan and Kunlun Ranges lie the much larger Tarim Basin and Taklimakan Desert. Most of the agricultural settlements in the XUAR are located around these two large desert basins. The snow-fed rivers which flow out of the high mountains supply water for numerous small pockets of irrigation in the valleys of the lower foothills and on the edges of the plains. In many places, the snow water is fed through underground channels over very long distances beneath the stony plains to minimise the loss of water due to evaporation and leakages. Underground irrigation canals have played a major part in the history of central Asia for thousands of years.

There is also a large, relatively fertile, triangular-shaped area lying between the northern spur of the Tianshan Range (the Borohoro Mountains) and the main Tianshan Range. This area is known as Yili Prefecture and it is the home of some of the best fine wool sheep in China. The oldest and best known of the four nationally recognised fine wool sheep studs, Gongnaisi, is located in Xinyuan County in eastern Yili Prefecture.

7.1 Identifying the Pastoral Areas of the XUAR

Yili Prefecture is one of 16 prefectural-level administrative units in XUAR (Table 7.1). Four of these are cities and the rest prefectures. One of the cities (Kuitun) and three of the prefectures (Yili, Tacheng, and Aletai) are administered as a kind of super prefecture under the title "Yilikazak Autonomous Prefecture".

Table 7.1 Number of County Level Administrative Units in the XUAR by Prefecture/City Together with Number of Counties Recognised as being Pastoral or Semi-Pastoral, 1989

Prefecture or cities	Total number of administrative divisions at the county level	Number of cities under the jurisdiction of prefecture	Number of counties	Number of autonomous counties	Number of districts	Number of counties designated as: Pastoral	Semi-pastoral
Urumqi C	8		1	-	7		1
Kelamayi C	4		-	-	4		
Shihezi C	-		-	-	-		
Tulufan P	3	1	2	-	-		
Hami P	3	1	1	1	-	1	2
Changji P	8	1	6	1	-	1	1
Kuitun C	1	1	-	-	-		
Yili P	9	1	7	1	-	4	1
Tacheng P	7	1	5	1	-	3	2
Aletai P	7	1	6	-	-	7	
Bertala P	3	1	2	-	-	1	2
Bayinguole P	9	1	7	1	-	1	3
Akesu P	9	1	8	-	-		2
Kezile	4	1	3	-	-	2	
Kashi P	12	1	10	1	-	1	1
Hetian P	8	1	7	-	-	1	
TOTAL	95	13	65	6	11	22	15

Source: Xinjiang Statistical Yearbook, 1990

There are 95 administrative divisions at the county level (Table 7.1). Thirty-seven of these are included in the 266 counties defined as pastoral or semi-pastoral counties by the Central government. The XUAR government also recognises the same 37 counties as being pastoral or semi-pastoral. Table 7.2 presents some characteristics of these counties. Obviously, some of the counties are extremely large, with eight being larger than Belgium which has an area of 30,513km^2. They are also sparsely populated, with 20 counties having fewer than 6 people/km^2 and seven of these having less than 1 person/km^2.

Hebukesaier, one of these large, sparsely populated pastoral counties, was surveyed. It covers an area of 30,400km^2 and has a population density of only 1.45 people/km^2 (Table 7.2). Hebukesaier is a Mongolian Autonomous County in the northern part of Tacheng prefecture which, as can be seen from Map 7.1, is one of the prefectures sharing the national Chinese border with Kazakhstan.

The other county in the XUAR which was surveyed, Cabucaer, is located in Yili Prefecture but it is not included in the list of 266 pastoral and semi-pastoral counties. Nevertheless, Cabucaer includes large pastoral areas and it is fairly representative of the Yili Prefecture. As can be seen from both Tables 7.1 and 7.2, there are four pastoral counties and one semi-pastoral county in Yili Prefecture which are in the list of 266 counties. Map 16.1 shows that three of these five counties have common borders with Cabucaer County.

7.2 Physical Details

7.2.1 Geography

As already pointed out, the XUAR is dominated by three major mountain ranges and two large desert basins. The total area of the XUAR is 1.6 million km^2 but little of it is habitable. The climate is extremely harsh and dry. The development of irrigated agriculture fed by the melting snow on the high mountain ranges has made it possible for increasing numbers of people to live in the XUAR. In the past four decades, a great deal of industrial as well as agricultural development has occurred. The XUAR has vast mineral resources (nickel, lead, copper and manganese) and three major oil fields. Many of these natural resources have yet to be developed and further exploration is being planned. For example, tenders have been called for expressions of interest from international companies in oil exploration in the Tarim Basin.

7.2.2 Population

In 1949, the population of the XUAR was 4.33 million. Forty years later, in 1989, it had reached 14.54 million. In 1949, 88% of the population were rural people, but by 1989 the corresponding proportion was 72%.

The increasing diversification of the XUAR economy is reflected in the declining proportion of the population classified as rural. In fact, on this indicator of development, the XUAR is a relatively highly-developed part of China since the proportion of the population designated as rural is well below the national average which was 81% in 1990.

Table 7.2 Some Characteristics of the Central Government Recognised Pastoral and Semi-Pastoral Counties in XUAR, 1990

Prefecture/city and county		Poverty status[1]	Population		Total area	Population density	Sheep numbers[2]		Wool production	Total arable land	Usable pastoral area
			Number of people	Prop. of minority nationality			Total all sheep	Fine and improved fine wool sheep			
			('000)	(%)	(km²)	(person/km²)	('000)	('000)	(tonne)	('000 mu)	('000 mu)
Pastoral counties											
Hami	- Yiwu	nsp	17	46.6	19,735	0.86	91.3	60.2	206	49	8,852
Changji	- Mulei	nsp	87	26.8	22,171	3.92	264.0	107.1	532	541	12,547
Bertala	- Wenquan	-	58	-	5,664	10.24	287.6	247.0	693	282	7,404
Bayinguole	- Hejing	-	135	41.6	39,686	3.40	625.1	126.1	798	234	24,469
Kezile	- Aheqi	nsp	31	89.0	12,189	2.54	143.1	83.1	255	66	8,331
	- Wuqia	nsp	38	72.1	19,200	1.98	158.1	79.2	330	58	11,392
Kashi	- Tashikuergan	nsp	25	-	25,000	1.00	61.4	8.4	52	57	5,958
Hetian	- Minfeng	nsp	28	-	54,236	0.52	147.8	29.1	148	67	11,790
Yili	- Xinyuan	-	216	59.7	6,446	33.51	446.3	357.9	1,562	581	7,868
	- Zhaosu	-	109	37.9	11,163	9.76	401.6	301.2	1,096	609	15,772
	- Tekeshi	-	115	41.8	8,352	13.77	402.9	322.3	1,129	292	10,135
	- Nileke	nsp	125	71.0	-	-	360.3	200.2	1,006	320	12,744
Tacheng	- Tuoli	nsp	7	66.3	20,097	0.35	289.5	129.0	661	228	19,115
	- Yumin	-	39	36.5	6,200	6.29	197.9	163.6	528	362	6,872
	- **Hebukesaier***	-	**44**	**65.9**	**30,400**	**1.45**	**235.7**	**188.7**	**490**	**92**	**23,559**
Aletai	- Aletai	-	143	39.0	11,354	12.59	255.7	78.9	760	288	19,541
	- Buerjin	-	56	55.0	10,540	5.31	177.5	177.5	375	212	12,274
	- Fuyun	-	8	-	32,155	0.25	309.7	309.7	644	202	36,115
	- Fuhai	nsp	5	43.9	31,842	0.16	155.5	149.4	364	374	35,358
	- Habahe	-	61	68.7	8,430	7.24	143.6	14.0	320	280	6,717
	- Qinghe	-	47	72.0	15,530	3.03	192.7	190.3	400	115	34,359
	- Jimunai	-	3	66.8	8,222	0.36	134.1	131.1	303	203	7,595
Total above 22 counties			1,397		398,610	3.50	5,481.4	3,454	12,652	5,512	338,767

Table 7.2 continued

Prefecture/city and county		Poverty status[1]	Population		Total area	Population density	Sheep numbers[2]		Wool production	Total arable land	Usable pastoral area
			Number of people	Prop. of minority nationality			Total all sheep	Fine and improved fine wool sheep			
			('000)	(%)	(km²)	(person/km²)	('000)	('000)	(tonne)	('000 mu)	('000 mu)
Semi-pastoral counties											
Urumqi City	- Urumqi	-	152	54.6	11,300	13.45	269.0	210.4	693	436	9,811
Hami	- Hami	-	248	33.8	85,035	2.92	189.7	65.0	302	202	11,550
	- Balikun	-	92	21.3	38,445	2.39	278.2	97.4	485	439	19,442
Changji	- Qitai	-	180	23.0	18,087	9.95	332.1	273.2	867	1,023	16,206
Bertala	- Bole	-	119	-	7,782	15.29	264.8	214.4	758	337	5,775
	- Jinghe	-	84	31.8	11,275	7.45	155.0	114.7	333	255	11,159
Bayinguole	- Weili	-	42	31.0	5,976	7.03	92.4	-	112	115	13,616
	- Qiemuo	-	44	85.9	138,278	0.32	202.8	2.2	203	126	27,301
	- Heshuo	-	41	49.0	12,769	3.21	81.4	68.3	159	174	5,897
Akesu	- Wensu	-	152	81.3	14,569	10.43	287.6	247.0	456	562	5,027
	- Shaya	-	164	87.0	31,972	5.13	187.2	13.5	175	567	3,182
Kashi	- Aketao	nsp	138	69.0	24,176	5.71	292.2	29.0	327	262	11,133
Yili	- Gongliu	-	123	67.3	4,513	27.25	286.2	198.7	1,001	368	4,519
Tacheng	- Tacheng	-	112	36.1	4,357	25.71	309.3	247.4	902	1,036	3,940
	- Emin	-	131	33.0	8,874	14.76	401.6	336.2	1,111	914	11,752
Total above 15 counties			1,822	n/a	417,408	4.37	3,629.5	2,117.4	7,884	6,816	160,310
Grand total above 37 counties			3,219	n/a	816,018	3.94	9,110.9	5,571.4	20,536	12,328	499,077
Total for all of XUAR			15,200	62.5	1,657,590	9.17	23,814.0	12,189.0	49,297	32,300	858,880

*This pastoral county was surveyed as part of ACIAR Project No. 8811
[1]Poverty status "nsp" means recognised as a poor county under the Central government poverty alleviation scheme. These counties receive special assistance from both the Central and XUAR governments. (See Section 3.4.3 for more details.)
[2]Sheep numbers are end-of-year data for 1989. In that year no distinction was made in the XUAR statistics between fine and semi-fine wool sheep. Therefore the number of "fine and improved fine wool sheep" also includes "semi-fine and improved semi-fine wool" sheep. In 1991, semi-fine and improved semi-fine wools represented 10% of the total XUAR flock (see Table 7.6).

The XUAR is home to a significant number of people belonging to more than a dozen minority nationalities. Most of these minority populations have more than doubled since 1949 but the 3.5 fold increase in the total population between 1949 and 1989 has been largely due to the immigration of Han people from elsewhere in China. Table 7.3 presents population data by ethnic group at roughly twenty year intervals.

Table 7.3 Population of the XUAR by Ethnic Group, 1949, 1970 and 1990.

Ethnic group	Population		
	1949	1970	1990
	('000)		
Han	291	3,867	5,647
Uygur	3,291	4,673	7,094
Kazak	444	616	1,115
Kirgiz	66	80	141
Hui	123	383	674
Mongolian	53	88	140
Xibe	12	21	34
Russian	20	1	1
Tajik	14	18	34
Uzbek	12	7	11
Tatar	6	2	4
Manchu	1	3	16
Daur	2	3	5
Others	1	13	65
Total population	4,333	9,775	15,200

The migration of Han people to XUAR began in the early 1950s but increased during the "Great Leap Forward" era (1958 to 1961). In 1949, there were less than 300,000 people of Han Nationality in XUAR. By 1957, there were 821,000 but by 1962 the Han population had increased to 2.077 million. The rapid inward migration of Han people continued throughout the 1960s and did not slow down until after the Cultural Revolution ended in the mid-1970s.

The location of the population by prefecture is shown in Table 7.4. The larger cities of Urumqi (Urumqi Prefecture), Changji (Changji), Yining (Yili), Akesu (Akesu), Kashi (Kashi), and Hetian (Hetian) clearly have a major influence on the prefectural population distribution.

7.2.3 Land Use

Virtually all the cropland in XUAR is irrigated to some degree. The major concentrations of irrigated agriculture tend to be in the more densely populated prefectures (Table 7.4). Urumqi, the capital of XUAR, is close to Changji Prefecture which has a large area of good quality cropland.

The XUAR is recognised as one of the top five "pasture provinces" in China. The total area of pasture in the XUAR is about 860 million mu, of which about 756 million mu is usable. The pasture areas are located in the three mountain ranges (Altay, Tianshan and Kunlun) and around the Junggar and Tarim Basins.

Table 7.4 Population and Land Use in the XUAR by Prefectural Administrative Unit, 1989

Prefecture/city	Population	Cropland	Land use — Usable pasture land						Total
			Summer pasture	Spring-autumn pasture	Winter pasture	Summer-autumn pasture	Winter-spring pasture	Four season pasture	
	('000)				('000 mu)				
Urumqi C	1,269	574.5	1,500	3,330	4,980				9,810
Kelamayi C	202	34.5	270	9,720	3,630				13,620
Shihezi C	536	-	-	-	-				-
Tulufan P	466	640.5	6,075	6,975	3,810				16,860
Hami P	404	697.5	4,665	18,480	16,710				39,855
Changji P	1,259	4,320.0	9,165	28,365	22,605				60,135
Yili P and Kuitun C	1,751	4,458.0	27,630	11,700	19,110				58,440
Tacheng P	768	4,008.0	17,910	29,025	37,395				84,330
Aletai P	503	1,612.5	51,210	71,460	29,475				152,145
Bertala P	310	843.0	7,440	8,280	8,610				24,330
Bayinguole P	824	1,509.0	24,075	8,175	8,400	37,410	19,875	31,110	129,045
Akesu P	1,642	4,704.0	13,515	8,910	8,805			19,770	51,000
Kezile P	366	601.5	4,980	600	2,760	18,195	9,225	1,215	36,975
Kashi P	2,679	5,887.5				9,510	8,655	15,660	33,825
Hetian P	1,342	2,359.5				19,320	17,805	9,225	46,350
TOTAL*	14,542	32,280.0	168,435	205,020	166,290	84,435	55,560	76,980	756,720

*Excludes PCC State farms for which the only comparable data available were for total cropland which was 13,920 thousand mu.
Source: Xinjiang Statistical Yearbook, 1990

Desert and semi-desert pastures (i.e. the stony desert areas, the semi-desert loess plains and the dry low hills) represent about 47% of the total pasture area. Lowland pasture (i.e. the good pastures on the better plains and on the low hills near the mountain ranges) make up another 24%. So-called middle pastures (which include high altitude summer pasture areas) are another 23%. Swampy pastures are rare and constitute only about 0.5% of the total. The remainder consists of various other pasture types. It is not easy to classify pastures in the mountains. The sunny sides (southern slopes) of the mountains are usually drier and hotter than the shady sides (northern slopes). The pasture is usually better on the northern sides of the ranges. However, the best mountain pastures are usually at the edges of forests and along creek beds and where there are springs and soaks.

As discussed in more detail in both the Cabucaer and the Hebukesaier County case studies (Chapters 16 and 17), there are commonly three types of pastures in XUAR from a grazing perspective: *winter* pastures, usually located near the agricultural areas or in the low to middle hills near the pastoral villages; *spring/autumn* pastures on the plains between the agricultural areas and the low hills and in the low hills; and *summer* pastures, usually located in the high altitude mountain areas. As shown in Table 7.4, some prefectures have substantial areas classified as *summer/autumn* pastures, *winter/spring* pastures and *four season* (or all year round) pastures. In these parts of the XUAR, the grazing pattern may differ somewhat from the norm just described.

Degradation of pastures is a major problem in the XUAR. As reported in the Hebukesaier case study, for example, many pastoral areas are seriously over-stocked (Sections 17.1.2 and 17.2.1). The XUAR, like the IMAR, has introduced regulations which are intended to be regional interpretations of the National Rangeland Law. These regulations came into effect on 1 September 1989. Under these by-laws, Pasture Utilisation Certificates are supposed to have been issued to all households with contractual rights to the use of pastures. These certificates (as in the IMAR and Gansu) specify limits on stocking rates for summer, winter, spring/autumn pastures, etc. However, at least in the two counties surveyed, Cabucaer and Hebukesaier, officials working for the AHB Pasture Management Stations do not check to ensure that farmers comply with the limits stated in their certificates.

Officially, it seems every effort is made to downplay the seriousness of rangeland degradation. In Cabucaer County, for instance, senior County government officials claimed that less than 5% of the pastureland was degraded. Yet the staff of the AHB Pasture Management Station provided data which demonstrated that average dry matter production from native pastures was now less than half what it had been in the 1950s (Section 16.1.2). Direct observations while travelling over a large part of Cabucaer County confirmed that most native pastures were degraded. The large expanse of spring/autumn pastures on the dry loess plains between the low hills and the agricultural areas along the Yili River were particularly badly affected, having been turned almost to desert in many places.

7.2.4 Herbivorous Livestock Numbers

Perhaps reflecting the extremely harsh environment and the consequential physical limitations this places on the raising of grazing animals, the numbers of herbivorous animals have not increased as rapidly in XUAR as elsewhere in China since 1949

(Table 7.5). Sheep numbers have expanded much faster than other types of grazing livestock so that the XUAR still has the largest sheep flock of any province/autonomous region.

Table 7.5 Long-Term Changes in Herbivorous Livestock Numbers in XUAR and All China, 1949 and 1990

Livestock type	XUAR			All China		
	1949	1990	Increase*	1949	1990	Increase*
	('000)	('000)	(%)	('000)	('000)	(%)
Large livestock	3,275	5,759	76	60,023	130,213	117
Sheep	6,640	23,814	259	26,221	112,816	330
Goats	2,071	4,494	117	16,126	97,205	503

*Calculated $[\frac{1990 - 1949}{1949}] \times 100$

Source: Collection of Statistical Data by Province (1949-1989)
Statistical Yearbook of China, 1991

While Table 7.5 presents the overall picture, the pattern in the long-term trends in sheep numbers and in total herbivorous animals (as measured by sheep equivalents) is shown in Fig. 7.1. Except for the Cultural Revolution period (1966 to 1976), there has been a consistent upward trend in both the sheep population and total sheep equivalents. The steady growth in the sheep flock since 1978 (when the XUAR had slightly fewer sheep than the IMAR) has allowed the XUAR to surpass the IMAR as the province/autonomous region with the largest sheep flock. By 1991 there were 3.6 million more sheep in the XUAR than in the IMAR.

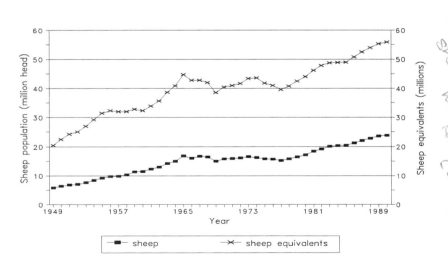

Fig. 7.1 Long-Term Trends in the Sheep Population and Total Sheep Equivalents in the XUAR

On the other hand, in the past the XUAR has always been the province/ autonomous region which the official statistics indicated had the largest number of fine and improved fine wool sheep. However, as discussed in Section 4.3.1, the XUAR statistics redefined almost 2.4 million (or approx. 20%) of the fine and improved fine wool sheep as being semi-fine and improved semi-fine wools in 1991. Consequently, the XUAR slipped slightly behind the IMAR in terms of number of fine and improved fine wool sheep. Despite the revision in the statistics, the data in Table 7.6 suggest that the number of sheep in this category probably increased at about the same rate as the expansion in the total flock during the 1980s. However, as mentioned in the footnote under Table 7.6, there is evidence that there was a sharp drop in fine and improved fine wool sheep numbers between 1990 and 1991.

Table 7.6 Number and Proportion of Total Sheep Represented by Each Major Category of Sheep in XUAR, 1981 to 1991

Year	Total sheep		Fine and improved fine wool sheep		Semi-fine and improved semi-fine wool sheep[1]		Other sheep	
	('000)	(%)	('000)	(%)	('000)	(%)	('000)	(%)
1981	18,408	100	8,637	46.92			9,771	53.08
1982	19,206							
1983	20,074							
1984	20,236	100	9,870	48.77			10,366	51.23
1985	20,371	100	10,158	49.87			10,213	50.13
1986	21,193	100	10,178	48.03			11,015	51.97
1987	21,963	100	10,776	49.06			11,187	50.94
1988	22,782	100	11,510	50.52			11,272	49.48
1989	23,537	100	11,821	50.22			11,716	49.78
1990	23,814	100	12,189	51.18			11,625	48.82
1991	23,706	100	9,245	39.00	2,380	10.04	12,081	50.96

Note: Year end data.
[1]Although the official data do not include this category until 1991, it seems that in 1990 there were 1,784.9 thousand semi-fine and improved semi-fine wools and 10,404 thousand fine and improved fine wool sheep in the XUAR. If this is correct, the number of fine and improved fine wool sheep in the XUAR declined by over 11% between 1990 and 1991.
Source: *China Agriculture Yearbooks*, Agriculture Publishing House, Beijing

7.2.5 Distribution of Herbivorous Livestock by Prefecture

The distribution of sheep and goats by prefecture in 1989 is presented in Table 7.7. These figures suggest that the total sheep flock is well dispersed, with 8 prefectures having in excess of 1 million head. However, the pure-bred and improved sheep are concentrated in five prefectures (Changji, Yili, Tacheng, Aletai, and Akesu). Of these, Changji, Yili and Tacheng are major fine wool growing prefectures. Aletai is a meat/coarse wool sheep area, and Akesu is a specialist semi-coarse wool and meat producing prefecture.

The percentage of prefectural flocks which is described as pure-bred or improved varies a great deal. Bertala, which is another fine wool growing area, has the highest proportion of pure-bred and improved sheep (81%). Of the prefectures with a significant number of sheep, Kashi with 10% pure-bred or improved has the lowest proportion of this type of sheep. Overall, half the sheep in XUAR are classified as pure-bred or improved.

Another significant difference between the major sheep raising prefectures (and one which is not clear from the cross-sectional data in Table 7.7), relates to the rate of expansion in the number of sheep and other herbivorous livestock since 1949. While sheep have increased faster than other grazing livestock in all prefectures, the difference has been especially great in Yili Prefecture where sheep numbers have quadrupled since 1949. The graphs in Fig. 7.2 compare Yili with Tacheng, another major sheep-raising prefecture. In 1949, both prefectures had sheep flocks of about the same size, but Yili now has 50% more sheep. Although the sheep flock has grown much faster in Yili than in Tacheng, the total number of grazing animals (as measured in sheep equivalents) has increased 2.3 times in both prefectures. Clearly, sheep are now much more numerous in Yili relative to other herbivorous animals than was the case in 1949.

All four graphs in Fig. 7.2 exhibit marked drops between 1968 and 1969. Livestock numbers declined sharply in 1969 owing to the worst excesses of the Cultural Revolution and bad weather conditions.

Table 7.7 Distribution of Sheep and Goats in XUAR by Prefecture/City, 1989

Prefecture/city	Sheep			Goats
	Total all sheep	Pure-bred and improved sheep[1]		
	('000)			
Urumqi C.	348	231	(66%)	108
Kelamayi C.	36	2	(6%)	5
Tulufan P.	678	-	(0%)	105
Hami P.	562	224	(40%)	285
Changji A.P.	1,692	994	(59%)	268
Yili P.	3,156	2,276	(72%)	141
Tacheng P.	2,062	1,622	(79%)	380
Aletai P.	1,450	1,068	(74%)	330
Bertala P.	707	576	(81%)	52
Bayinguole P.	1,607	434	(27%)	393
Akesu P.	1,964	948	(48%)	988
Kezile P.	766	219	(29%)	344
Kashi P.	3,209	311	(10%)	504
Hetian P.	2,515	539	(21%)	192
PCC[2]	2,785	2,383	(86%)	201
Total	23,537	11,827	(50%)	4,296

Note: End of year data.
[1]Includes "fine wool" and "improved fine wool" sheep, "semi-fine wool" and "improved semi-fine wool" sheep and all other pure-bred and improved sheep. The percentage in brackets is the proportion these sheep represent of the total sheep flock in the respective prefecture/city.
[2]Production and Construction Corp - see Section 7.4 below for details.
Source: Xinjiang Statistical Yearbook, 1990

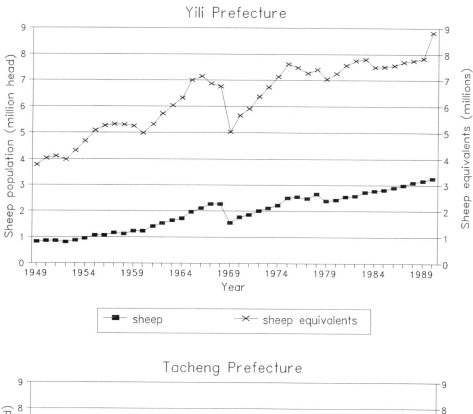

Fig. 7.2 Long-Term Trends in the Sheep Population and Total Sheep Equivalents in Yili and Tacheng Prefectures of the XUAR

7.2.6 Location of Pastoral Production

While the distribution of sheep and goats by prefecture shown in Table 7.7 provides some indication of the location of pastoral production within the XUAR, the output data in Table 7.8 presents a more detailed picture.

Yili Prefecture (see Map 7.1) is clearly the dominant pastoral prefecture in terms of the quantity of beef and raw wool produced while it ranks second to Kashi in mutton and goat meat output. As mentioned earlier, one of the case study counties (Cabucaer) is in Yili Prefecture. The next most important wool-growing prefecture after Yili is Tacheng and the second case study county in the XUAR (Hebukesaier) is in this prefecture.

As mentioned in the previous section, Changji, Yili and Tacheng are major fine and improved fine wool growing prefectures. Furthermore, as discussed below in Section 7.4.2, a high proportion of the fine and improved fine wool sheep raised by the Production and Construction Corp (PCC) are also in these three prefectures. The data in Table 7.8 indicates that Changji, Yili, Tacheng and the PCC together contributed almost 60% of the total raw wool output of the XUAR in 1987.

Table 7.8 Output of Major Pastoral Products in the XUAR by Prefectural Administrative Unit, 1990

Prefecture/city	Beef	Mutton and goat meat	Raw wool (1987)[1]
		(tonne)	
Urumqi C.	1,450	3,209	808
Kelamayi C.	102	248	46
Tulufan P.	1,006	4,521	1,060
Hami P.	679	3,130	1,232
Changji A.P.	3,451	10,702	4,170
Yili P.	15,431	20,199	9,255
Tacheng P.	4,869	12,351	5,238
Aletai P.	10,872	16,828	2,834
Bertala P.	1,696	3,915	1,650
Bayinguole P.	2,876	10,347	2,464
Akesu P.	4,541	11,898	2,996
Kezile P.	3,363	6,304	1,221
Kashi P.	11,885	22,333	3,131
Hetian P.	4,037	17,084	2,352
PCC[2]	4,564	14,468	7,752
Total	70,822	157,537	46,209

[1]Raw wool production by prefecture for 1990 is unavailable. The data in the table is for 1987.
[2]Production and Construction Corp - see Section 7.4.
Source: *Xinjiang Statistical Yearbook, 1991*. Statistical Press of China, Beijing

7.3 Administration of Agriculture and Animal Husbandry in the XUAR

Agriculture and animal husbandry activities in the XUAR are grouped into two distinct networks. That is, there are two vertical hierarchies responsible for administering agricultural and livestock production.

The first is the usual provincial or autonomous region government bureaucracy. As in the IMAR, Gansu and most other provinces/autonomous regions, the Agricultural Commission of the XUAR has an Agricultural Department/Bureau and an Animal Husbandry Department/Bureau. These departments/bureaus have corresponding units at the prefectural and county level. The XUAR Animal Husbandry Bureau is structured in much the same way as similar bureaus in other provinces/autonomous regions. That is, it has Pasture Management Stations, Animal Improvement Stations, and so forth at the provincial, prefectural and county level. Like the Animal Husbandry Bureau in the IMAR, it also has responsibility for an Academy of Animal Sciences. The Xinjiang Academy of Animal Sciences, as it is called, consists of five divisions: Institute of Animal Sciences; Grassland Institute; Information Institute; Veterinary Institute; and Sheep Breeding Centre. The (National) Sheep Breeding Centre was established in 1991 as part of the AIDAB funded China-Australia Sheep Research Project.

The second vertical bureaucracy concerned with agricultural and animal husbandry production in the XUAR is the Xinjiang Production and Construction Corp.

7.4 Xinjiang Production and Construction Corp

The Xinjiang Production and Construction Corp (PCC) has its headquarters in Urumqi. The administration of the PCC is more or less independent of the XUAR government. The agricultural and animal husbandry activities of the PCC are under the control of the Xinjiang General Bureau of State Farms and Land Reclamation. This Bureau functions independently of (and in competition with) both the Agricultural Bureau and the Animal Husbandry Bureau under the Xinjiang Agricultural Commission. There are similar PCC systems in three other provinces in China (Yunnan, Hainan and Heilongjiang).

7.4.1 A Brief History of the Xinjiang PCC

After entering the XUAR in 1949, the People's Liberation Army (PLA) was charged with frontier defence and other military tasks and with becoming self-sufficient in grains and other food-stuffs. In February 1951, the Xinjiang Military Region (XMR) administration introduced a three-year program for developing agricultural and industrial production in the XUAR. By 1953, 27 army reclamation farms and 42 industrial enterprises (coal mining, steel making, textiles, etc.) had been established and it was decided to divide the PLA in XUAR into two parts: a national defence troop and a production troop. The production troop formally became the PCC of the XMR in August 1954 after the establishment of the PCC had been ratified by the Military Commission of the Central Committee of the Communist Party of China.

The period 1955 to 1966 was one of great expansion for the Xinjiang PCC. The gross value of agricultural and industrial output increased 11.5 times relative to 1954. By 1966, the PCC sector represented 25% of the whole XUAR economy.

During the Cultural Revolution, the Xinjiang PCC went through a period of great social upheaval and economic recession. Between 1966 and 1974, the PCC population increased by 772,000 (or 52%) while the gross output value in the PCC sector increased by only 7% and the agricultural gross output value actually declined by 9%. By 1974, the PCC administration had accumulated a large deficit.

The XMR administration was abolished in April 1975 and the PCC organisation became responsible to the General Bureau of Reclamation for the XUAR in May 1975. Despite these administrative changes, the period 1975 to 1977 was a difficult period for the PCC. Funds were not available for investment in capital construction and there was a major shortage of production inputs. The PCC continued to generate a large deficit.

In 1978 the State Council included the Xinjiang PCC in the group of reclamation organisations which received essential inputs directly as part of the State Plan. Consequently, the Xinjiang PCC entered a new period of development. The structure and management systems of the PCC were adjusted and the household production responsibility system was introduced for agricultural production on PCC State farms. By 1981, the PCC was once again self-sufficient and it operated with a net profit of ¥21 million in that year.

Further major changes occurred in December 1981. In order to encourage PCC reclamation activity in the XUAR, the Central Committee, the State Council and the Military Commission of the Central Committee, decided to strengthen the Xinjiang PCC organisation. The PCC was restructured into the current ten agricultural divisions, three farm management bureaus and one engineering and construction division. The reconstructed PCC then embarked on a new era of reclamation and construction.

The position of the Xinjiang PCC was further strengthened in May 1990. The State Council agreed to include the Xinjiang PCC plan for economic and social development in the State Plan for all departments under the State Council. This decision gave the PCC the authority to implement its plan on an equal footing with governments of provinces and autonomous regions. The PCC became a "state within a state" in the XUAR.

The economic size of the Xinjiang PCC relative to the whole economy of the XUAR is indicated by the data in Table 7.9. The PCC population in 1990 represented only 14% of the total for the XUAR but the PCC employed more than one-third of all workers in state-owned enterprises. The PCC also generated almost one-fifth of the XUAR GNP and 28.5% of the gross value of agricultural output; and it produced 17% of raw wool, 26% wool fabrics, and 50% of hand knitting wool yarn. Clearly, the PCC is a major economic force in the XUAR and, in particular, it has a major interest in wool growing and processing.

Table 7.9 Some Major Indices of the Importance of the Xinjiang PCC in Relation to the XUAR Economy, 1990

Index	Unit	PCC	XUAR	PCC as a proportion of XUAR (%)
Population	1000 p.	2,143.5	15,155.8	14.1
Staff & workers of state-owned units	1000 p.	935.2	2,616.0	35.7
Gross national product (GNP)	Mill. ¥	4,546.0	23,969.0	19.0
Total output value of society	Mill. ¥	10,059.0	43,914.0	22.9
National income	Mill. ¥	4,147.0	19,441.0	21.3
Gross output value of industry & agric. In which:	Mill. ¥	8,277.0	34,677.0	23.9
Industry	Mill. ¥	4,445.0	21,209.0	21.0
Agriculture	Mill. ¥	3,832.0	13,468.0	28.5
Output of grains	1000 t	1,566.3	6,768.9	23.1
Output of cotton	1000 t	194.8	468.8	41.6
Output of aquatic products	1000 t	8.2	23.2	35.3
Livestock population. In which:	1000 h	3,311.0	34,964.0	9.5
pig	1000 h	324.0	897.1	36.1
sheep & goats	1000 h	2,729.4	28,308.1	9.6
Output of meats	1000 t	47.1	304.6	15.5
Output of wool	1000 t	8.5	49.3	17.2
Output of cow milk	1000 t	43.5	308.1	14.1
Output of eggs	1000 t	15.4	63.0	24.4
Output of wool fabrics	1000 m	3,220.0	12,303.7	26.2
Output of hand-knitting wool yarn	1000 t	1.1	2.2	50.0
Gross output value of construction	Mill. ¥	964.0	3,951.0	24.4
Total value of retail commodities	Mill. ¥	1,970.0	11,594.0	17.0

Source: Statistical Summary of Xinjiang Production and Construction Corp, 1991, pp.14-15

7.4.2 Sheep and Wool in the PCC System

The organisational structure put in place in 1981 still exists. Thus, as already mentioned, the PCC consists of 11 divisions and three farm management bureaus (which are a little below divisional status). These units are listed in Table 7.10, which also indicates the city (or large town) in which each unit has its headquarters and the prefecture. Beneath each division are regiments, beneath each regiment are companies, etc. The PCC retains a traditional military hierarchical organisational structure, although it is now entirely devoted to production, construction and other commercial operations.

As can be seen from Table 7.10, all PCC Divisions and Bureaus raise sheep and goats, even the No. 1 Engineering and Construction Division. In addition, there are some other units directly under the control of the PCC, but not in the formal divisional/bureau structure, which also have sheep and goats. Most of the sheep and goats (73%) in the PCC system are fine wool or improved fine wool sheep. The prefectural distribution pattern for PCC sheep and goats shown in Table 7.10 is similar to the pattern for non-PCC sheep (see Table 7.7). Yili Prefecture has easily the largest number of PCC sheep and almost all of these sheep are fine or improved fine wools. The PCC also has a large number of this category of sheep in the Changji/Shihezi/Kuitun/Bertala/Tacheng group of prefectures.

Table 7.10 Distribution of Sheep and Goats in the Xinjiang PCC System by Division/Bureau and Prefecture/City, 1989

Division/bureau and location of headquarters	Prefecture or city	Sheep				Total goats
		Total all sheep	Fine wool and improved fine wool	Semi-fine wool and improved semi-fine wool sheep	Other sheep	
		------------------ (no.) ------------------				
No. 1 Agric Div - Akesu	Akesu	155,445	56,953	9,300	89,192	9,164
No. 2 Agric Div - Korla	Bayinguole	192,572	146,721	45,851	0	1,214
No. 3 Agric Div - Kashi	Kashi	226,145	21,028	--	205,117	103,446
No. 4 Agric Div - Yining	Yili	434,625	427,192	3,534	3,899	4,849
No. 5 Agric Div - Bole	Bertala	169,375	169,375	--	0	2,106
No. 6 Agric Div - Wujaiqu	Changji	320,584	236,901	46,373	37,310	25,338
No. 7 Agric Div - Kuitun	Kuitun	220,764	205,730	7,478	7,556	5,709
No. 8 Agric Div - Shihezi	Shihezi	329,716	317,573	12,143	0	5,214
No. 9 Agric Div - Emin	Tacheng	231,952	230,340	1,612	0	5,508
No. 10 Agric Div - Beitun	Aletai	168,083	121,543	25,237	21,303	4,924
Urumqi Farm Man Bur	Urumqi	84,035	70,527	13,067	441	4,652
Hami Farm Man Bur	Hami	150,408	104,125	14,213	32,070	25,918
Hetian Farm Man Bur	Hetian	73,725	53,679	20,046	0	2,615
No.1 Eng & Const Div	Urumqi	8,834	6,704	2,130	0	20
Units directly under PCC		18,401	17,593	304	504	308
Total All PCC		2,784,664	2,185,984	201,288	397,392	200,985

Source: Statistical Yearbook of Xinjiang Production and Construction Corp. 1990, pp.162-164

Interestingly, more than 7% of the sheep are classified in Table 7.10 as semi-fine and improved semi-fine wools. As discussed earlier, until 1991 the official statistics suggested there were no sheep of this type in the XUAR. Clearly, historical data series on sheep in the XUAR are not reliable in this regard.

The PCC State farms permit households living on the farm to keep private sheep. The data in Table 7.10 includes these private sheep. While no precise details were available, it was suggested that about half the sheep in the PCC system were now privately owned.

Similarly, it was not possible to get data on how much wool was produced by privately-owned sheep. The total production for each PCC Division/Bureau shown in Table 7.11, therefore, includes wool grown by both privately-owned and PCC owned sheep. Of the total clip of 8,891 tonne in 1989, 89% was graded as fine or improved fine wool. As would be expected given the distribution of sheep shown in Table 7.10, Yili and Shihezi are the most important wool-growing areas. Almost 40% of the fine and improved fine wool produced in the PCC system came from these two prefectures/cities in 1989.

7.4.3 Other PCC Activities

Apart from the State farms and other commercial undertakings (such as the large wool-processing mill in Shihezi City), the PCC operates the Shihezi Agricultural

University and an affiliated training college, and the Xinjiang Academy of Agricultural Reclamation Sciences which has a major compound in Shihezi City.

The PCC originally established the August 1st Agricultural College in Urumqi. However, this College (or "University" as it should now be termed) has been transferred to the control of Urumqi Prefecture.

The Xinjiang Academy of Agricultural Reclamation Sciences is a parallel institution to the XUAR AHB's Academy of Animal Sciences. The Academy of Agricultural Reclamation Sciences is formally a regiment under the No. 8 Agricultural Division of the PCC which has its headquarters in Shihezi City. It was established in 1979, having evolved from the old PCC Agricultural Research Institute which dated from 1959. In 1992, the Academy consisted of eight research departments or institutes including an Animal Husbandry and Veterinary Research Institute. A new Special Products Research Centre has recently been established to commercialise the research results of the Academy so that it can become more financially independent.

Table 7.11 Distribution of Wool Production in the Xinjiang PCC System by Division/Bureau and Prefecture/City, 1989

Division/bureau and location of headquarters	Prefecture or city	Total all wool	Fine and improved fine wool	Semi-fine and improved semi-fine wool	Other wool
		------------------ (tonne) ------------------			
No. 1 Agric Div - Akesu	Akesu	328	162	23	143
No. 2 Agric Div - Korla	Bayinguole	638	563	75	--
No. 3 Agric Div - Kashi	Kashi	268	79	--	189
No. 4 Agric Div - Yining	Yili	1,543	1,533	8	2
No. 5 Agric Div - Bole	Bertala	577	577	--	--
No. 6 Agric Div - Wujaiqu	Changji	930	797	97	36
No. 7 Agric Div - Kuitun	Kuitun	788	766	22	--
No. 8 Agric Div - Shihezi	Shihezi	1,627	1,565	62	--
No. 9 Agric Div - Emin	Tacheng	831	809	22	--
No. 10 Agric Div - Beitun	Aletai	467	382	63	22
Urumqi Farm Man Bur	Urumqi	276	248	22	6
Hami Farm Man Bur	Hami	383	278	31	74
Hetian Farm Man Bur	Hetian	156	115	41	--
No.1 Eng & Const Div	Urumqi	34	28	6	--
Units directly under PCC		45	44	1	--
Total All PCC		8,891	7,946	473	472

Source: Statistical Yearbook of Xinjiang Production and Construction Corp, 1990, pp.169-170

The Special Products Research Centre has developed a number of new consumer products which can be manufactured by the Academy. For example, in 1992 the Academy dairy factory had just begun to manufacture and market an "ice-cream sandwich" developed by the Special Products Research Centre. This delicious new confectionery consisted of a small block of ice-cream between two wafer-like biscuits. It was cleverly and attractively packaged in a sealed plastic wrapper. It was said to cost the Academy about ¥0.4 to manufacture while the prefectural Price Bureau had set a wholesale price of ¥0.6. The retail market was a free market and the product was selling well to consumers at around ¥1.0. The "ice-cream sandwich" was only one of several new entrepreneurial products and activities being developed by the Academy.

The Academy in Shihezi is the centre of the PCC research and development activities supporting the large PCC State farm network. For example, the Animal Husbandry and Veterinary Research Institute within the Academy, plays an active role in the PCC fine wool sheep improvement program (see Section 7.5.2). It also contributes to the training programs for sheep and wool technicians conducted by the PCC Sheep and Wool Training Institute located on Ziniquan State Farm and Sheep Stud.

7.5 Livestock Improvement Programs

Both the XUAR AHB and the PCC networks are engaged in animal improvement programs designed to upgrade the quality of livestock in the XUAR. In the case of sheep, there are a number of separate programs designed to build on the existing sheep resources of the XUAR.

As Table 7.7 shows, at the beginning of the 1990s there were about 24 million sheep in the XUAR (including the sheep in the PCC system). In general terms, these sheep could be classified along the following lines:

- Fine wool sheep   - 8 million
- Improved fine wool sheep   - 4 million
  Fine wool and improved fine wool sheep are raised mainly in six prefectures (Yili, Tacheng, Bertala, Changji, Hami, and Urumqi).
- Meat breeds of sheep   - 4 million
  These sheep are mainly Altay (Fat-tail) Sheep raised in the Aletai Prefecture. They are primarily meat sheep but they also produce coarse wool.
- Kazak Sheep   - 5 million
- Mongolian Sheep   - 0.4 million
- Bashbay Sheep   - 0.25 million (raised in one county only)
  The last three breeds are mutton and coarse wool producing breeds.
- Hetian Sheep   - 1.6 million
  These sheep produce semi-coarse wool and mutton. They are raised in the Hetian area around the southern edge of the Tarim Basin.
- Xinjiang Lamb Pelt Sheep   - 1 million
  These sheep are a Chinese selection from the Russian Kalequer breed. They are renowned for black lamb skins.

7.5.1 Fine Wool Sheep Improvement within the XUAR AHB Network

The breeding of fine wool sheep in China began in 1935 at what is now Nan Shan State Farm and Sheep Stud which is located 75km by road south of Urumqi in the foothills of the Tianshan Range. In 1939, the best sheep and breeding technicians moved to what is now called Gongnaisi State Farm and Sheep Stud in Xinyuan County of Yili Prefecture. In the 1950s and 1960s, Gongnaisi became the main parent fine wool sheep stud for the XUAR and, in many respects, for the whole country. Nowadays it is one of the four nationally recognised fine wool sheep studs in China. Gongnaisi is the home of the Xinjiang Fine Wool Sheep. Since 1985, the best sheep at Gongnaisi have met the breed standard for the Chinese Merino.

The XUAR AHB fine wool sheep improvement program is based on Gongnaisi (the only first-level stud) and eight second-level studs. The eight second-level studs and the prefectures in which they are located are:

> Cabucaer* (Yili)
> Houcheng (Yili)
> Tacheng (Tacheng)
> Nan Shan* (Urumqi)
> Fukang (Changji)
> Weunqian (Bertala)
> Bouhu* (Bayinguole)
> Baicheng* (Akesu)

Unlike the other seven studs which are all under prefectural AHBs, Nan Shan is controlled by the Xinjiang Academy of Animal Sciences. As already mentioned, the Academy is under the provincial AHB. Nan Shan, Cabucaer and Tacheng studs are described in considerable detail in Longworth and Williamson (1993).

There is a general consensus of opinion that Cabucaer, Nan Shan, Bouhu and Baicheng (the four studs marked with an asterisk above) are the better studs. In the past, all eight second-level studs have drawn on Gongnaisi for genetic material and they would all claim to raise Xinjiang Fine Wool Sheep. In fact, some of these studs (such as Cabucaer and Nan Shan, for example) now have a significant number of sheep which are classified as Chinese Merinos. However, in recent years there seems to have been a strong desire "to go it alone". Indeed, the importation of new blood from Gongnaisi or from any other Chinese stud seems to be frowned upon. On the other hand, almost any merino ram imported from Australia is used as widely as possible.

Xinjiang Fine Wool rams from these studs are dispersed to county-level "ram breeding" stations, or to township or village-level "ram care" stations, or even to individual households assigned to undertake ram care responsibilities. The county-level ram breeding stations act more or less as third-level studs since they breed rams for the "ram care" stations. The ram care stations supply the AI centres with rams during the mating season and provide approved rams for natural mating. Of course, there are also a significant number of locally bred rams used for natural mating and probably even at the AI centres. Nevertheless, to the extent that farmers use the AI services available, the breed improvement program could be most effective.

During the commune/collective era (1956 to 1978), the XUAR government implemented a policy of steadily increasing the number of fine and improved fine wool sheep by requiring the communes to meet specified planning targets in this regard. Since the breakup of the communes, it has been much more difficult to implement such a policy. Indeed, officials from the XUAR AHB Animal Improvement Station reported that there was a serious genetic regression in a significant proportion of the Xinjiang fine wool and improved fine wool flocks after the individual households purchased the sheep from the collectives (usually in 1984). The officials of the XUAR SMC and many other people interviewed in the XUAR made the same point. Given the freedom to choose, many farmers did not mate their ewes to the fine wool rams available through the AI centres. They consider fine wool sheep are not well suited to the harsh environment and, in particular, require too much supplementary feeding over the long winter. In pastoral areas, supplementary feedstuffs (especially grain) are in critically short supply and, if available, are too expensive to feed to sheep.

The official AHB position is that the genetic regression problem is principally concentrated in the private small-holder sector. Although all State farms in the AHB network implemented a kind of household production responsibility system in 1984/85, the managers of the State farms still control the key sheep production decisions.

According to the official view, these managers are conforming to government policy. However, State farm managers and leaders of sheep production teams on State farms who were interviewed, expressed the same negative views about fine wool sheep as the private farmers.

These "grass roots" reservations about the wisdom of continuing to expand fine wool production in Xinjiang are at odds with official policy. For example, the 8th Five Year Plan (1986 to 1990) for the XUAR called for a marked increase both in the absolute output of fine and improved fine wool and in the proportion of the clip which was wool of this type. Motivated by these planning targets, the XUAR government took action in 1988 to protect the integrity of the fine wool sheep improvement program in both the small-holder and the State farm sectors. Twenty-seven counties in Yili, Bertala, Tacheng, Changji, Urumqi and Hami Prefectures (the so-called Xinjiang fine wool base) were set targets which called for an increase in their fine wool flocks. The county governments were told to achieve these targets by ensuring that a specified percentage of eligible ewes were mated by AI to fine wool rams. When interviewed in 1992, the provincial-level Animal Improvement Station officials claimed that these administrative controls were combating the genetic regression problem and, in most areas, flocks were once again being upgraded.

As reported in the Hebukesaier County case study (Section 17.2.3), the real situation at the grass roots may be rather different to the official perception in Urumqi. In Hebukesaier, which is one of the counties for which AI targets have been set, the responsible officials were not attempting to enforce the targets. Indeed, the County Animal Improvement Station could not supply even an estimate of the number of ewes being mated by AI.

The officials in Hebukesaier County gave the impression they felt there had been an over-emphasis placed on upgrading for fine wool production. Even in Urumqi, senior people in the AHB recognised that fine wool sheep had their limitations and that perhaps there had been too much weight given to fine wool breed improvement over the last three decades.

7.5.2 Fine Wool Sheep Improvement within the PCC Network

Ziniquan is the parent stud for the PCC fine wool sheep improvement program in the XUAR. It is located in Shawan County of Tacheng Prefecture in the foothills of the Tianshan Range south of Shihezi City (Map 7.1). The sheep at this stud, it is claimed, have been developed independently of Gongnaisi. Ziniquan is the home of the Xinjiang Fine Wool Sheep (PCC Type) and one of the four nationally recognised fine wool sheep studs. For a history of the breeding program (including the critical role played by Australian merino blood), details of the four bloodlines, a description of the sheep, and other details about Ziniquan, see Longworth and Williamson (1993).

Until 1990, rams from Ziniquan were sold to three daughter studs in the PCC network. These three daughter studs supplied 33 commercial wool-growing State farms. All of these State farms are located in the Shihezi/Kuitun area. The three daughter studs are operated by PCC Regiment No. 124 (located near Kuitun) and Regiments No. 142 and No. 147 (located near Shihezi). Ziniquan, the three daughter studs and the 33 commercial State farms are all in PCC Divisions No. 7 (Kuitun) or No. 8 (Shihezi).

In the Kuitun/Shihezi area, there are about 500,000 fine wool sheep (Table 7.10) of which 150,000 to 200,000 would be adult ewes. The three daughter studs have a total of 15,000 adult ewes (No. 142 has 5,000 ewes, No. 124 has 6,000 ewes, No. 147 has

4,000 ewes). For many years these three daughter studs and Ziniquan have operated a joint sheep breeding and assessment program.

Beginning in 1990, the PCC extended the Ziniquan network to four new PCC Divisions: No. 4 (Yining), No. 5 (Bole), No. 9 (Emin) and No. 2 (Korla). There are as yet no second-level (or daughter) studs in these four new divisions. However, there are some divisional-level ram breeding State farms which could be upgraded. The plan is to establish five new daughter studs - one each in the Bole, Korla and Emin areas and two in Yili Prefecture (Yining area). The expanded Ziniquan network will include a total of almost 100 commercial fine wool growing State farms within the six PCC Divisions. [Each farm is a separate Regiment under the respective Division.]

Of the 10 PCC divisions concerned with agricultural and animal husbandry, the four not now included in the Ziniquan fine wool production network are not fine wool growing divisions. No. 1 (Akesu) and No. 3 (Kashi) are interested primarily in semi-fine wool sheep. No. 10 (Beitun) is concerned with breeding fat tail mutton sheep. No. 6 (Wujaiqu) is close to Urumqi and concerned more with pigs and dairy cattle.

The six divisions now in the Ziniquan network have over one million fine wool sheep of which 40% would be adult ewes. The remainder are replacement ewes and rams (yearlings), wethers, and breeding rams.

The PCC State farms in the Ziniquan network have formed an association called "The Association for PCC Type of Xinjiang Fine Wool Sheep". Amongst other things, this association sets the prices for each Ziniquan ram which is for sale within the PCC system. The Association also organises a joint assessment of the sheep on Ziniquan. Usually the chief breeding specialists from the other divisional studs come to Ziniquan to make these assessments.

7.5.3 Breed Improvement Programs for Meat and Coarse Wool Sheep

In 1987, the XUAR AHB launched a meat and coarse wool sheep improvement program. Initially it was to be concentrated in three prefectures (Aletai, Kashi, and Hetian). Since 1987, pure-bred sheep have been imported from Kazakhstan (Altay Fat-tails) and from Australia and New Zealand (Lincoln, Dorset and Suffolk) to up-grade the local breeds in these three prefectures.

A fourth prefecture, Akesu, has been designated as a specialist semi-coarse wool growing area. Pure-bred sheep have been imported from Kazakhstan and elsewhere to upgrade the semi-coarse wool sheep in Akesu Prefecture.

Since the opening up of the live sheep export market in Kazakhstan, there has also been increasing interest in improving the meat sheep in other prefectures such as Yili, Changji and Bayinguole.

The emerging new meat sheep production system involves mating local Altay or Kazak ewes to Suffolk or Dorset rams by AI for a late-winter/early-spring lambing. The lambs are fed supplements from 15 days of age until the end of May. They are then transferred to summer pasture. Once the lambs reach 30kg liveweight (usually around 5½ months of age), they are marketed and slaughtered. It was anticipated that over 100,000 ewes would be mated to pure-bred meat breed rams in 1992.

Suffolk crosses grow faster than Dorset crosses and they can graze a wider selection of plant species. However, they usually have black heads and legs and the black wool is not popular with the textile mills. All the F_1 must be slaughtered.

The Dorset crosses do not grow as fast as the Suffolks but they are faster growing than the native meat sheep. Their wool is better than that of the Suffolks (white and around 56/58 in spinning count). Because the wool is acceptable, it may be possible to keep some female F_1 sheep for further breeding. This option is popular with the farmers.

Even though Yili Prefecture has the second largest area of summer pasture (Table 7.4) and some of the best summer pastures in the XUAR, and is close to the Kazakhstan border, there are not many Altay or Kazak ewes in Yili from which to breed meat sheep. Officially, therefore, Yili is expected to remain a major fine wool growing prefecture. Meat-sheep raising will not replace fine wool production. On the other hand, even Cabucaer State Farm and Sheep Stud, the second-level fine wool stud owned by the Yili prefectural AHB, has moved into meat-sheep raising in a relatively big way. For more details on the new meat-sheep raising enterprise at this stud, see Longworth and Williamson (1993). If the live sheep export market continues to grow, more and more of the fine wool growers in Yili Prefecture will switch to meat sheep production. Not only are meat sheep more profitable but also the quick turn-off production system outlined above matches the feed requirements of the sheep to the natural pasture growth cycle much better than fine wool growing production systems.

7.6 Xinjiang Animal Husbandry Industrial and Commercial Company

The XUAR Animal Husbandry Bureau (AHB) has a commercial arm called the Xinjiang Animal Husbandry Industrial and Commercial Company (XAHICC). This commercial company was established in 1980 but did not really commence activities until 1984. It is vertically connected to the China AHICC which is one of the commercial companies controlled by the Ministry of Agriculture in Beijing.

Names of Chinese organisations are often translated in a number of different ways. Hence, the China AHICC may have a number of other English names such as "China General Corporation of Animal Husbandry Industry and Commerce". This national commercial organisation controlled by the Ministry of Agriculture is engaged in a large number of commercial joint ventures related to animal husbandry, feedstuffs, grass seed production, and apiculture. These joint ventures involve mainly local Chinese partners but participation by foreign investors is also being encouraged. Since the China AHICC is backed by the Ministry of Agriculture, it would seem to be an attractive joint venture partner for a potential foreign investor.

The XAHICC initially set up two subsidiary companies: The Animal Husbandry Industrial and Commercial Trading Company which has responsibilities for buying and selling animal products such as wool and hides (and some agricultural products); and The Animal Husbandry Industry Breeding and Livestock Company which trades in breeding livestock (and semen) as well as livestock for fattening and/or slaughter.

In 1986, the XAHICC purchased a major tourist hotel in Urumqi called the Tian Shan Hotel. A further subsidiary was established to operate this venture. About the same time, a fourth subsidiary called the Animal Products Trade Centre was created to provide wholesaling services and to operate a retail department store. The Animal Products Trade Centre plans to develop a wholesale market for raw wool in Urumqi. In 1988, a fifth subsidiary known as The Animal Medicine and Machinery Company was established.

In addition to these five specialised subsidiary companies, the XAHICC has a 25% interest in a joint venture called the Hua Xin AHICC. The other partners are the China AHICC and the Urumqi Agricultural Committee (which belongs to the Urumqi City government). The Hua Xin Company operates a number of large enterprises in the XUAR producing eggs and chickens.

Companies similar to the XAHICC exist at the provincial level in all provinces/regions in which animal husbandry is a major activity. They are vertically connected with similar commercial units at the prefectural level and even down to the county level in some areas. For example, in the XUAR, there are AHB trading companies in Aletai and Yili Prefectures but not in Tacheng prefecture nor in either of the counties surveyed (i.e. Cabucaer and Hebukesaier). In the absence of these relatively new commercial institutions, livestock farmers and collectives must trade with the traditional government monopolists, the SMCs and the Food Companies.

During the "wool war" period, these AHB companies became major competitors for the Supply and Marketing Cooperatives (SMCs), even in provinces or prefectures where the wool market was officially closed and the SMCs were supposed to have a monopoly over wool purchasing. With the opening of the wool market in the XUAR in 1992, the trading company subsidiary of the XAHICC is likely to expand its wool trading activities at the expense of the SMCs.

7.6.1 Wool Trading Activities

Since about 1986, State farms in the XUAR have not been compelled to sell all their wool to the SMCs. Most larger State farms (i.e. county- and prefectural-level farms) outside the PCC network now either sell directly to mills or to the XAHICC Trading Company. In the past, the trading company has bought very little wool from PCC State farms, township level State farms, village collectives or individual farmers. However, with a completely free market from 1992, this situation could change.

In 1990, the XAHICC Trading Company purchased about 4,500 tonne of greasy wool. Purchases dropped to 3,000 tonne in 1991 due to the depressed state of the market. The total output of fine wool in the XUAR, excluding the PCC network, is 6,000 to 8,000 tonne per year. The Trading Company therefore handles about half this segment of the wool trade.

Unlike the SMCs which buy on a greasy weight basis, the Trading Company usually buys and sells on a clean wool basis. The wool is tested for yield by the Fibre Testing Bureau, and on-sold to mills without regrading or repacking. The grading done on the State farms is acceptable because the graders are trained by the XUAR AHB.

Some State farms find it convenient to deal through the Trading Company rather than sell direct to the mills because the Company has better information about the state of the wool market not only in the XUAR but in other provinces as well. The mills buy from the Trading Company because it can offer larger lines of wool of each grade. That is, the Company may combine lots of wool of similar grade from different sources before selling to processors. The Company is said to have good relations with a wide range of mills throughout China.

Before the XUAR wool market was opened to free trade in 1992, the Trading Company appeared to be paying State farms the official Price Bureau established "instructive prices" for wool landed in Urumqi and selling to the mills with a fixed 25% mark-up. The 25% represented a 10% product tax payable to the government of the county from which the wool came, 5% handling costs, and 10% trading profit. There was clearly room for negotiation in relation to both the buying and selling prices. With free trade from 1992 onwards, the marketing margin may narrow somewhat.

7.6.2 Livestock Trading Activities

In 1992, the market for livestock in the XUAR was declared open and the XAHICC Livestock Company was completely free to compete with the Food Company of the Ministry of Commerce (now known as the Ministry of Domestic Trade). Prior to 1992, the State established delivery quotas for livestock. For example, in 1991 all State farms were required to deliver to the Food Company 10% of the livestock they wanted to sell. They were paid State list prices for these animals. In earlier years, the delivery quotas were much larger proportions of the stock being sold. With the decline in the importance of compulsory delivery quotas, the number of livestock available for purchase by the XAHICC has increased rapidly in recent years.

At the same time, major new export markets for live animals have developed in Kazakhstan and the other Central Asian republics. Until 1959, livestock were often exported to Kazakhstan and other parts of the USSR. During 1988, after an almost thirty-year gap, the XUAR AHB re-established relations with the Department of Agriculture in Kazakhstan. In 1989 and 1990, the XUAR sent Xinjiang Fine Wool sheep and Chinese Merinos (Xinjiang Type), and some stud cattle (both beef and dairy breeds) to Kazakhstan. It imported in exchange some semi-fine wool breeding sheep, some semi-coarse wool sheep and some Kalequer (Black Lamb Pelt) Sheep as well as some dual-purpose cattle.

In 1991, the XUAR exported about 140,000 live animals to Kazakhstan in exchange for agro-chemicals (fertilisers) and some steel. Of the 140,000 animals, about 112,000 were fat sheep for slaughter, 20,000 were cattle for slaughter and the remaining 8,000 head were horses. In 1992, it was expected that about 360,000 live animals would go to Kazakhstan from the XUAR, of which about 300,000 would be sheep and 60,000 large animals. In return, the XUAR AHB would receive fertiliser, phosphorus, wood and perhaps motor trucks. The AHB could use these materials or sell them to other units.

These trade deals with Kazakhstan have been organised by the XAHICC. In 1991, the XAHICC paid extraordinarily high prices for good quality fat sheep. Farmers interviewed in Cabucaer County of Yili Prefecture, for example, reported selling sheep to the XAHICC for ¥182 per head.

The XUAR had six open border posts with Kazakhstan in 1992 and there were plans to open the border in a number of other places. The opportunities for future trade with Kazakhstan and the other four new Central Asian republics (Kirghizstan, Tadzhikistan, Uzbekistan and Turkmenistan) as well as with Mongolia, the Russian Federated Republics and Afghanistan (now the civil war is more or less over) are improving. Urumqi has an international airport with weekly flights to Tashkent, the capital of Uzbekistan and on to Istanbul in Turkey. The railway line to the Kazakhstan border was completed in 1990 and it now links with the old USSR rail network at Alataw Shankou north of Bole, the capital of Bertala Prefecture. This new line, therefore, offers a second railway link from China to Europe. It is also shorter in distance, if not time, than the traditional Trans-Siberian/Mongolia route from Beijing to Europe.

7.6.3 Import/Export of Stud Sheep

The XAHICC has been the vehicle by which the XUAR AHB has purchased Australian merino rams. The XAHICC in turn relies on the Canton Animal Development Company to arrange the transportation of these sheep. Most of the rams imported from Australia by the XAHICC have come from Western Australia. The officials being interviewed said there was general dissatisfaction with many of the rams purchased from Australia.

Another role of the XAHICC is to act as an agent for individuals, State farms, or various levels of government wishing to purchase Xinjiang Fine Wool or Chinese Merino (Xinjiang Type) rams for use in other provinces. Buyers from outside the XUAR are not required to buy through the XAHICC and may approach the studs directly if they wish. Purchasers who use the XAHICC as an agent must pay between 2% and 4% of the purchase price as commission.

7.7 Other Economic Issues

Ten of the 37 pastoral and semi-pastoral counties in the XUAR are designated for assistance under the Central government poverty alleviation program (Table 7.2). While one of these counties is in Yili Prefecture and another is in Tacheng Prefecture, the counties surveyed in these prefectures (Cabucaer and Hebukesaier respectively) are not classified as poor counties.

Since sheep are the dominant grazing animals, the prices of wool and mutton have a major impact on incomes in the pastoral areas of the XUAR. Although the markets for wool and meat were both declared completely free markets in 1992, the traditional SMC marketing channels and official pricing policies (especially with regard to premiums and discounts) for wool will continue to influence the prices paid for wool. This will be especially true in relation to the private smallholder sector. On the other hand, the larger-scale State farms have gradually broken away from the SMC since 1986.

official influence on wool price

As discussed earlier (Section 7.6.1), about half the wool produced on non-PCC State farms is handled by the XAHICC. Some of this wool is now sold by auction. Many State farms also sell directly to mills. The State farm sector, therefore, has already embraced a more sophisticated approach to wool marketing and is in a good position to take full advantage of the more open trading opportunities offered by the free market. Nevertheless, the XUAR is economically isolated from the rest of China and government attitudes to wool prices are likely to continue to be important, even under a so-called free market system. Past official pricing policies, therefore, will remain relevant for some time into the future.

7.7.1 Rural Income Trends and Future Sustainable Development

The available average net income per capita data for the rural households in the two case study counties are plotted in Fig. 7.3. There is a striking difference between the trends in nominal and real income in both cases. Nominal incomes rose dramatically during the 1980s, especially in Hebukesaier. This increase is in sharp contrast with the situation during the 1970s which, as demonstrated in Fig. 7.3 for Hebukesaier, was a period when incomes stagnated or declined.

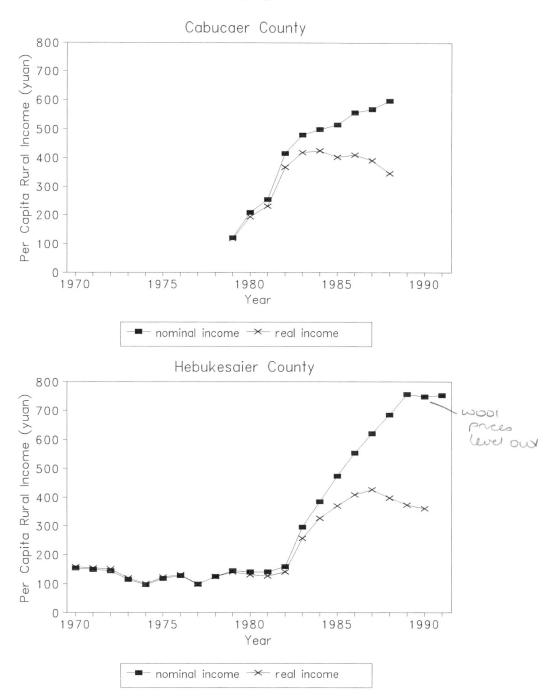

Fig. 7.3 Nominal and Real Average Net Rural Income Per Capita for Two Case Study Counties in XUAR

Nominal incomes rose much more in Hebukesaier County than in Cabucaer. As explained in the case studies (Chapters 16 and 17), these two counties are very different in terms of the make-up of their rural populations. In Hebukesaier, pastoral and semi-pastoral households represent almost half the rural population. On the other hand, only 8% of households in Cabucaer are classified as pastoral. Consequently, the trend in nominal rural incomes in Hebukesaier was much more influenced by the improvement in the prices paid for pastoral commodities (especially wool and mutton) during the 1980s than was the case for Cabucaer. This conclusion is supported by the levelling off of nominal incomes in Hebukesaier in 1990 and 1991 when wool prices declined.

The average nominal net income per capita for rural households in the XUAR in 1990 was ¥622. On this basis, rural incomes were about average in Cabucaer and well above average in Hebukesaier. However, as pointed out previously in relation to remote isolated counties, life is not easy in places such as Hebukesaier. A higher than average nominal income does not readily translate into an above average standard of living.

The falling real incomes in both counties must be of great concern to the XUAR government because it is likely to be a general phenomenon in the pastoral areas. As suggested in relation to a similar situation in the IMAR, there is virtually no capacity to increase further the production of pastoral commodities in order to boost pastoral incomes. In fact, as discussed in the case studies in Chapters 16 and 17, the current overgrazing and consequent degeneration of native pasture will inevitably lead to a contraction in output of pastoral products. Further exploitation of the natural pastures in the pastoral areas of the XUAR is not a sustainable approach to further economic development in these parts of the region.

7.7.2 Wool Pricing in the XUAR

When in 1985 the Central government delegated to provincial-level governments the power to set prices for wool and/or to open the wool market, the XUAR government did not elect to allow a free market to develop for wool. In the case of the XUAR, unlike the situation in the IMAR, there were no contiguous wool-growing areas in other provinces which had adopted a free marketing system for wool. Consequently, the XUAR government was more or less able to retain control over the marketing of wool in the 1985 to 1991 period. In 1992, the XUAR government declared the wool market in XUAR "open" for the indefinite future. All prefectures and counties accepted this decision and, in theory at least, there is now a free market for wool in the XUAR.

The official State prices for wool in XUAR were established by the XUAR Price Bureau between 1980 and 1991. Prior to the establishment of the Price Bureau in 1980, the setting of State prices was the responsibility of the XUAR State Planning Commission. Irrespective of which organisation was charged with setting wool prices, the mechanism for establishing the premiums and discounts for each grade remained essentially unaltered from 1970 to 1991.

These price relativities were constant and were almost the same as the quality differentials defined in the National Wool Grading Standard and discussed in Section 4.5.2. The grade differentials relative to "Improved fine and semi-fine wool - Grade I" which applied in the XUAR were:

Fine and semi-fine wool
- Special grade   124%
- Grade I   114%
- Grade II   107%

Improved fine and semi-fine wool   - Grade I   100%
                                               - Grade II   75%

Coarse wool   [no fixed quality differential, price varied with market conditions]

The only difference between the above grade differentials and the relativities established in the National Wool Grading Standard was that "Improved fine and semi-fine wool - Grade II" is set at 91% in the National Standard (compared with 75% above).

The National Standard has "Improved fine and semi-fine wool - Grade I" as the base type and grade (i.e. with a relativity of 100%). The prices of XUAR were set using the same base until 1986. From 1986 to 1991, the base grade became "Fine and semi-fine wool - Grade I", with an assigned base value of 100%. The other prices were established relative to this new base on the same proportional quality differentials as applied before 1986.

While the quality differentials (premiums and discounts) remained constant in proportional terms between 1970 and 1991, the absolute level of prices changed significantly both before and after 1985. Table 7.12 shows that prices jumped in the late 1970s and again between 1984 and 1985. There was also a big increase in 1986. However, the base grade in 1986 was a better grade than the base grade in 1985 and part of the jump in the prices between 1985 and 1986 indicated in Table 7.12 reflects this change.

Table 7.12 Administratively Determined Prices for the Base Grades of Wool in the XUAR, 1970 to 1991

Year	Price for improved fine and semi-fine wool - Grade I [i.e. *the base price* (100%)]
	[¥/kg (greasy)]
1970 to 1977	Around 2.0
1978 to 1980	3.4
1981 and 1982	3.5
1983 and 1984	3.8
1985	4.5
	Price for fine and semi-fine wool - Grade I [i.e. *the base price* (100%)]
1986	7.5
1987	8.5
1988 (the main "Wool War" year in XUAR)	13.5 (the price reached ¥28/kg in some areas!)
1989	12.00 to 15.00
1990	10
1991	8.76
1992	(free market)

Source: XUAR Price Bureau

Prior to 1992, the SMC was expected to pay the official State purchase price for all wool bought to satisfy the State purchase quota. This quota, of about 40,000 tonne, was set each year until 1991. The SMC did not pay any premium for over-quota wool in most years. However, in 1988 especially, the SMC was forced to increase its prices for all wool to compete with the "illegal" traders (including other government units as well as private individuals) operating at that time.

Prices rose to sensational levels in 1988 because the "wool war" had spread to the XUAR. The official State prices remained high in 1989 despite the collapse of free market prices in China. The downward adjustment in State prices continued in 1990 and 1991. It is understood that, in general, the so-called free market in XUAR in 1992 did not generate prices for the bulk of the clip which were significantly different from the 1991 State prices, although discounts for lower grades of wool increased.

In 1991, perhaps as a transitional step towards the opening of the market in 1992, prefectural administrations were for the first time allowed some flexibility in the setting of wool prices in their area of jurisdiction. They could vary prices ±10% around the provincial Price Bureau established prices. In some cases, even counties were given permission to adjust wool prices in this way.

Until 1992, formal permission was required from the XUAR government to take wool out of Xinjiang. That is, buyers of wool (even if they bought from the SMC) were required to have special permission to export wool to other provinces. There was also a 20% "export tax" on the value of the wool, payable before permission to export wool was granted. This tax was called a "pasture construction tax" and it was intended that the revenue collected would be used to improve pastures in XUAR. Since there is only one practical route out of the XUAR (through Gansu Province), it was easy for the government to police these limitations on the export of wool. From 1992 onwards, the export tax has been lifted and no special permission is required to take wool out of the XUAR. However, all buyers of wool must still pay the 10% product tax to the county governments. It is widely perceived that, under free market conditions, it will be extremely difficult for county governments to monitor wool sales and to collect the 10% product tax.

As already mentioned, the XAHICC Animal Products Trade Centre plans to develop a centralised wholesale market for wool to facilitate the assembly of larger lots, to provide bench-mark prices, and generally to facilitate efficient wool trading. Private traders will be encouraged to use this market and they will be required to pay the AHB a 1% commission on all sales.

Plate 1

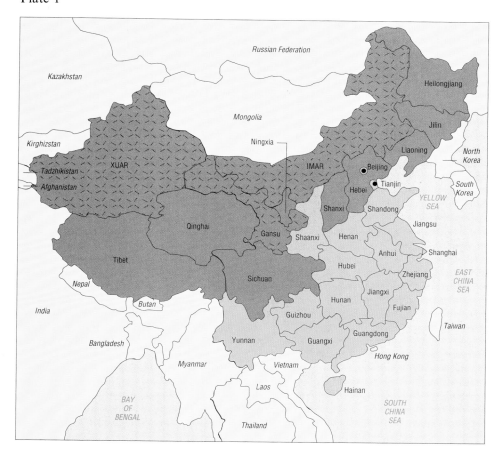

Geographic Location of the 12 Pastoral Provinces of China

Plate 2

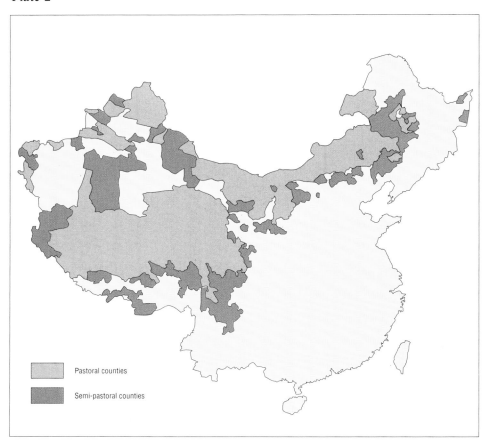

Geographic Location of the Pastoral and Semi-Pastoral Counties of China

Plate 3

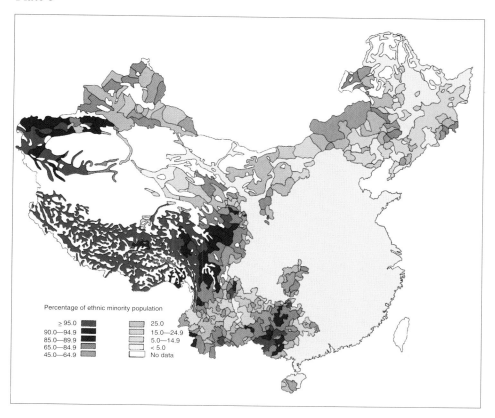

Proportion of the Population Belonging to Minority Nationalities

PART III

COUNTY CASE STUDIES

Chapter Eight

Balinyou County (IMAR)

Balinyou County (or Banner) is the central county of the five northern counties in Chifeng City Prefecture (Map 8.1). It lies in the Eastern Grassland area of the Inner Mongolia Autonomous Region (IMAR). In many respects, Balinyou is a typical county in the pastoral region of China (Liu, 1990). Although pasture degradation due to overgrazing has been recognised as a major problem since the early 1960s, herbivorous livestock numbers have remained at unsustainable levels and the natural pastures have continued to deteriorate.

With almost 45% of the population belonging to the Mongolian Nationality and other non-Han nationalities representing an additional 1%, Balinyou has the highest proportion of minority people of all the counties in Chifeng City Prefecture.

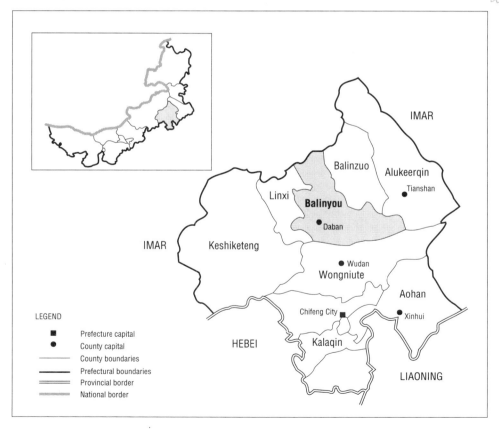

Map 8.1 Balinyou County in Chifeng City Prefecture of IMAR

8.1 Resource Endowment

Despite extensive degradation, the rangelands remain the chief natural resource in Balinyou County.

While not especially distant from Beijing in a geographic sense, Balinyou is a relatively remote and isolated county. In economic terms, it is a poor county and it is included in the poverty alleviation scheme for backward counties operated by the Central government.

8.1.1 General Features

The average annual rainfall at Daban (the county capital) is 349.46mm. However, the distribution of this rainfall is highly seasonal. Furthermore, there is considerable variation from year to year as illustrated in Table 8.1. The lowest rainfall year in the period 1959 to 1990 was 1989 (120mm) while the highest was 1959 (548mm). Not only was 1989 the worst year on record but 1988 was the second worst. The drought of 1988/89 was easily the most serious in living memory.

In Balinyou County, the average length of the warm and cold seasons is 160 and 205 days respectively. The winters are extremely cold and summers are very hot.

Underground water accessed by open wells is one of the major sources of water for livestock in Balinyou County. A high incidence of drought conditions and the extreme cold in winter make this underground water most valuable.

Table 8.1 Annual Rainfall at Daban the Capital of Balinyou County, 1959 to 1990

Year	Annual rainfall (mm)	Year	Annual rainfall (mm)
1959	547.6	1975	244.6
1960	314.9	1976	258.3
1961	407.6	1977	341.0
1962	363.6	1978	229.7
1963	456.8	1979	477.7
1934	368.2	1981	287.4
1965	275.9	1982	316.0
1966	444.2	1983	403.2
1967	330.4	1984	305.2
1968	206.0	1985	416.3
1969	530.5	1986	511.9
1970	327.5	1987	384.3
1971	349.1	1988	158.6
1972	289.7	1989*	120.0
1973	277.2	1990**	430.0
1974	459.8		

* estimate
**provisional estimate
Source: Balinyou County Animal Husbandry Bureau

8.1.2 Land Use

About 78% of the land area in Balinyou is described as usable pasture while as little as 2.6% is suitable for agricultural production.

- Total land area   14,460,000mu   (or 10,256km^2)
  - usable rangeland   11,287,000mu
  - forestry land   2,087,000mu
  - arable land   376,000mu
- Total degraded land area (1980)   10,234,000mu
  - heavily degraded   2,885,000mu
  - medium degraded   3,649,000mu
  - lightly degraded   3,700,000mu

The only publicly available data on the overall extent of rangeland degradation in Balinyou are the 1980 figures presented above. However, a report prepared by Wang (1989) for the county government demonstrates that the situation has become progressively worse since 1980. Some of the points made by Wang are:

• The rate at which pasture land became degraded in the 1980 to 1988 period (i.e. 206,000mu per year on average) was almost double the average annual rate between 1964 and 1979 (i.e. 115,000mu per year).

• The area described as "sandy waste" land (i.e. all of the heavily degraded pasture and some of the worst of the medium degraded areas) increased from 2.9 million mu (or 26% of the usable pasture area) in 1980 to 5 million mu (or 44% of usable pasture) in 1988.

• The average quality of the remaining natural pasture has also declined, with annual plants replacing the more durable and productive perennial species. At the same time, the proportion of leguminous plants has declined and weed species have become more common.

• Average annual dry matter production from the natural pastures has fallen progressively from 180kg of DM per mu in 1958, to 170kg in 1962, to 133kg in 1980, and by 1988 it was only 107kg of DM per mu.

• Mobile and semi-mobile sand dunes covered an area of 1.744 million mu in 1988 which was nearly seven times the area of dunes recorded in 1958.

• The area of degraded pasture land has continued to increase despite substantial investment in improved pastures. (The average area sown to semi-artificial and artificial pasture each year between 1972 and 1988 was 168,000mu.)

• Overstocking is the obvious direct cause of pasture degradation. On the basis of aggregate estimates for the whole county, the number of herbivorous animals exceeded the theoretical carrying capacity by 8% in 1963, 45% in 1980, and 79% in 1988.

• Animal productivity has declined significantly. For example, the annual meat production per head for local breeds of cattle, sheep and goats in the 1970s was estimated to be only about two-thirds of that achieved in the 1950s. Although breed improvement since the 1970s may have reversed this trend to some extent, average meat and fibre production per head per year remain well below potential levels.

8.1.3 Population

- Total population (1988)   154,987   (100%)
    - rural   131,950   (85.14%)
    - urban   23,037   (14.86%)
- Ethnic composition (1988)
    - Han   83,740   (54.03%)
    - Mongolian   69,472   (44.82%)
    - Other   1,775   (1.15%)
- Population density (1988)   15.11 people/km^2

8.1.4 Administrative Structure

The general administrative structure of Balinyou County is set out in Table 8.2. About one-third of the rural households are classified as pastoral households. There are eight State farms running sheep in Balinyou County, of which five are agricultural and animal husbandry farms, and three are forestry farms which also run sheep.

Table 8.2 Administrative Units in Agricultural, Pastoral and Semi-Pastoral Areas of Balinyou County, 1989

Classification	Rural population	No. of townships	No. of natural villages	No. of administrative villages	No. of households
Agricultural	75,000	n/a	278	70	22,000
Pastoral	59,000	n/a	266	74	12,000
Semi-pastoral	0	0	0	0	0
Total	134,000	19	544	144	34,000

Source: Balinyou County Animal Husbandry Bureau

8.1.5 Livestock Population

The major grazing animals are cattle, sheep and goats, although there are also significant numbers of horses, mules and donkeys. The numbers of livestock in June 1990 were:
- cattle   117,100
- goats   506,300
- sheep   390,100
- total sheep equivalents   1,639,000

The long-term trends in total sheep equivalents and sheep numbers are presented graphically in Section 5.2.5 as Figure 5.1. The impact of the severe drought in 1988 and 1989 on the total stocking rate in sheep equivalents is obvious. It is also noteworthy that the total number of grazing animals as measured by sheep equivalents peaked in the mid-1960s but sheep numbers continued to rise until 1983. As discussed in Section 5.2.5, the decline in sheep numbers since the introduction of the household production responsibility system in the early 1980s, reflects the relative unprofitability of sheep compared with other grazing animals, particularly goats.

8.2 Improvements to Production Conditions in the Pastoral Systems

8.2.1 Improved Pastures

• Area of pasture land fenced: It was claimed that a total of 3.48 million mu is currently fenced in the county. About 30% of this enclosed pasture land is cutting land. Perhaps 55% of the area currently enclosed was fenced prior to the introduction of the household production responsibility system.
• Area of pasture sown by aircraft: The figure given was 580,000mu sown by aircraft since 1979. Prior to 1979, no pasture had been sown by aircraft in the county.
• Sand control and other artificial pasture establishment: It was stated that 770,000mu had been sown with artificial pasture, both for sand control and for increasing the productivity of pre-existing pasture land. Of this 770,000mu, greater than 300,000mu was of the second type (i.e. genuine artificial pasture land or improved pasture, as distinct from sand control).
• The IFAD North China Pasture Development Project funded the development of 244,200mu of irrigated pastures in Balinyou County (Brown and Longworth, 1992a). This significantly increased the area of improved pasture available in the county.

8.2.2 Pasture Management Programs

One hundred per cent of the grazing land in Balinyou County is said to be contracted out. [For details of these contracts see the Certificate of Contract presented as an Appendix.] However, despite having comprehensively implemented the land contract system, approximately 70% of contract grazing land within Balinyou is still grazed in common. In this regard, contracting of grazing land *per se* has proven relatively ineffectual in preventing degradation of pasture lands in this county. The Pasture Management Station indicated that farmers were unwilling to take advantage of the institutional changes embodied in the land contract system because of severe shortages of capital required for fencing contract land and the relatively small areas of land contracted out. This latter problem refers principally to the absolute size of contract land areas. It is not due to the fragmentation of the areas under contract. Very little of the contract grazing land in Balinyou County is fragmented in as far as each household is concerned. Most farmers are reported to have no more than two or three pieces of grazing land, and fragmentation presents few problems with regard to water supplies and fencing costs.

The most important policy approach to increasing investment in pastures has been the introduction of a "user pays" system. This approach to collecting funds to finance pasture construction was tried in Bayantala sumu of Balinyou in 1987. As discussed in Section 5.5.2, by 1990 it was being applied in a number of counties including Balinyou, and by 1992 half the herdsmen in the IMAR were supposedly paying a pasture use fee. The money collected in this way was being channelled back into pasture development projects.

Other measures being taken by the Balinyou County government to speed up the rate of pasture improvement include:

- aerial spot planting of pasture land;
- encouraging individual and collective units to fence and improve the pasture land; and
- encouraging collective units and private households to invest in improved pastures.

The county government is requesting the Agricultural Bank of China to make credit available to farmers. In addition, the government is encouraging households to take advantage of the low interest loans being made available.

In an effort to encourage households to fence pasture lands, the Balinyou County government, in conjunction with the IMAR government, is running an experiment which makes interest free loans available to households. These loans are available to cover part of the cost of fencing small areas of pasture land under the family pasture fencing program.

The move towards fencing small areas of grassland represents a significant policy shift away from large-scale, more extensive type, fencing operations such as those conducted in the collective era and shortly afterwards.

Officials interviewed at the Balinyou County AHB suggested that of the 170,000mu (approx.) of semi-artificial and artificial pasture sown each year, about 80,000 to 100,000mu was artificial pasture. The theoretical life of this pasture is 5 to 8 years, but because of overgrazing in late spring and winter, most artificial pastures are resown after only about 3 years. Usually these artificial pastures are sown on areas of relatively good native pasture land. Logically, some of the best native pasture land is likely to be ploughed up and sown to artificial pasture if maximum productivity from the artificial pastures is to be achieved. Therefore, there is not a net addition of 80,000 to 100,000mu of new pasture each year. Most of this area will involve either replanting previously sown pasture or the destruction of better quality native pasture.

While the above measures are designed to increase investment in fencing and planting of pastures, the Balinyou County government has also introduced policies aimed directly at protecting the existing natural pastures. These include:

(a) Placing a limit on the number of livestock each household is permitted to graze on their contracted pasture land.

In 1989 the Balinyou County government introduced an indicative limit of 1 sheep equivalent per 6mu of pasture land (Liu, 1990). It was possible for township administrations to vary the limit according to local conditions and the productivity of different classes of pasture land. For example, the two townships in Balinyou which were surveyed elected to impose different stocking rate limits: Bayantala accepted the 1:6 ratio but Shabutai imposed a 1:7 limit. Officials from the AHB are required by the county government to check stock numbers at the end of each year. Herdsmen found in breech of their allowed number of livestock (in sheep equivalents per mu) are supposed to be forced to comply with the regulatory limits through the sale or slaughter of an appropriate number of livestock. When the officials were questioned as to whether or not the policy of enforcing compliance with the regulatory limits had actually been implemented within the county, the answer appeared to be that it had not! The officials questioned indicated that they were trying to "persuade" herdsmen to comply with the strategy. At the same time, the

county government was also introducing a second policy aimed at encouraging herdsmen to raise the quality of their livestock.
(b) Requiring herdsmen wanting to increase their sheep numbers beyond the present 1 sheep equivalent per 6mu, to increase the productivity of their pastures (through improved pastures, etc.) before being allowed to do so.
As an example, one farmer was said to have invested ¥20,000 in improved pastures and has subsequently been allowed to increase his sheep numbers significantly.
(c) Discouraging goat production.
Goats are considered more damaging to rangeland than either sheep or cattle. The government is planning to introduce the following three measures to discourage goat production.
   (i) Goats are to be reclassified from being equivalent to 1 sheep unit to 2 sheep units for the purpose of determining the allowable stocking rate.
   (ii) The annual pasture use fee for goats is to be doubled. Currently, the pasture use fee for 1 sheep equivalent is ¥0.2 per year. For goats, the pasture use fee will be increased to ¥0.4 per goat per year.
   (iii) The tax on goats is to be increased. Currently, the tax levied for sheep and goats is ¥0.3 to ¥0.4 per head per year respectively. In the future, this tax may be tripled or perhaps even quadrupled for goats.

Another interesting initiative was that during the very dry year of 1989, there was considerable official encouragement given for herdsmen to sell livestock out of the county. The aim was to encourage sheep to be moved from the purely pastoral areas to the cropland areas, and even as far as to Beijing City. It was claimed that 36% of the total number of sheep in the county were sold and moved outside the county in 1989.

The county officials were questioned about whether the maximum carrying capacity of the county had been established in order to stabilise degradation and overstocking at current levels, or to reduce overstocking and hence allow the pastures to recover. The answer seemed to imply that the aim was to stabilise degradation at current levels and at the same time increase the overall production capacity of the system through the sowing of more pasture land to artificial and semi-artificial pasture, thus easing the stocking rate burden on the native pasture.

8.2.3 Improved Sheep and Wool Production

- *Sheep production systems*

Apart from the eight State farms, sheep production is generally undertaken on a small scale by households under the household production responsibility system. As discussed by Liu (1990, p.101), Balinyou County has also pioneered the "family farm" concept. The objective has been to create large-scale (by Chinese standards) family farms to take advantage of economies of scale and to improve overall efficiency. A significant number of these large-scale farms have been established in Balinyou County.

The annual cycle of events for a sheep-herding farmer is largely dictated by the seasons of the year. Most sheep in Balinyou are raised under a pastoral production system. The lambs are born in late winter (April); the sheep are shorn in June; they are taken to summer pastures for the months of July–August–September; in late

September they are brought back close to the farm house and village; and for most of the winter period (November to May) they are shedded at night in barns adjoining the family house.

In Balinyou County, the distance between the winter, spring, summer and autumn pastures seldom exceeds 10km. Usually, each household (or group of households) can practise a type of rotational grazing by moving their flock from one area of pasture to another as the seasons change.

- *Supplementary feeding*

During the long winter period, ewes, replacement weaners, and rams are fed some grain (usually maize) and native or improved pasture hay. The quality of the hay is low because it is usually cut after the onset of seed production when the content of energy and protein is at relatively low levels. Rarely are wethers provided with grain or hay supplements. Crop residues such as maize and sorghum stalks are usually reserved for draught animals such as donkeys, mules, horses and oxen.

Pasture hay is conserved by traditional cut and carry techniques utilising designated areas of pasture land known as "cutting land". Typically, irrigation and fertiliser (including animal dung which is used as a source of fuel for cooking and heat) is not applied to cutting land in pastoral areas. Consequently, the average yield of hay is relatively low and varies tremendously with rainfall.

Livestock in Balinyou County are confined around the clock for a total of no more than ten full days per year in most areas. Sheep are, however, partially confined for at least seven months of the year (usually from November to May). Partial confinement is restricted to the hours between dusk and dawn. During the day, the farmer (or a hired labourer) shepherds the sheep on nearby "winter" pastures. Snow on these pastures is said to be rare and in any event the sheep are able to continue grazing by scraping the snow aside with their hooves. Only in extreme circumstances when the snow partially thaws and subsequently refreezes, are the sheep not able to graze. In these circumstances, the frozen snow forms an icecap over the pastures, effectively preventing the sheep from grazing. Among Mongolian pastoralists, this event is widely referred to as a "black disaster".

Farmers have the opportunity to purchase hay in times of drought or when on-farm production is not sufficient to keep sheep alive over the long winter period. Hay may be purchased privately in the "free market" or, in especially serious circumstances, the county government may arrange purchases of hay from distant counties on behalf of local farmers/herders.

Farmers also have the opportunity to purchase grain for their livestock. Normally, the free market is the only available source of grain in pastoral areas. In agricultural areas, grain may also be available through the State grain distribution system. In the past, if grain has been available from the State, the price has been well below the free market price. Recent Central government policy changes mean that less subsidised grain will be available in the future.

Not only has low priced government grain not been available in pastoral areas but also the free market price of grain is usually higher than in agricultural areas. Consequently, most farmers in pastoral areas use only a limited amount of grain for supplementary feeding purposes. Even in the agricultural areas as the cost of grain has increased (due to the phasing out of low priced government grain) relative to the price of wool, less grain is being fed to sheep during the winter.

As frequently mentioned by managers of wool textile mills, adequate winter feeding is essential if wool quality is to be maintained. Recent changes in policy in China which increase the cost of grain for livestock feeding in winter could, therefore, have a substantial adverse effect on the quality of wool produced.

- *Use of artificial insemination*

Around 40% of mature ewes in Balinyou in 1990 were artificially inseminated. In terms of individual villages, the percentage of ewes inseminated varied from zero to 100%. Villages which made little or no use of artificial insemination were located in the more remote parts of the county. The provision of AI services in these areas is extremely difficult and costly.

Data in Table 8.3 indicate a general decline in the application of AI in Balinyou County. This trend raises serious concerns about the capacity of the AHB to maintain present levels of breed improvement. The decline in the use of AI reflects, in part, the increased costs of providing AI services since the break up of the communes. However, it also reflects the reality that given the freedom to choose, herdsmen are deciding not to breed fine wool sheep. Similar conclusions are reached in Chapter 17 in relation to Hebukesaier County in Xinjiang. To the extent that these findings can be applied to other pastoral areas, they suggest that it will become increasingly difficult to expand the output of quality fine wool in the pastoral region of China.

Table 8.3 Percentage of Ewes Artificially Inseminated in Balinyou County, 1980 to 1990

Year	Percentage of ewes artificially inseminated
	(%)
1980	56.9
1981	57.3
1982	57.6
1983	44.8
1984	43.9
1985	45.2
1986	36.1
1987	47.0
1988	50.3
1989	n/a
1990	40.0

Source: Animal Husbandry Bureau, Balinyou County

The system of artificial insemination used in Balinyou County follows closely the national system set up by the central level Animal Husbandry and Veterinary Sciences Department within the Ministry of Agriculture in Beijing. Essentially the system is designed to make efficient and low cost use of a relatively small number of high quality rams available to the Animal Improvement Stations at the county level. The system used in Balinyou County is based on two ram holding/breeding stations at the county level. All rams and ewes used to breed rams at these two stations are purchased from approved studs, in particular Aohan Sheep Stud in Aohan County and Haoluku Stud located in Keshiketeng County. Aohan Stud is the home of Aohan Fine Wools and Haoluku Stud breeds Salisi dual-purpose sheep as well as Xinjiang Fine Wools (Longworth and Williamson, 1993).

Of the total number of purchased rams at the two stations, 10% are said to be of special grade, 70% are said to be of grade I standard, and the remaining 20% are said to be of grade II standard. Around 70% of the rams held at the two ram breeding/holding stations are bred at the stations. Rams owned by the stations are made available for sale or rental directly to households charged with the responsibility of keeping a flock of high quality rams in each village. These households are under no obligation to purchase rams from the ram breeding/holding stations, and may instead purchase directly from approved sheep stud farms. The ram holding or ram care households are, however, prohibited from breeding rams either for their own use or for sale. They are also prohibited from purchasing rams that are of sub-standard quality. The ram holding household in each village typically maintains a total flock of between 20 and 30 quality rams.

The actual task of performing the artificial insemination is left to a village level technician. A charge of ¥1 per successful conception is levied on the households owning the ewes inseminated. This cost is said to be sufficient to cover the AI technician's salary, as well as providing sufficient remuneration for the ram holding household. The artificial insemination is carried out routinely once a year for a specified time period, usually between 50 and 60 days. Ewes which fail to fall pregnant following artificial insemination are mated naturally using only rams of approved quality (i.e. usually those rams owned by the ram holding household).

Households which do not have access to artificial insemination services are permitted to use natural mating on condition they use rams of approved quality. It is extremely difficult for the county AHB officials to "police" this policy. Most rams used for natural mating in the more remote areas are raised by the farmers themselves or purchased from other farmers. Some of the old AI rams are also made available to farmers for natural mating.

The Animal Improvement Station is currently experimenting with the use of frozen semen technology as a means of making more efficient use of quality rams. By efficient use, it is meant that semen can be stored outside the normal AI period employed in the county. There are said to be a number of problems, however, with the use of frozen semen. These problems include the requirement for much more expensive equipment and the reduced rates of conception compared with that achieved using fresh semen. The conception success rate obtained from two attempts using warm semen is said to be 90%. For frozen semen, the rate falls to between 75% and 80%. Around 4,000 ewes are inseminated with frozen semen each year in Balinyou County.

- *Sheep breeds and wool quality*

Officially, 95% to 98% of the sheep in Balinyou County are supposed to have been bred to Aohan Fine Wool rams for several generations. However, as discussed in Section 5.3.2, less than 2% of the sheep in Balinyou are classified as genuine fine wool sheep. On the other hand, as the data in Table 8.4 demonstrates, about 65% of the wool purchased by the Balinyou Supply and Marketing Cooperative (SMC) is graded as "fine wool" and almost all the wool purchased is either improved fine wool or fine wool.

Table 8.4 Proportion of Each Grade of Wool Purchased by the Balinyou SMC, 1989

Wool type	Grade	Proportion
		(%)
Fine wool	– Special Grade	0.0
	– Grade I	5.0
	– Grade II	60.0
Improved fine wool	– Grade I	30.0
	– Grade II	4.0
Local/coarse wool		1.0
Total		100.0

Source: Balinyou County Supply and Marketing Cooperative

- *Animal productivity*

In the 1950s, the wool produced per sheep was roughly 1kg per head of raw wool from the local breeds. In the 1960s, it began to increase and reached 1.5kg per head of raw wool by the late 1960s. In the 1970s, the proportion of improved sheep rose quite rapidly and wool yields also rose, reaching around 2.5kg per head towards the end of the 1970s. By 1982, the average had reached 3.4kg per head from the improved sheep. Since then, there has been relatively little improvement in the yield per head of sheep of the improved type. In 1991, the improved fine wool sheep averaged 3.5kg of greasy wool per head.

It was stated that the best fine wool stud rams yielded 7.5 to 10kg of raw wool per head, which had a clean scoured yield of about 45% on average.

- *Specific sheep breeding programs*

Sheep improvement began in the 1960s on a relatively small scale. Initially, Russian merinos were introduced to upgrade the local sheep, and accounted for 60% of the improved breeds. In the late 1960s, additional merinos were introduced from Russia and from East Germany.

In the mid to late 1970s, the sheep improvement program was boosted by importing fine wool sheep from Xinjiang Province. In the late 1970s, the Aohan Fine Wool Sheep was developed and is now the most popular sheep breed in Balinyou County. The Aohan Fine Wool Sheep is a dual-purpose sheep, useful for both meat and wool. There have also been some Australian merino rams imported to Balinyou County but these sheep are not considered well suited to the environment in this part of China.

Officials of the AHB are of the opinion that a dual-purpose sheep (meat and wool) is preferable to one bred primarily for wool production. The main reasons offered for this preference are:
– they are easier to manage; [Fine wool sheep require higher quality feed and more precise animal husbandry management. For example, the last pure–bred Australian merino in Balinyou county died in 1989 despite good care being provided (it was fed hay, green feed and carrots, and it was kept warm).]
– many households have meat quota obligations to the government;

- the consumption of mutton is extremely popular in Mongolian households; and
- the Balinyou Supply and Marketing Cooperative (SMC) has both meat and wool processing capacity to satisfy.

County AHB officials reported two major problems with the wool being grown in Balinyou at present.

(1) The length of the locally produced wool is considered by processors to be too short. In view of this concern, the AHB officials have forecast an increase in the use of Aohan Fine Wool rams.

(2) The grease level of local wool, currently around 7%, is considered to be far too low. Despite this view, officials of the AHB indicated that there were no concrete plans to increase the level of wool grease, for while it may be important to increase the amount of wool grease, the local officials were concerned to avoid excessive amounts of grease in the wool. If the grease level in the wool is too high, too much sand will be retained in the wool. If the grease level is too low, the wool will be too open to dirt, dust and seeds. [Excessive dust etc. results in the wool fibre being weakened and easily broken which leads to a low clean yield.]

At the township (sumu) and village (gaca) level, there are technicians in charge of improving the sheep and goat herds. Officially, farmers are said to be favourably disposed to breed improvement. However, there must be a reluctance to continue upgrading the sheep to pure fine wools because there are so few sheep of this type in Balinyou after more than fifteen years of the upgrading program.

8.2.4 Improved Goat Breeds

In the mid to late 1970s, a major improvement program for goats began using frozen semen and artificial insemination. The main breed used to upgrade the goats has been the Liaoning white cashmere goat, and roughly 50% of all goats are now improved breeds. The Liaoning goats were used to develop dual-purpose goats (useful for meat and cashmere).

In the last five years, there has been a significant increase in the cashmere yield per head. In the early 1980s, average cashmere production was 150g per head per year, but more recently it has reached 250g per head per year.

When it was suggested that this may be related to the very good seasons in 1990 and 1991 rather than a genuine increase in genetic potential for higher productivity, the AHB officials claimed that cashmere yields rose by between 70 and 80g per head per year, between 1989 (poor season) and 1991.

8.2.5 Improved Cattle Breeds

New breeds of cattle together with artificial insemination were first introduced into Balinyou County in the 1960s. The breeds introduced were Shorthorns and the Three River breed of cattle. In the 1970s, the program was altered to include the Simmental breed from Russia. By the 1980s, the improvement program for cattle was entirely dependent upon the use of the Simmental breed.

8.3 Socio-Economic Development

8.3.1 Income Growth

As with most other parts of China, nominal incomes in Balinyou increased markedly in the early 1980s. Although nominal average per capita net income in Balinyou continued to rise after 1983, real incomes plateaued and then declined (Table 8.5). The severe drought in 1988 and 1989 (see the rainfall data in Table 8.1) was a major factor depressing pastoral household incomes and hence average rural incomes in these two years.

The per capita income data in Table 8.5 demonstrates that net rural incomes in Balinyou County have generally been a little above the net incomes for the whole county. That is, rural incomes are a little higher than urban incomes. Within the rural sector, pastoral households usually have higher net incomes than agricultural households. Nevertheless, the average net per capita nominal income for rural households in Balinyou (¥491 in 1990) remains well below both the average for the IMAR (¥607 in 1990) and the national average (¥630 in 1990).

Table 8.5 Average Net Income Per Capita in Balinyou County, 1970 to 1989

Year	Net income per capita for rural population		Net income per capita for whole population	
	Nominal	Real*	Nominal	Real*
	----- (¥/year) -----		----- (¥/year) -----	
1970			69.21	71.5
1971			63.00	65.6
1972			53.32	55.7
1973			70.93	73.6
1974			46.77	48.3
1975			61.00	62.8
1976			56.00	57.5
1977			59.00	59.4
1978	74.9	74.9	61.00	61.0
1979	93.3	91.5	78.56	77.0
1980	168.0	155.4	69.10	63.9
1981	203.0	183.4	101.00	91.2
1982	243.0	215.4	153.40	136.0
1983	310.0	270.7	224.69	196.2
1984	312.0	265.1	262.00	222.6
1985	342.0	267.0	275.45	215.0
1986	382.0	281.3	243.65	179.4
1987	432.0	296.5	291.00	199.7
1988	479.0	277.5	474.00	274.6
1989	376.0	184.9	473.00	232.6
1990	491.0	236.4	457.00	220.0

*Deflated by "Overall Retail Price Index" in State Statistical Bureau of the PRC, *China Statistical Yearbook 1991*, p.199.
 The base year is 1978.
 Source: Balinyou County Animal Husbandry Bureau

8.3.2 Changes to Commodity Prices

Before 1959, there was no plan controlling the sale or purchase of wool and the price was essentially determined by the free market. After 1959, wool was subject to a purchase quota or plan quota, with only small residual amounts being sold on the free market. The wool plan was determined by the prefectural government and the quotas were based on historical average yields which were adjusted for seasonal factors.

The prices paid for wool were administratively determined and remained unchanged for long periods prior to 1985. The official State purchase prices in Balinyou from 1975 to 1991 are shown in Table 8.6. In every year except 1989 the premiums and discounts relative to improved fine wool of grade I were determined by the quality differentials written into the National Wool Grading Standard. (See Section 4.5.2.) In 1989, the price for fine wool of special grade was 138% above the price for the base grade while the quality differential which determined the premium for this kind of wool in all other years was 124%.

In 1985, the wool market was formally declared "open" by the Central government but each province could decide whether to open their own wool market. The IMAR provincial government did not open the wool market in 1985 but in 1986 it was decided to allow prefectural administrations to open their wool markets if they wished. Chifeng City Prefecture did not declare the wool market open and so the Balinyou SMC remained the only "official" buyer of wool in Balinyou County. The IMAR government eventually declared the wool market in the whole of the IMAR to be a free market in 1992.

Free market wool prices in China increased sharply after 1985 until the second half of 1989 when they collapsed. The prices paid by the Balinyou SMC and set out in Table 8.6 lagged behind the free market. These prices were administratively determined by the Chifeng City prefectural Price Bureau and were supposed to apply in all the counties in the prefecture.

Compulsory purchasing quotas were introduced for mutton, beef and goat meat from 1963 although state prices had been in place since 1958. The markets for these meats were opened in 1991.

Table 8.6 Prices Paid by the SMC for Wool Purchased from Farmers in Balinyou County, 1975 to 1991

Wool type	Grade	1975 to 1980	1981 to 1984	1985	1986	1987	1988	1989	1990	1991
						(¥/kg greasy)				
Fine wool	– Special Grade	5.63	5.28	6.08	6.60	7.13	9.51	12.08	8.71	7.90
	– Grade I	5.31	4.86	5.59	6.06	6.56	8.74	10.00	8.03	7.26
	– Grade II	4.86	4.56	5.24	5.69	6.16	8.21	9.40	7.53	6.82
Improved fine wool	– Grade I	4.54	4.26	4.90	5.32	5.75	7.67	8.78	7.04	6.37
	– Grade II	4.13	3.88	4.46	4.84	5.24	6.98	7.98	6.40	5.80
Local wool		3.20	2.32	3.20	3.60	3.78	4.22	8.94	6.40	5.80

Source: Balinyou County Supply and Marketing Cooperative

8.3.3 Timing of Introduction of the Household Production Responsibility System

The household production responsibility system was introduced in a series of steps commencing in 1979/80 with the assignment of livestock to the households (Liu, 1990, pp.93-94). By 1984, it had been converted to a double contract system involving both livestock and pastures. Pasture land was allocated to the households on the basis of the number of people and the number of livestock held by the household at the time the double contract system was introduced in 1984.

8.3.4 Level of State Assistance

In 1988, the Central government declared Balinyou County to be a poverty county within the pastoral region of China. To qualify for this status, a county must be recognised as either being a pastoral or semi-pastoral county and have an average net per capita income for three years (from 1984 to 1986) of less than ¥300.

Funds allocated for poverty alleviation in Balinyou County are coordinated through the county government. Funds are distributed as loans via the Agricultural Bank of China. In 1990, a total of ¥1.7 million was allocated to Balinyou County for the purpose of poverty alleviation. Of these funds, 70% was lent to households engaged in animal production. Officials of the Agricultural Bank of China also pointed out that of the total ¥1.7 million, approximately ¥1.5 had been used for pasture construction. These funds for poverty alleviation are made available at an interest rate equal to approximately half that of the going commercial rate in China. In this regard, there appears to be an implicit recognition that without substantial discounting of commercial interest rates on loans for rangeland improvement, individual households would invest much less than is socially desirable, owing to both the uncertainty which surrounds current tenure arrangements and the low commercial rates of return available from these investments (see Williamson and Longworth, 1993). The remaining funds allocated for poverty alleviation were used within the county as follows: agricultural production (10%), fisheries (11%) and forestry (9%). The ¥1.7 million level of funding had been fixed for three years (i.e. from 1988 to 1990). In 1990, there was uncertainty about whether or not the amount would remain unchanged for the 1991 year.

The criteria used to determine which households actually receive the funds are not clear. What is certain, however, is that the households must have the capacity to repay the loan and that the net per capita income of the household must be less than ¥300. Clearly with criteria such as these, the poverty alleviation funds are not targeted to households which would be regarded in a true sense as being poverty-stricken. Rather they appear to be targeted at households which are perhaps above an absolute level of poverty but are still below the relative levels of poverty recognised for such counties by the Central government. An allocation policy of this type has obvious implications for the equitable distribution of capital within the pastoral region.

In another move designed to stimulate greater farm level investment, the county government, in conjunction with the IMAR provincial government, is providing households with grants for fencing small areas of pasture land. Under this policy introduced in 1988, households which choose to fence a minimum of 10mu of grassland are eligible for a grant of up to ¥500 from the county government. One of the problems with the program is the very limited amount of funds available. In

1990, for example, a total of ¥350,000 was made available for grants to households under this scheme. In addition, a further 1,000 households were said to be eligible for interest-free loans to cover part of the cost of fencing small areas of pasture land. These loan funds were said to be part of the provincial poverty alleviation program.

8.3.5 Development of Pastoral Product Processing

The IFAD North China Pasture Development Project in Balinyou County provided loans for machinery purchases to establish processing facilities. There were 298 such loans made under the IFAD Project between 1980 and 1988 (Brown and Longworth, 1992a). As one of the counties participating in the IFAD Project, Balinyou was able to obtain IFAD funds to assist with the establishment of a wool scouring plant. Unfortunately, however, this may not have been a wise investment for the reasons discussed in detail in Brown and Longworth (1992b).

8.3.6 Growth in Output of Major Pastoral Commodities

The major pastoral commodities produced in Balinyou are wool, cashmere, beef, mutton and goat meat. Some data are available for wool output since 1978 (Table 8.7).

Table 8.7 Production of Raw Wool in Balinyou County, 1978 to 1992

Year	Raw wool (tonne)	Year	Raw wool (tonne)
1978	478.05	1986	937.00
1979	585.95	1987	864.00
1980	679.95	1988	970.00
1981	656.85	1989	950.00
1982	797.55	1990	950.00
1983	1,104.00	1991	n/a
1984	1,161.50	1992	1,250.00
1985	1,073.00		

Source: Balinyou County Animal Husbandry Bureau

Wool production increased substantially from 1978 to 1984 but then appeared to have plateaued at a level of output below the peak reached in 1984 until the record clip of 1992. The new record wool output achieved in that year reflected the extremely favourable wool growing seasons in 1990 and 1991, which were years of well above average rainfall in Balinyou County (Table 8.1).

Taking into account the current levels of productivity per animal and the progressively worsening pasture degradation, there is little prospect of further significant increases in the output of wool in Balinyou County unless the prices for wool increase substantially relative to meat and cashmere prices thus inducing a change in the product mix. Indeed, since the current number of herbivorous animals exceeds the long-term sustainable stocking rate and wool production is not especially profitable, it is possible that wool production could decline as pastures become progressively more degraded.

Appendix 8
Certificate of Contract for Pasture Land use with Payment

Issued by the People's Government of Balinyou County

Side A: Land Owner (gaca)
Side B: Land Contractor (households)

The contract is signed following the negotiations between Side A and Side B.
    Side B contracts in pasture land of (how many) mu from Side A. Of the total pasture land area contracted in .... contains (how many) mu of essential pasture land for livelihood; (how many) mu of responsibility pasture land; (how many) mu of grazing pasture land; (how many) mu of cutting pasture land.

Payment standard:
    (how many) yuan/mu for grazing land
    (how many) yuan/mu for cutting land
    (how many) yuan/sheep equivalent

[handwritten: payment for land & sheep]

Term of the Land Contract:
    (how many) years from date, month, year to date, month, year

1. The rights and obligations of Side A and Side B

For Side A: Pasture land is owned by Side A. Side A is entitled to coordinate and approve pasture land sub-contract; to command and organise pasture programming, construction and treatment; to punish those who break the regulation; to approve the utilisation of the fund for pasture construction.
    Side A has the obligation to provide Side B with production technology and information services; to assist pasture management sectors in pasture management; to protect the legal interests of Side B; to be responsible for filling out the records of pasture utilisation with payment.

For Side B: Side B has the right to construct and rationally use the pasture land contracted in; to enjoy all kinds of services provided by State governments and collectives; to monitor pasture management sectors and their staff.
    Side B has the obligation to cooperate with pasture management sectors to conduct all kinds of work regarding pasture management; to make efforts to contract and rationally utilise the contracted pasture land and abide by the collective plan and contribute labour forces for collective-planned pasture construction; to pay the pasture land use fees according to regulations; to meet the purchase quota for animals and the purchase quotas for animal products set by governments both quantitatively and qualitatively.

2. Breach – Contract Responsibility of Side A and Side B
    If Side A does not fulfil its (any) obligations in the contract, Side B is entitled to ask Side A for its requirements and Side A should meet the requirements by regulations concerned.
    If Side B does not fulfil its (any) obligations in the contract, Side A is entitled to warn and punish Side B, even withdraw the pasture land from those who break the contract very seriously.

[handwritten: (ie no punishments for gaca)]

    This contract is triplicated, Side A, Side B and township government keeps one copy each.

Signature of Side A: ....
                                                                                                       date   month   year

Signature of Side B: ....
                                                                                                            date   month   year

Monitoring Unit: ....
                                                                                                              date   month   year

Table 1 Stocking Number by End of June

Stocking number by end of June	Year					
	1990	1991	1992	1993	1994	1995
Goats:						
sheep equivalents						
young goats						
adult goats						
Sheep:						
sheep equivalents						
young sheep						
adult sheep						
Cattle:						
sheep equivalents						
young cattle						
adult cattle						
Horses:						
sheep equivalents						
young horses						
adult horses						
Camels, mules, donkeys:						
sheep equivalents						
young animals						
adult animals						
Total sheep equivalents:						
Table filled by:						
Table checked by:						

Table 2 Contracted Pasture Land Area, Pasture Production, Hay Storage and Stocking Numbers in Cold Season

	Year					
	1990	1991	1992	1993	1994	1995
Contracted pasture land area						
Sub-contracted pasture land area (in)						
Sub-contracted pasture land area (out)						
Actual pasture land area in use						
Pasture production in a normal year						
Stocking number in a normal year						
Because of the change in the weather:						
increased pasture production						
decreased pasture production						
Increased pasture construction area:						
pasture area newly fenced						
planted pasture artificially						
Examined:						
annual pasture production						
Hay storage in cold season:						
hay						
silage						
others						
total storage						
Proper stocking numbers in cold season						
Table filled by:						
Table checked by:						

Table 3   Year End Stocking Number and Overstocking Number

	Year					
	1990	1991	1992	1993	1994	1995
Stocking number and sheep equivalents:						
Goats:						
sheep equivalents						
young goats						
adult goats						
Sheep:						
sheep equivalents						
young sheep						
adult sheep						
Cattle:						
sheep equivalents						
young cattle						
adult cattle						
Horses:						
sheep equivalents						
young horses						
adult horses						
Camels, mules, donkeys:						
sheep equivalents						
young animals						
adult animals						
Total stocking sheep equivalents						
Proper stocking number in cold season						
Stocking situation:						
overgrazed						
undergrazed						
Table filled by:						
Table checked by:						

Table 4   Payment Collection

	Year					
	1990	1991	1992	1993	1994	1995
Payment for pasture land use:						
Contracted pasture land area						
Cutting land area:						
area						
payment						
Grazing land area:						
area						
payment						
Sheep equivalents by end of June:						
area						
payment						
Estimated total payment						
Payment free or decreased:						
natural disasters						
others						
Payment for overgrazing:						
overgrazed sheep equivalents						
payment						
Additional payment for goats:						
goat number at end of June						
payment						
Total payment collection						
Table filled by:						
Table checked by:						

Chapter Nine

Wongniute County (IMAR)

Wongniute is another county in the Eastern Grassland area of the Inner Mongolia Autonomous Region (IMAR). It lies in the centre of Chifeng City Prefecture to the south of Balinyou and Alukeerqin counties. The eastern part of Wongniute is north of Aohan and borders the Zhelimu Prefecture of IMAR (Map 9.1). Before the Cultural Revolution (1966 to 1974), the western and eastern parts of Wongniute were two separate counties. This historical division is still evident in the organisation of some of the government agencies serving the county.

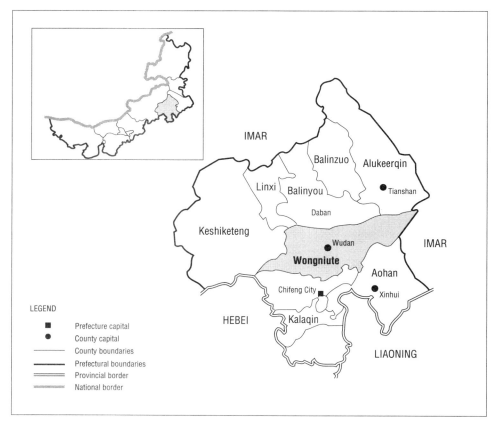

Map 9.1 Wongniute County in Chifeng City Prefecture of IMAR

9.1 Resource Endowment

All levels of government designate Wongniute as a pastoral county. The Central government also classifies Wongniute as a Minority poverty county. The western half of the county has most of the population and much of the agricultural activity. The eastern part of the county is more sparsely settled and depends heavily on pastoral pursuits.

9.1.1 General Features

A major river flows from west to east through Wongniute County and there are mountains to the west. The altitude ranges from 300m to 2,025m above sea level.

Wongniute is characterised by an inland climate with many windy days and with large amounts of sand and dust being blown about. The average annual rainfall is 367mm but as the data in Table 9.1 shows, there is considerable variation from year to year. The rainfall data in Table 9.1 is for Wudan, the county capital, which is in the climatically more favoured western half of Wongniute. The eastern border regions include parts of the Keerqin Desert (mentioned in regard to northern Aohan) and these areas are much drier.

While the data in Table 9.1 do not necessarily support the argument, county officials in Wongniute seemed convinced that rainfall in the county had declined and become more variable over the last four decades. They claimed that, in the case of Wudan for example, in the 1950s, annual rainfall was 400mm to 413mm; in the 1960s it was 300 to 380mm; in the 1970s it ranged from 320 to 350mm; and in the 1980s it was only 200 to 300mm. Furthermore, it was claimed that a greater proportion of the rain than in the past was now falling in winter. Winter rainfall is not considered as useful as summer rainfall, since the weather is too cold for grass to grow.

perception of climate change

Table 9.1 Annual Rainfall at Wudan the Capital of Wongniute County, 1969 to 1990

Year	Annual rainfall	Year	Annual rainfall
	(mm)		(mm)
1969	329.0	1980	289.9
1970	269.4	1981	241.7
1971	293.6	1982	347.9
1972	300.1	1983	355.6
1973	503.3	1984	403.3
1974	397.6	1985	455.7
1975	371.8	1986	519.3
1976	387.1	1987	425.2
1977	372.9	1988	228.6
1978	416.0	1989	261.9
1979*	487.0	1990**	410.0

* estimate
**provisional estimate
Source: Wongniute County Animal Husbandry Bureau

9.1.2 Land Use

While Wongniute and Balinyou are roughly the same size in total land area, there is four times as much cropland in Wongniute. On the other hand, Wongniute has less usable pasture land, all of which is classified as degraded to some extent.

- Total land area   17,710,000mu   (or 11,807 km^2)
- Total usable land area   14,640,000mu
- Total rangeland   10,420,000mu
  - usable pasture   8,830,000mu
  - "cutting" land   600,000mu
- Arable land   1,655,000mu
- Forestry land   3,790,000mu
- Total degraded land (1990)   9,043,000mu
  - heavily degraded   1,167,000mu
  - medium degraded   2,853,000mu
  - lightly degraded   5,023,000mu

Wongniute County encompasses five different classes of pasture. These are:
(1) dry mountain tableland pasture – located mainly in the mountains in the western part of the county and subject to very low temperatures in winter;
(2) dry hill pasture land – located in the foothills of the mountains in the western part of the county;
(3) plains pasture land – located in the middle of the county and consisting mainly of marshland soils, salty soil, sandy soil and alluvial soil;
(4) semi-desert plain sandy pasture land – occupying the majority of the eastern part of the county; and
(5) high-moisture, low-lying land – located in the eastern part of the county in a triangle of two converging rivers.

The major type of pasture land which is badly degraded is the semi-desert plain sandy pasture. In 1965, the total area of degraded pasture was 6.2 million mu. By 1976, the degraded pasture area had expanded to 7.4 million mu; and by 1990 the corresponding figure was 9.043 million mu, which represents 87% of the total rangeland in Wongniute. Rangeland degradation is a most serious problem in Wongniute County because almost all of the usable pasture land is degraded, at least to some extent. It is clearly the view of the county government that the primary reason for the increasing grassland degradation is the excessive number of herbivorous livestock. As demonstrated in Fig. 5.1 in Chapter 5, this is not a new problem. Indeed, the total sheep equivalents have tended to decline since the early 1960s.

9.1.3 Population

- Total urban and rural (1989)
  - total   424,933   (100%)
  - rural   392,458   (92.36%)
  - urban   32,475   (7.64%)
- Ethnic composition (1989)
  - Han   368,365   (86.69%)
  - Mongolian   51,896   (12.21%)
  - Tibetan   4,672   (1.10%)
- Population density (1989)   35.991 people/km^2

9.1.4 Administrative structure

Table 9.2 provides information on the number of townships, villages and households in the rural sector of Wongniute. Less than 20% of households are classified as pastoral.

There are five large State farms located in the county.

Table 9.2 Administrative Units in Agricultural, Pastoral and Semi-Pastoral Areas of Wongniute County, 1989

Classification	No. of townships	No. of villages	No. of households
Agricultural	17	185	58,000
Pastoral	10	98	16,000
Semi-pastoral	4	55	14,000
Total	31	338	88,000

Source: Wongniute County Animal Husbandry Bureau

9.1.5 Livestock Population

- Livestock numbers (mid-year 1990)
  - cattle   120,000
  - horses   40,000
  - donkeys   35,000
  - mules   15,000
  - camels   1,500
  - sheep   400,000
  - goats   240,000
  - pigs   130,000

Long-term data on total sheep equivalents and sheep numbers are presented graphically in Figure 5.1. As in the case of Balinyou County, the total sheep equivalents peaked in the early 1960s. However, while the overall trend in stock numbers since 1970 has been upward in Balinyou, the trend has been downward in Wongniute since about 1965. On the other hand, sheep numbers expanded rapidly until 1983 but, as in Balinyou, they have fallen significantly since that year.

9.2 Improvements to Production Conditions in the Pastoral Systems

9.2.1 Improved Pastures

- Total area of improved pasture by type
  - improved pasture   2,600,000mu
  - fenced pasture   1,600,000mu
  - semi-artificial pasture   500,000mu
  - artificial pasture   320,000mu

There have been large annual increases in both improved pasture and fenced pasture since 1985 (Table 9.3).

Table 9.3 Annual Increase in Area of Improved Pasture and Fenced Pasture in Wongniute County, 1985 to 1990

Year	Increase in improved pasture area	Increase in fenced pasture area
	('000 mu/year)	
1985	100	100
1986	310	150
1987	440	200
1988	240	200
1989	250	210
1990	350	180

Source: Wongniute County Animal Husbandry Bureau

- *Specific pasture improvement programs*

Pasture land improvement began in 1949 and has accelerated since 1979. The major effort has been directed at the development of artificial and semi-artificial pastures. In Wongniute County, semi-artificial pastures are classified as aeroplane-planted pastures. Artificial pastures are classified as those pastures that are planted by ground-based machinery or by hand.

The county government has established a special program to improve a specific amount of pasture land each year. Each individual township is given a target. The township governments then decide which pasture areas should be improved.

The county government distributes rewards in the form of money and encouragement plaques to farmers who have improved their pasture land. It also supplies investment funds, some of which have a non-repayment period, provides credit to farmers at low interest rates, and supplies wood, steel and cement posts to farmers for the purpose of fencing improved pastures.

The Wongniute County government has also developed irrigation systems for cropland production and encouraged the planting of trees by individuals and collective units.

The IFAD North China Pasture Development Project (1980-1988) facilitated the development of 100,000 mu of irrigated pasture in Wongniute (Brown and Longworth, 1992a).

- *Specific pasture management programs*

The county government's concerns over the increasing level of grassland degradation have led to the introduction of policies aimed at controlling animal numbers.

In 1984, the county government introduced an absolute ceiling on the stocking capacity for the county as a whole. This ceiling was pegged at 1.19 million sheep equivalents and has remained unchanged up to the present. After the introduction of this policy, total sheep equivalents in the county increased from 1.21 million in 1984, to 1.25 million in 1989. It is apparent from these figures that county policy aimed at pegging the total level of stock equivalents at around 1.19 million has not been entirely successful. It is another issue, however, whether even the formal ceiling of 1.19 million sheep equivalents is a realistic and sustainable long-term carrying capacity for the county as a whole. Given that the area of degraded pasture land is increasing significantly every year, it would appear that the actual maximum sustainable stocking rate is falling and would be below 1.19 million sheep equivalents.

The county government attempts to ensure that the aggregate ceiling on stock equivalents is met by allocating permissible stock equivalent quotas to each production team in the county. Production teams which breach their allocated stock equivalent limit are fined and/or warned of impending fines. It was said that different administrative areas within the county have different policies with regard to the enforcement of allocated stocking limits. In general, the number of sheep equivalents each household is permitted to keep is determined according to the amount of pasture land contracted or used in a grazing-in-common situation by the household.

The county government has also chosen to introduce a system of "user pays" for grasslands as a second major policy measure aimed at reducing rangeland degradation. The introduction of this policy is only a recent phenomenon. It was introduced in two townships (sumus) in the pastoral area of the county in 1988. The user-pays system as it operates in these two pastoral sumus requires households to pay a fee of ¥.5 per mu of cutting land contracted and a fee of ¥.1 to ¥.2 per mu of grazing land contracted. The sumu surveyed in 1989 (Bahantala Sumu) had just introduced the user-pays system.

In semi-pastoral areas, the user-pays system is an even more recent innovation. It was reported that in May of 1990, only one village in the semi-pastoral areas of the county had experimented with the user-pays system. The system being used in this village is fundamentally different from the user-pays system being used in pastoral areas of the county. Households are being charged on a per sheep unit basis rather than on a per unit of contracted land basis. This is presumably necessary because only a very small proportion of grazing lands in the semi-pastoral areas have as yet been contracted out to individual and/or joint households.

A user-pays system based on charging farmers for the use of a given land area raises the fixed cost faced by that particular pastoral household. A user-pays system based on charging the household according to sheep units, however, raises the marginal cost of running one more sheep. Therefore, in theory, levying the charge per head should lead to more rational stocking rates and to the raising of sheep which are more productive on a per head basis.

9.2.2 Improved Sheep and Wool Production

- *Sheep production systems*

In general terms, there are three rather different sheep production systems in Wongniute County.

In the pastoral areas of the county which occupy 5.3 million mu of rangeland, farmers usually operate as teams typically consisting of from 3 to 10 households. These teams are called "duguilongs" by people of the Mongolian Nationality. Although each household owns its own flock, most of the pasture land is allocated to the production team and it is grazed in common. Each individual household owns a flock of between 100 and 300 sheep and goats. The opportunities for agricultural production or for any form of off-farm work are extremely limited. The sheep are grazed all the year except for 30 to 50 days in winter when the snow is deep and it is too cold to let the sheep out of their sheds.

In the semi-pastoral areas (1.2 million mu of rangeland), flock sizes are smaller, many households have non-pastoral based sources of income, and supplementary feed is more readily available for the sheep. While still a very significant household activity, sheep-herding is less important and tends to be delegated to older or younger members of the family.

In the agricultural areas where there are 3.9 million mu of rangeland, sheep-raising is often only a side-line activity. The sheep are kept in pens for much more of the time and each household owns a relatively smaller number of sheep (usually less than 25 head). The management and herding of the sheep are given a relatively low priority.

- *Proportion of sheep artificially inseminated*

The county government has attempted to ensure that at least 90% of the total number of mature ewes in the county are artificially inseminated. The actual proportion of mature ewes artificially inseminated is said to be something in the order of 91% to 92%. This percentage varies substantially between sumus and ranges from zero to 100%. In very remote areas of the county, there are quite significant numbers of households which do not use AI. These households are reported to use quality Aohan Fine Wool rams for natural mating purposes owing to the lack of AI services.

- *Animal productivity*

The average yield of greasy wool per adult sheep in the county in 1989 was 5.74kg, with a range of yields from 3.5kg per head to 11kg per head. This was said to be the extreme range; a more typical production range is 4.5kg to 6.5kg of greasy wool per head per year.

The average clean wool yield as determined by random sampling is around 45% and ranges from 34% to 51%. The clean wool yield for fine wool produced in the eastern areas of the county (38-40%) is lower than the county average, while wool produced in the western areas of the county is above average.

- *Specific sheep breeding programs*

Between 1952 and 1980, merino rams were introduced into Wongniute County from the Soviet Union. After 1980 and up to 1985, fine wool breeds from Xinjiang were crossed with the Aohan Fine Wools. A total of 15 Australian rams arrived in the county between 1985 and 1989. Ten of these rams were introduced in accordance with the Joint Agricultural Cooperation Agreement between Australia and China, and five were imported by the Government of IMAR. The major characteristics of the Australian merino of interest are the wool length and the high clean yield. Since 1984, 60 Chinese Merinos have also been brought into the county. As well, some Polwarth x Chinese Merino crosses have been introduced from Zhelimu Prefecture. All these sheep breeds are discussed in detail in Longworth and Williamson (1993).

The Animal Improvement Station, as the name suggests, is responsible for ensuring the improvement of animal breeds in Wongniute County. In this connection, the Station is responsible for administering an extensive artificial insemination program throughout the county, aimed at improving sheep breeds. Artificial insemination is not a compulsory management practice in Wongniute County. Technicians skilled in the technique of artificial insemination are located in every township and village in the county. However, technicians are not always available at the team or duguilong level.

Artificial insemination has been used in Wongniute County since 1952. However, only since 1985 has it become popular. Artificial insemination is being used as an effective means of utilising a small number of high quality fine wool rams to inseminate and upgrade the great majority of the sheep flock in Wongniute County. Semen is collected between March and June and stored as frozen semen for use later in the insemination period. In 1987, one high quality fine wool ram was used to inseminate 3,850 ewes. The official record for artificial inseminations from one ram in one year is 5,162 ewes.

The AI period for the county as a whole, extends for 6 to 7.5 months, which is considerably longer than in other counties in Chifeng City Prefecture. Typically, the AI period lasts for 2 to 2.5 months. Wongniute County, as mentioned above, is unusual in that it has essentially three different production systems, namely a pastoral production system, a semi-pastoral production system, and an agricultural production system. Because of broad differences in each of these production systems, the timing of AI is also different.

The majority of AI undertaken in the county is conducted using warm semen technology. The effectiveness of this technology, as indicated by the level of conception achieved, is as follows:
- 75% of inseminated ewes conceive after one attempt;
- 90% of ewes conceive after two attempts; and
- 95% to 98% of ewes conceive after three attempts.

For those ewes that do not conceive after three attempts, natural mating is used to achieve conception.

It is County government policy that only high quality fine wool rams are used for the purpose of AI. There are essentially two sources of high quality rams available to village-level and/or township-level artificial insemination centres.

The first source is the Animal Improvement Station itself, which normally has about 60 high quality rams available for rental to artificial insemination centres. Of the 60 rams owned by the Animal Improvement Station in 1989, 46 were rams which had met the Chinese merino breed standard from Gadasu Stud Farm in Zhelimu Prefecture. The remaining 14 rams were Australian merinos introduced from Western Australia between 1984 and 1989.

The distribution of the high quality rams owned by the Animal Improvement Station is largely done through a system whereby the township-level Animal Improvement Station submits an artificial insemination plan to the county-level Animal Improvement Station. Contained within this plan are details relating to which villages require rams and when these rams are required. Villages classified as sheep specialisation areas are given priority for the allocation of high quality rams owned by the Animal Improvement Station. Rams allocated/rented to a particular village are cared for during the AI period (usually 2 to 2.5 months in any particular place) by the village-level AI technician.

Approximately 100,000 mature ewes in the County are inseminated using the 60 high quality rams owned by the Animal Improvement Station. This high use ratio of rams to mature ewes is largely made possible by the very broad differences in the production systems within the county.

The second principal source of high quality rams is the Baiyinhua State Farm. As discussed in detail in Longworth and Williamson (1993), this State farm is a county-level stud recognised as producing quality Aohan Fine Wool sheep. The

Baiyinhua State Farm is closely associated with the Aohan Stud Farm in neighbouring Aohan County.

In total, there are around 1,000 high quality rams owned by township and village-level AI centres. Most of these rams are obtained from the Baiyinhua State Farm. These rams are used to inseminate around 60,000 adult ewes per year. In addition, some of these rams may also be made available to more remote areas where AI is not available and quality rams are required for natural mating purposes.

The efficiency of artificial insemination has also been enhanced by lagging conception time according to location. In the western areas of the county, sheep are artificially inseminated from July to September. This allows for a winter lambing. In the eastern areas of the county, sheep are artificially inseminated from October to November, thus allowing for an early-spring lambing.

Officials of the Wongniute Animal Husbandry Bureau report that there are no problems with the practice of moving outside the traditional reproduction cycle of conception in autumn and lambing in spring. Scientists have found that lambs born in the spring are by the following winter almost as large as lambs born in the previous winter. Because the ewes, after mating, are in better condition, lambs born in winter (typically in the agricultural areas) are usually larger than lambs born in spring. A spring lambing (typically in the pastoral areas) usually results in small lambs because the ewes must carry their lambs through the winter. However, lambs born in spring in pastoral areas have access to high quality green feed immediately following birth. As a result, they grow rapidly and tend to "catch-up" to older winter-born lambs.

- *Supplementary feeding*

Animal scientists attached to the Baiyinhua State Farm recommend the following supplementary winter-feeding regime for sheep which are allowed to graze/forage during the day.

Ewes and replacement weaners — hay 1.5kg per day
— grain feeding : November/December 0.1kg per day
: January/March 0.15 to 0.25kg per day
: April 0.15 to 0.2kg per day
: May 0.1kg per day

Wethers — hay/grain fed January/April at half the rate for ewes and weaners

When sheep are confined for 24 hours owing to excessive snowfalls or high winds, all sheep should receive 4kg of hay per head and the amount of grain as indicated above.

- *Specific animal health programs*

Officials claimed that the genetic potential of the fine wool sheep in Wongniute is being realised. The reasons given for this conclusion include:
(i) the use of greenhouse technology and soil brick houses to keep sheep warm in winter;
(ii) better drenching, dipping and disease prevention;
(iii) supplies of grain from the crop areas to animals in the pastoral areas of the county; and
(iv) significant levels of investment in infrastructure such as fencing and improved pastures.

9.2.3 Improved Goat Breeds

Twenty-five thousand goats in Wongniute County have been improved for cashmere production. This represents just over 10% of the total goat population.

9.2.4 Improved Cattle Breeds

Sixty-eight thousand cattle or more than half of the cattle population are of improved breeds.

9.3 Socio-Economic Development

9.3.1 Income Growth

Average nominal per capita net incomes increased more than threefold between 1978 and 1989 in Wongniute County (see Table 9.4). Real incomes, on the other hand, increased by only about 50%. The gap between rural and county-wide average incomes was not great in 1978 and there does not seem to have been any trend for the gap to widen. Nonetheless, the data for 1989 listed below Table 9.4 suggest that incomes in pastoral areas were substantially above the average for rural areas while households in semi-pastoral areas enjoyed about the average rural income. This is a relatively recent phenomenon. Prior to the boom in wool prices in 1986, pastoral households generally had lower incomes than other rural households. The additional information available in regard to pastoral households in Wongniute (Table 9.5) suggests that net incomes in the pastoral areas increased remarkably between 1985 and 1989.

Table 9.4 Average Net Income Per Capita in Wongniute County, 1978 to 1989

Year	Rural areas		County-wide	
	Nominal	Real*	Nominal	Real*
	------ (¥/year) ------		------ (¥/year) ------	
1978	116.2	116.2	128	128.0
1979	155.1	152.1	165	161.8
1980	190.0	175.8	203	187.8
1981	268.2	242.3	274	247.5
1982	248.2	220.0	262	232.3
1983	279.7	244.3	293	255.9
1984	294.1	249.9	317	269.3
1985	366.1	285.8	391	305.2
1986	254.0	187.0	299	220.2
1987	302.0	207.3	340	233.4
1988	303.0	175.6	359	208.0
1989	376.0	184.9	373	183.4

*Deflated by "Overall Retail Price Index" in State Statistical Bureau of the PRC, *China Statistical Yearbook 1991*, p.199.
The base year is 1978.
Net nominal rural income per capita in 1989 — pastoral areas ¥456
— agricultural areas ¥303
— semi-pastoral areas ¥385
Source: Animal Husbandry Bureau, Wongniute County

Perhaps the most important statistics in Table 9.5 are in the final two columns. Households in the pastoral area have demonstrated their willingness to increase investment in their pastoral activities as their incomes rise.

Table 9.5 Income, Expenditure, and Investment Statistics for Pastoral Households in Wongniute County, 1983 to 1989

Year[1]	Total gross income[2]	Total production costs, depreciation and taxes[2]	Total net income[2]	Total capital expenditure[3]	Total capital expenditure as a percentage of total net income
	------------- (million yuan) ---------------				(%)
1983	8.05	1.28	6.77	1.21	17.87
1985	15.77	3.93	11.84	3.15	26.60
1989	48.00	14.00	34.00	19.20	56.47

[1]For statistical purposes, the total pastoral population increased substantially from 38,000 to 59,000 people between 1983 and 1985. This increase occurred in 1984 when a number of previously agricultural villages were redefined as being pastoral villages.
[2]See Section 2.5.4 and Table 2.2 for definitions.
[3]Capital expenditure includes investments in fencing, shed building, pasture construction, and other on-farm capital items.
Source: Economic Management Station, Wongniute County

9.3.2 Changes to Commodity Prices

In Wongniute, as elsewhere in the pastoral region of China, the overwhelming factor involved in the improvement of pastoral incomes in the second half of the 1980s in nominal terms was the substantial increases in the prices paid for pastoral products. However, as real incomes have fallen since 1985, the increases in prices have not been sufficient to compensate for inflation.

Tables 9.6 and 9.7 present details of the prices for wool (by grades), beef, mutton and cashmere. It is particularly interesting that the official State prices for wool in Wongniute were not always the same as in Balinyou especially in 1987. On the other hand the premiums and discounts in all years and for all kinds of wool were determined according to the quality differentials in the National Wool Grading Standard.

Table 9.6 Prices Paid for Raw Wool (by grades) in Wongniute County, 1987 to 1990

Wool type	Grade	1987	1988	1989	1990
		----------- (¥/kg greasy) ------------			
Fine wool	– special grade	5.28	9.51	10.88	8.71
	– grade I	4.86	8.74	10.00	8.03
	– grade II	4.56	8.21	9.40	7.53
Improved fine wool	– grade I	4.26	7.67	8.78	7.04
	– grade II	3.88	6.98	7.98	6.40
Local wool		n/a	n/a	n/a	n/a

Source: Wongniute County Supply and Marketing Cooperative

Table 9.7 Prices Paid for Beef, Mutton and Cashmere in Wongniute County, 1985 to 1990

Year	Beef	Mutton	Cashmere
		(¥/kg)	
1985	2.00	2.10	32.0
1986	3.00	2.70	48.0
1987	4.50	4.08	67.2
1988	5.60	6.00	130.3
1989	5.20	4.80	147.7
1990	3.60	4.60	116.0

Source: Wongniute County Animal Husbandry Bureau

9.3.3 Timing of Introduction of the Household Production Responsibility System

As elsewhere in the pastoral region, the sheep were distributed first and later the land was contracted out.

The system of contracting out grazing land to individual households or to small groups (teams) of households on a joint contract basis was introduced in the pastoral areas of Wongniute County in the beginning of 1984. By the end of that year, 100% of the cutting and grazing land in the pastoral areas of the county had been contracted out. In the semi-pastoral areas of the county, the contracting out of cutting and grazing lands was commenced in 1989. By the end of 1990, 100% of these lands was contracted out to individual or joint households. In the agricultural areas of the county, none of the cutting or grazing lands had been contracted out to households before 1991. However, when interviewed in mid-1991, the staff at the Pasture Management Station indicated that the County government was planning to commence contracting out grazing land to households in the agricultural areas in the second half of that year.

9.3.4 Level of State Assistance

As mentioned earlier, Wongniute County is recognised by the Central government as being a poverty county within the pastoral region of China. The definition of a poverty county in the pastoral region is a pastoral or semi-pastoral county with an average net income of less than ¥300 per capita per annum determined for the three years 1984 to 1986.

A county classified as a poverty county in the pastoral region is able to use a large proportion of its poverty alleviation funds for animal husbandry purposes. For more detail on the fiscal arrangements for poverty counties, see the comments on this topic in the Aohan and Alukeerqin case studies (Chapters 10 and 11).

9.3.5 Development of Pastoral Product Processing

The Wongniute County Supply and Marketing Cooperative (SMC) was planning to commence construction of a wool scouring plant in 1992. According to the SMC officials, the main reasons for establishing the plant are:
(i) wool mills prefer to buy scoured rather than raw wool; [In 1990, when some

SMC officials went to Shanghai and to Wuxi City in Jiangsu Province, and received a request for clean scoured wool. No request, however, has been received for clean scoured wool from the mills in Chifeng City.] and
(ii) scoured wool can reduce transport costs.

Until 1990, the Wongniute County SMC was not allowed to sell wool to any other prefecture or province. All wool had to be delivered to Chifeng City SMC which would then decide the fate of the wool. Alukeerqin County, on the other hand, was not forced to sell all of its raw wool to Chifeng City SMC because it could supply to either Chifeng City or Zhelimu prefectural SMCs. With the opening of the wool market throughout IMAR in 1992, these formal restrictions will be relaxed.

The SMC officials argued that good-quality scoured wool could be sold at premium prices. Some of the profits could then be paid back to farmers in the form of rewards. These rewards were specifically called subsidies and not price increases even though they were really the latter. The SMC may also pass back some of the premiums for good-quality wool by purchasing top-grade rams for the township and village AI centres.

9.3.6 Growth in Output of Major Pastoral Commodities

The available data for wool and cashmere production are presented in Table 9.8.

Table 9.8 Wool and Cashmere Production in Wongniute County, 1978 to 1988

Year	Wool	Cashmere
	(tonne)	(tonne)
1978	918.4	
1979	1,027.7	
1980	1,194.8	
1981	1,156.7	
1982	1,459.4	
1983	1,662.1	30.5
1984	1,564.5	32.0
1985	1,289.3	35.0
1986	1,415.4	41.0
1987	1,354.0	46.0
1988	1,814.1	n/a

Source: Wongniute Animal Husbandry Bureau

Wool production increased up to 1983 and then declined until the peak year of 1988. Cashmere output increased by more than 50% between 1983 and 1987. As discussed in Section 5.2.5 of Chapter 5, the swing towards raising cashmere goats rather than sheep reflects the marked shift in the relative prices of wool and cashmere in favour of cashmere in the second half of the 1980s (Table 9.7).

The potential to increase further the total number of herbivorous animals in Wongniute is extremely limited. Indeed, the total grazing capacity of the county is falling. While it is technically feasible to substantially increase the area of artificial pasture, such an approach is unlikely to be economically justified. Therefore, if wool production is to increase, this can only occur as a result of wool-growing sheep replacing other herbivorous animals and/or by increasing the output of wool per head of sheep. Neither of these two possibilities is likely to lead to any marked increase in wool production in the foreseeable future.

Chapter Ten

Aohan County (IMAR)

Aohan County is situated on the southern fringe of the Eastern Grassland area of the Inner Mongolia Autonomous Region (IMAR). It is the most south-easterly county in Chifeng City Prefecture. Aohan County has extensive common borders with Liaoning Province to the south and with Zhelimu Prefecture of IMAR to the north east (Map 10.1).

The Aohan Fine Wool breed of sheep has been developed at Aohan State Farm and Sheep Stud which is located roughly in the middle of the county. As explained in Longworth and Williamson (1993), this breed is one of the major provincially recognised fine wool sheep breeds in China.

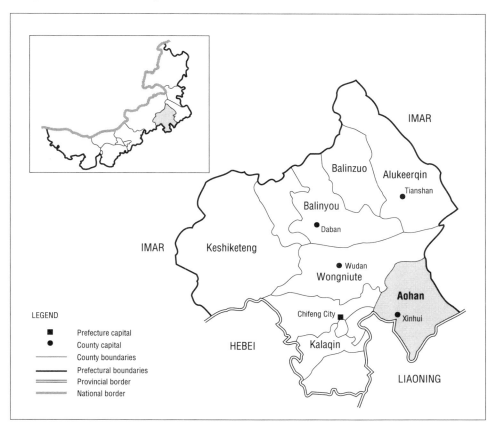

Map 10.1. Aohan County in Chifeng City Prefecture of IMAR

10.1 Resource Endowment

Aohan County is officially classified by the Central government as a semi-pastoral county. Furthermore, as demonstrated in Table 5.2 of Chapter 5, Wushen is the only county in the IMAR which produces more wool than Aohan. Nevertheless, Aohan is widely recognised in the IMAR as an agricultural county. The agricultural areas in the southern parts of the county are relatively well developed but most of Aohan is poor sandy semi-desert grazing country. In economic terms, Aohan is the poorest county surveyed. It is also the most densely populated county surveyed. Both the Central government and the IMAR government recognise Aohan as a poverty county.

10.1.1 General Features

Aohan County is very dry with a rainfall in the order of 200 to 350mm. The county is typified by a harsh environment and extreme wool growing conditions. The northern part of the county borders on a large degraded area of natural pasture which is now known as the Keerqin Desert.

As already mentioned, Aohan Fine Wool State Farm and Sheep Stud is located roughly in the middle of the county. The physical features of the environment at the Stud would be fairly typical of the whole county. The Stud Farm is located at 400–500 metres above sea level and has a rather severe climate with daily temperatures which average 5–8°C. The highest average daily temperature is about $36°C$ and the lowest about -29.5°C. The average rainfall is 250–350mm and most of the rain falls in the July–September period. Annual evaporation is 2,040–2,530mm. Spring is often very windy with more than 100 days of severe wind per year. The strongest wind experienced on the farm is a class 8 wind (i.e. the strongest kind of wind measured in China and said to be 18–28 metres per second). The soil is a loose sandy loam of the loess type, and is rather poor and very low in organic matter. There is no river in the main farm area. Underground water of a reasonable quality occurs at 4–6 metres under much of the farm but some of it is as far as 80 metres down. Much of the native pasture is salt-affected and not of very high quality.

Rainfall is lower in the northern part of Aohan County than at the Stud Farm but is higher in the southern part of the county. The official rainfall records provided by the County AHB for Xinhui (the county capital) are presented in Table 10.1. Over the period 1957 to 1990, annual rainfall at Xinhui varied from a low of 217mm (1980) to a high of 595mm (1978).

10.1.2 Land Use

Aohan has only about half the usable pasture area of Balinyou but the cropping area is even larger than in Wongniute.

- Total land area   12,160,000mu   (8,294km^2)
- Total rangeland area   5,240,000mu
  - usable pasture   4,450,000mu
  - cutting land   n/a
- Total arable land   1,875,000mu
- Total forestry land   3,830,000mu
- Proportion of pasture land which is degraded:
  - light   7.31%
  - medium   45.69%
  - heavy   29.36%

Table 10.1 Annual Rainfall at Xinhui the Capital of Aohan County, 1957 to 1990

Year	Annual rainfall (mm)	Year	Annual rainfall (mm)
1957	399.8	1974	458.8
1958	310.1	1975	410.8
1959	480.4	1976	480.1
1960	301.7	1977	595.1
1961	382.3	1978	471.1
1962	531.3	1979	497.1
1963	357.4	1980	217.6
1964	491.2	1981	394.3
1965	459.1	1982	284.5
1966	404.0	1983	389.5
1967	313.1	1984	493.0
1968	364.9	1985	521.4
1969	487.7	1986	554.8
1970	402.4	1987	513.7
1971	334.5	1988	311.5
1972	317.1	1989	359.1
1973	325.9	1990*	390.0

*provisional estimates
Source: Animal Husbandry Bureau, Aohan County

10.1.3 Population

- Total, urban and rural population (1989)
  - total population   526,480 (100%)
  - rural   495,815 (94.18%)
  - urban   30,665 (5.82%)
- Ethnic composition (1989)
  - Han   500,926 (95.15%)
  - Mongolian   19,551 (3.71%)
  - other   6,003 (1.14%)
- Population density (1989)   63.47 people/km^2

10.1.4 Administrative Structure

In Aohan County there are 30 townships including Aohan Stud Farm (Table 10.2). Only one township and seven villages are classified as purely pastoral. The 1,479 households which are described for statistical purposes as being pastoral households constitute only a little over 1% of all rural households in the county. Administratively, the Stud Farm is a semi-pastoral township responsible for six villages.

Table 10.2 Administrative Units in Agricultural, Pastoral and Semi-Pastoral Areas of Aohan County, 1989

Classification	No. of townships	No. of villages	No. of households
Agricultural	13	156	63,937
Pastoral	1	7	1,479
Semi-pastoral	16	160	55,005
Total	30	323	120,421

Source: AHB, Aohan County

10.1.5 Livestock Population

It was estimated that there were about 565,600 sheep equivalents of grazing animals in Aohan in mid-1990. These animals were mainly sheep, cattle and goats. There were a relatively large number of horses, mules and donkeys as well, together with eight camels.

- Livestock numbers (mid-year 1990)
    - cattle   21,600
    - goats   11,500
    - sheep   402,700
    - total sheep equivalents   565,600

The long-term trends in total sheep equivalents and sheep numbers are presented graphically in Figure 5.1 of Chapter 5. Sheep numbers have more than doubled since 1960 but, as in Balinyou and Wongniute, there has been a downward trend since the introduction of the household production responsibility system in the early 1980s. Total stock numbers (measured in sheep equivalents) have steadily declined since about 1968, reflecting the gradual reduction in carrying capacity as the pastures become more and more degraded.

10.1.6 Flock Sizes

County-wide data on sheep flock sizes were available for 1990. This information is presented as Table 10.3. Although the total number of households in Table 10.3 is less than the total rural households shown in Table 10.2, the data in Table 10.3 implies that about half the rural households in Aohan keep some sheep. Most households keeping sheep have less than 10 head. On the other hand, there were 6,600 households with flocks in excess of 45 sheep.

Table 10.3 Distribution of Sheep-Raising Households in Aohan County by Flock Size, 1990

Flock size	Households	
(no.)	(no.)	(%)
0	54,000	49.0
0 to 10	34,100	31.0
11 to 15	5,500	5.0
16 to 25	5,500	5.0
26 to 45	4,400	4.0
> 45	6,600	6.0
Total	110,100	100.0

Source: Aohan County Animal Husbandry Bureau

10.2 Improvements to Production Conditions in the Pastoral Systems

10.2.1 Improved Pastures

As elsewhere in China, reliable data on the extent of pasture improvement was difficult to obtain for Aohan County. The Policy Research Division of the Aohan County government provided the following data for 1989:
- total area of improved pastures   962,320mu
- area of improved pasture fenced   400,000mu
- area of improved pasture aerial seeded   500,000mu

10.2.2 Pasture Improvement Programs

The great majority of pasture improvement in Aohan County is undertaken by village collective units. In 1989, a total of 150,000mu of aerial seeded pasture was planted by collective units in the County. The County government provides these units with a subsidy equal to ¥1 for each mu of aerial seeded pasture.

In addition, the government is attempting to encourage individual households to plant pastures by allocating certain areas of agricultural land for pasture planting. This land is planted to pastures for a period of two to three years and then returned to agricultural production. It would seem that this is a mechanism more for imposing compulsory crop rotation practices than for encouraging the planting of pastures. This conclusion is based upon evidence that the returns per mu using agricultural land for cropping purposes are much higher than the returns using the same land for pastoral purposes.

10.2.3 Aerial Seeding of Pasture/Tree Planting

One large area of aerial seeded pasture was inspected closely. The paddock concerned seemed to be a remarkably successful example of aerial seeding. It had previously been essentially wasteland consisting of mostly sand dune and salt-affected wadi areas. In fact, surrounding this paddock and across the other side of the road, there were still large areas of salt-affected wadi, and in some cases, sand dunes. This paddock had been aerial sown in June and July 1988 and in October 1989 it carried a remarkable body of feed. It had been decided by the local village collective which contributed to the cost of the aerial seeding that this paddock would not be used even for "cutting land" for two years. After that, if the sown pasture had become sufficiently well established to stabilise the sandy areas then this area would be used for cutting hay. Aerial seeding is supposed to be cheaper than surface seeding for improved pastures.

Tree planting in Aohan County has been rather extensive. This County is acknowledged as being the best of all counties in the Chifeng City Prefecture with respect to the area planted to trees.

10.2.4 Supplementary Feeding

In winter, sheep are kept alive by feeding them grain, crop residues, hay and leaves collected during the autumn from planted poplar trees.

The problem of "broken" wool (i.e. wool with a weak spot in the fibre) was previously common in Aohan County. However, in recent years sheep have received better quality feed in winter, and according to AHB officials it is now rare to find a break in the wool. While this was the official view, tender wool was observed on many sheep in the county.

10.2.5 Sheep Production Systems

In this County, the average stocking rate on natural pasture is one sheep to 5mu. In the agricultural areas, flock sizes are small and the sheep are shedded/yarded for long periods. In the drier parts of the county, flocks are larger but sheep are still shedded at night for about 100 days or more during winter.

10.2.6 Improved Sheep and Wool Production

- *Proportion of sheep artificially inseminated*
  In 1989, according to the county officials, 81.56% of all mature ewes in the County were artificially inseminated. The percentage of households using artificial insemination in that year was 88.37%.

- *Sheep breeds*
  Eighty percent of all sheep in the county had Aohan Fine Wool blood in 1990.

- *Specific sheep breeding programs*
  The Aohan Fine Wool Sheep is the product of a local breeding program aimed at increasing the quantity and quality of wool produced per sheep (Longworth and Williamson, 1993).
  The responsibility for the improvement of sheep breeds in Aohan County rests with the Animal Improvement Station which operates under the Animal Husbandry Bureau. The Animal Improvement Station is undertaking sheep improvement through the combined use of high quality rams and artificial insemination.
  The use of poor quality rams for natural mating or artificial insemination is prohibited in Aohan County. Further, the use of good quality rams for natural mating, without prior recourse to artificial insemination, is also supposed to be prohibited. This would account for the relatively high percentage of households in the county using AI.
  In 1986, the Aohan county government introduced a regulation prohibiting individuals from owning rams, except with the express permission of the county Veterinary Station. The terms of the regulation are:
- The Veterinary Station will approve a ram only if the ram has been selected for breeding by a state breeding farm. For example, rams rejected from Aohan Stud Farm are not considered acceptable.
- Farmers given permission to keep rams are not allowed to use the rams for natural mating without first trying AI.

In principle, therefore, it is compulsory to mate all ewes to approved high quality fine wool rams by AI. However, there is no specific penalty associated with failure to obey this regulation. Administrative methods whereby Animal Husbandry Bureau officials "talk" to offenders are used to convince farmers of the errors of their ways.

The AHB officials insist that regulations are necessary because the price differences between the different grades of wool are insufficient to encourage farmers to strive for improvements in wool quality. In addition, without regulation, officials expressed concern that wool quality may actually decline. This problem is more serious in agricultural, *vis-à-vis* pastoral areas, for two reasons:
- artificial insemination costs tend to be higher in agricultural areas. Savings accruing from economies of size in pastoral areas would, therefore, tend to reduce the cost of artificial insemination in these areas; and
- the number of sheep per household in agricultural areas is relatively low (nine sheep per household in the areas surveyed). Under these conditions, the aggregate net return to each household resulting from improvements in wool quality is much less important in relation to total household income than in the pastoral areas. The incentives for individual households to improve wool quality in agricultural areas is, therefore, much less than in pastoral areas. Furthermore, households in agricultural areas are likely to put more emphasis on the meat producing capabilities of their sheep than would be the case in pastoral areas.

In apparent contradiction of county government policy relating to restrictions on the sale of quality breeding stock, the Aohan Stud Farm has been encouraging farmers who purchase Aohan ewes to purchase so-called "quality" Aohan rams. The quality Aohan rams purchased, however, are in fact said to be rams rejected for sale as breeding stock. This practice was reported both by farmers and AHB officials. It appears that there is an apparent reluctance on the part of the Stud Farm to castrate these rams. Many farmers within Aohan county are reported to be unaware of this somewhat unscrupulous practice. In one instance, a farmer being interviewed openly and happily spoke of his purchase of a quality Aohan ram. To his surprise and dismay, he was promptly informed of the true state of affairs by the local Animal Husbandry Bureau official present at the interview. Upon learning of the true situation, the farmer offered to have the ram castrated immediately. The local Animal Husbandry Bureau official reported that Veterinary Stations in all townships of Aohan have been told to castrate free of charge any rams found to be second quality Aohan Stud Farm rams.

Artificial insemination was said to be compulsory in the County because:
(a) it was difficult to control the use and distribution of good quality rams; and
(b) it was difficult to improve sheep breeds throughout the County. It was said that there is a severe shortage of good quality rams available in the County.

The efficient and widespread use of artificial insemination has been made possible through the establishment of a county-wide breeding and distribution system for high quality rams. Under this system, high quality rams are bred on the Aohan Stud Farm and distributed through two main channels:
• First, rams produced by the Aohan Stud Farm may be sold directly to breeding stations located at the county and township levels. There are currently four breeding stations at the county level and 23 breeding stations at the township level. Only Grade I rams are purchased by the breeding stations from the Aohan Stud Farm.

Township-level stations have the option of either purchasing rams from the Aohan Stud Farm or renting rams from the county-level ram breeding station. As the name suggests, ram breeding stations at either the county or township level are in fact able to breed their own rams. In order to do so however, they must obtain the sire rams and the dam ewes from the Aohan Stud Farm. It is prohibited for the stations to breed rams for the purpose of artificial insemination using sires and dams bred at the breeding

stations. Approximately 50% of township-level breeding stations rent rams from county-level breeding stations. About 60% of the rams used for AI in the county are held at the county-level and township-level breeding stations.
- The second major channel used to distribute high quality rams from the Aohan Stud Farm involves the direct sale of rams to collectives and specialist households which operate AI stations. These stations charge about ¥2 to ¥3 per successful mating per year. This fee is designed to cover the costs of equipment and the buildings together with the salaries of the technicians who carry out the AI.

There is currently some limited use of frozen semen in Aohan County. It was said that each year around 3,000 ewes were inseminated using frozen semen. The level of success using frozen semen is considerably lower than that using fresh semen. For example, the chances of achieving conception using frozen semen after two attempts is said to be around 50%. This contrasts sharply with a 95% success rate obtained using warm semen after two treatments. Another problem with using frozen semen is the much greater cost associated with capital equipment required.

10.2.7 Wool Output

The average annual output of greasy wool in Aohan County is about 1,750 tonne. The percentages of each grade of wool produced are:

Fine wool	–	special grade	10%
	–	grade I	30%
	–	grade II	50%
Improved fine wool	–	grade I	10%
	–	grade II	0%

The distribution of the wool clip by grades suggests that the general quality of the sheep in Aohan is considerably better than in Balinyou. Since 40% of the clip is fine wool special grade or grade I (in Balinyou virtually none of the wool was of these grades), the proportion of sheep which are genuine fine wools must be reasonably high, perhaps approaching 40%.

10.3 Socio-Economic Development

10.3.1 Income Growth

In common with the rest of China, Aohan has experienced substantial growth in nominal incomes since the early 1980s. However, as the data in Table 10.4 demonstrate, real incomes have fallen since the peak in 1985. The data also suggest that the gap between rural and urban incomes in Aohan has narrowed dramatically during the 1980s. [That is, the county-wide data shown in the last column in Table 10.4 is the average for both rural and urban incomes.]

Despite the improvement, net per capita rural incomes in Aohan County remain the lowest of the four counties surveyed in Chifeng City Prefecture. Indeed, as can be seen in Fig. 5.2 of Chapter 5, average rural incomes per capita are much higher in the two much less densely populated pastoral counties (Balinyou and Alukeerqin) than in Aohan. Since incomes in these other counties are below average in relation to both the IMAR and China as a whole, it is easy to see why Aohan County, where incomes are even lower, is one of the poorest counties in the whole country.

Table 10.4 Average Net Income Per Capita for Aohan County, 1970 to 1989

Year	Rural net income per capita		County-wide net income per capita	
	Nominal	Real*	Nominal	Real*
	------- (¥/year) -------		------ (¥/year) ------	
1970	49.0	50.6		
1971	35.0	36.5		
1972	15.0	15.7		
1973	43.0	44.6		
1974	33.0	34.1		
1975	42.0	43.3		
1976	55.0	56.5		
1977	40.6	40.9		
1978	51.5	51.5		
1979	40.7	39.9		
1980	35.0	32.4		
1981	43.0	38.8	129	116.5
1982	73.0	64.7	148	131.2
1983	136.0	118.8	196	171.2
1984	172.0	146.1	233	198.0
1985	250.0	195.2	294	229.5
1986	149.0	109.7	295	217.2
1987	212.0	145.5	250	171.6
1988	225.0	130.4	260	150.6
1989	214.0	105.2	220	108.2

*Deflated by "Overall Retail Price Index" in State Statistical Bureau of the PRC, *China Statistical Yearbook 1991*, p.199.
The base year is 1978.

10.3.2 Changes to Commodity Prices

As already mentioned, AHB officials feel that the price difference between fine and coarser wools is insufficient to provide an incentive for farmers to shift production towards fine quality wool. The management of the county SMC agreed with this viewpoint. They felt that the price differences between wool grades should be increased to encourage the production of finer wool.

After 1985, when the Central government decided to allow provincial governments to "open" the market for wool, wool marketing became "chaotic" in Aohan County. Officially, the IMAR government had declared that the wool market would not be a free market in IMAR in 1985 and, when given the opportunity, the Chifeng prefectural administration also refused to open the wool market in the 1986 to 1988 period. Liaoning Province, on the other hand, had opened its wool market from 1985. Since Aohan has a substantial border with Liaoning Province, private buyers from Liaoning became active in Aohan. Aohan, therefore, was one of the counties particularly affected by the "wool war" which occurred between 1986 and 1988 (Watson and Findlay, 1992).

Despite the illegal competition from across the county border, the Aohan SMC continued to purchase most of the wool in Aohan County in the 1986 to 1988 period.

The prices paid were supposed to be the same as in the other counties in Chifeng City Prefecture (see Table 8.6 in Chapter 8) because the prefectural Price Bureau was the authority charged with setting wool prices. In the "wool war" era (1986 to 1988), the actual prices paid by the Aohan SMC were often above the official prices. In 1989, the free market prices for wool declined sharply in the second half of the year. Nevertheless, the SMC was forced to purchase wool at the (high) official prices throughout 1989.

Details of the actual prices paid grade by grade over the 1985 to 1990 period are not available. However, the average prices paid for wool and cashmere are set out in Table 10.5.

Table 10.5 Average Prices Paid by the SMC for Wool and Cashmere in Aohan County, 1985 to 1990

Year	Wool	Cashmere
	---------- (¥/kg greasy) ------------	
1985	5.34	31.08
1986	6.06	48.00
1987	6.56	70.00
1988	8.74	120.00
1989	10.87	140.00
1990	8.03	140.00

Source: Aohan County Supply and Marketing Cooperative

Although Aohan is not a significant production area for cashmere, the price of cashmere rose much more than wool and there would have been a marked incentive for farmers to expand cashmere production at the expense of wool. The prices paid for other pastoral products have also increased greatly since 1983 as shown in Table 10.6. The relative profitability of fine wool production has probably declined, especially in agricultural areas.

Table 10.6 Prices Paid for Beef, Mutton and Goat Meat in Aohan County, 1978 to 1990

Year	Beef	Mutton	Goat meat
	-------------------- (¥/kg) --------------------		
1978	1.44	1.62	1.44
1979	1.44	1.62	1.44
1980	1.44	1.62	1.44
1981	1.44	1.62	1.44
1982	1.44	1.62	1.44
1983	1.44	1.62	1.44
1984	2.00	1.62	1.44
1985	2.20	2.40	2.40
1986	3.40	3.00	3.00
1987	3.40	3.60	3.60
1988	4.80	5.00	5.00
1989	5.60	5.80	5.80
1990	5.60	6.00	6.00

Source: Aohan County Food Company

10.3.3 Timing of Introduction of the Household Production Responsibility System

The household production responsibility system developed in Aohan County in three stages. The first stage began in 1979 with the introduction of a group responsibility system based on production targets and work points. The second stage occurred in 1982 when the animals were sold to individual or joint households. The final stage was the allocation of the land and pastures to the households. The allocation of agricultural land and cutting land was completed rather quickly. However, large areas of rangeland are still grazed in common.

10.3.4 Level of State Assistance

Aohan County is recognised at the national level as being a general "poverty county". The definition of a general poverty county is a county with an average net per capita income of less than ¥150 in 1985.

As a poverty county, the county government is entitled to special assistance from the Central government. Assistance provided by the government is mostly through the Agricultural Bank of China (ABC) and is in the form of a limited amount of subsidised credit. In 1990, a total of ¥4 million in subsidised credit was made available to collective units and individual households. Of this, ¥700,000 was said to be set aside for the development of sheep production in the County. The amount of subsidised credit made available by the ABC to the County is said to vary from one year to the next. For example, in 1990 the amount of subsidised credit available to households was ¥2 million. In 1991, the amount available for lending to households was only ¥1.4 million. It was claimed these funds were lent to poor households at one-tenth the normal rate of interest.

Aohan County also receives State assistance in the form of fiscal transfers from other counties and levels of government defined as having a surplus of tax revenue over expenditure. The total size of these transfers is fixed over a five-year term. In 1990, the total value of the transfer was said to be ¥22 million. The actual size of the fiscal transfer is determined according to the size of the county budget deficit for each of three years prior to the year under review. Currently, Aohan County government expenditures (including Central government grants) exceed county revenues by about 50%.

There is little published information on the detailed fiscal arrangements at the county level. These arrangements vary from province to province and even between prefectures within the same province. For information on Chinese fiscal arrangements down to the province level, see World Bank (1988).

10.3.5 Development of Pastoral Product Processing

Aohan County has built a new wool scouring plant and opened a leather factory. The wool washing factory was due to commence production in 1990. It has a capacity to wash around 3,000 tonne of greasy wool annually. This is about double the present output of greasy wool in Aohan County. The wool washing factory in Aohan was built to add value to the wool produced in the county and provide further income for the county. Brown and Longworth (1992b) raise serious doubts about the economic value of such scouring plants. On the other hand, officials of the County SMC claim that the sale of raw wool, by itself, is not profitable. The 1989 gross value of one tonne of greasy wool was ¥13,000 per tonne, compared to the gross value of one tonne of clean wool of ¥52,000. Effectively, therefore, washing wool added a gross value of about ¥13,000 per tonne, given that about 3 tonne of greasy wool is required to make 1 tonne of clean wool. The County SMC officials also claim that the transport costs associated with raw wool *vis-à-vis* clean wool are 30% higher. The actual cost of transporting raw wool to Chifeng City by truck, a distance of 118km, is ¥0.086 per kg.

The Aohan SMC also operates a food-processing factory which makes flavourings for food and a knitting factory which produces knitted goods.

10.3.6 Growth in Output of Major Pastoral Commodities

The only data available on trends in the output of products derived from grazing animals are presented in Table 10.7. Wool production has declined since 1984, reflecting the fall in sheep numbers referred to earlier.

The degraded state of the pastures in the pastoral areas of Aohan suggests that it will be difficult to increase significantly the output of wool and other products derived from grazing animals. Indeed, the level of output of wool, in particular, could fall as the prices of other commodities such as mutton, beef and cashmere increase faster than wool.

In the agricultural areas, there may be some potential to raise more sheep but again the trends in the relative prices of mutton and wool suggest that farmers will prefer to raise sheep for mutton rather than for fine wool. Furthermore, mutton-type sheep are better suited to the production systems employed in the agricultural areas.

Table 10.7 Wool and Cashmere Production in Aohan County, 1983 to 1987

Year	Wool	Cashmere
	------------- (tonne) -------------	
1978	969	
1979	1,120	
1980	1,220	
1981	1,270	
1982	1,420	
1983	1,820	1.5
1984	2,075	1.5
1985	1,700	n/a
1986	1,750	2.0
1987	1,650	2.0
1988	2,100	n/a
1989	1,800	n/a

Source: Aohan County Animal Husbandry Bureau

Chapter Eleven

Alukeerqin County (IMAR)

Alukeerqin, like Balinyou and Wongniute, is part of the Eastern Grasslands area of the Inner Mongolia Autonomous Region (IMAR). It is the most north eastern county in Chifeng City Prefecture (Map 11.1). It has an extensive eastern border with Zhelimu Prefecture of IMAR. Tongliao, the capital of Zhelimu, is the "natural" large commercial centre serving Alukeerqin. Furthermore, although under the prefectural administration in Chifeng City, Alukeerqin is in the same economic zone as Zhelimu Prefecture. Thus for both geographic and economic reasons, Alukeerqin is not as closely tied to Chifeng City as the other counties in the Chifeng City Prefecture.

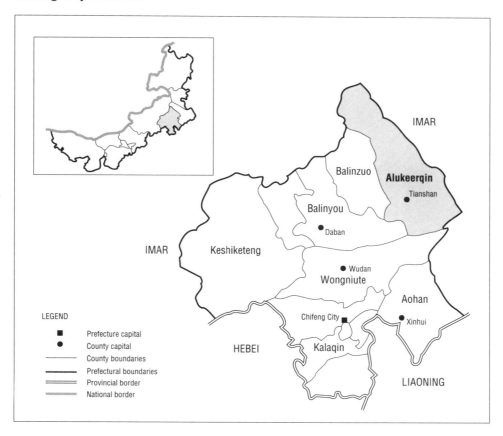

Map 11.1 Alukeerqin County in Chifeng City Prefecture of IMAR

11.1 Resource Endowment

In terms of topography and general resource endowments, Alukeerqin is similar to Balinyou. The rangelands are the chief natural resources available and the living standards (income) of up to three-quarters of the population are dependent upon pastoral products, at least to some extent. All levels of government designate Alukeerqin as a poverty pastoral county.

11.1.1 General Features

The official average annual rainfall at Tianshan, the county capital, is 353mm but it varies a great deal as illustrated by the data in Table 11.1. The lowest recording in Table 11.1 was 176mm in 1973 and the highest was 551mm in 1985.

The environment is harsh. Living conditions for people and livestock vary greatly during the year from being extremely cold in winter to windy and dusty in spring and very hot in summer.

Sheep are unable to utilise the pasture in the rough mountainous regions of the county. Goats are largely grazed in these areas.

Table 11.1 Annual Rainfall at Tianshan the Capital of Alukeerqin County, 1968 to 1991

Year	Annual rainfall	Year	Annual rainfall
	(mm)		(mm)
1968	228.0	1980	328.0
1969	328.8	1981	363.2
1970	303.6	1982	290.9
1971	266.2	1983	383.0
1972	343.2	1984	378.8
1973	176.0	1985	550.9
1974	314.8	1986	494.3
1975	305.3	1987	441.0
1976	364.3	1988	380.4
1977	266.5	1989	239.1
1978	487.0	1990	313.0
1979*	412.6	1991**	350.0

* estimated value
**provisional estimates
Source: Alukeerqin County Animal Husbandry Bureau

11.1.2 Land Use

Of the four counties surveyed in Chifeng City Prefecture, Alukeerqin has the biggest total area and the largest amount of usable pasture. While the area suitable for cropping in Alukeerqin is more than double the arable land in Balinyou, it is still much less than in either Wongniute or Aohan.

- Total land area   21,415,000mu   (or 14,277km^2)
- Total usable land area   20,614,900mu

- Total rangeland (pasture) 16,183,100mu
  – useful pasture land 15,138,000mu
- Arable land 981,000mu
- Forestry land 2,976,700mu
- Total degraded pasture land (1990) 10,452,000mu

There are five recognised classes of pasture/grassland in Alukeerqin County:
   (a) mountain and plain pastures
   (b) dry hill pastures
   (c) sandhill and sandy land pastures
   (d) riverland and lowland pastures
   (e) desert pastures.

The main soil types associated with the different classes of grassland are:
   (a) desert grassland – sandy soils
   (b) dry hill grassland – dark brown soils
   (c) lowland plain grassland – swamp soils alongside river alluvial plains.

Pasture degradation is now a major problem in Alukeerqin with almost two-thirds of the total pasture land degraded. Roughly half the affected area is heavily degraded, 40% is medium degraded and around 10% is lightly degraded using the following definitions.

Heavily degraded grassland   A grassland is said to be heavily degraded if it has the following characteristics:
   (a) previously dominant perennial species of grasses have been replaced by less productive annual species;
   (b) vegetation coverage per unit area of grassland has fallen below 30%;
   (c) the average height of grasses per unit area of grassland is < 10cm; and
   (d) the dry matter production is < 50kg per mu per annum.

Medium degraded grassland   A grassland defined as being medium degraded has the following characteristics:
   (a) vegetation coverage per unit area of grassland which ranges between 31% and 50%;
   (b) the average grass height per unit area of grassland is around 10cm; and
   (c) average dry matter production is between 50 to 100kg per mu per annum.

Lightly degraded grassland   A grassland defined as being lightly degraded is said to have the following characteristics:
   (a) the grass species which originally dominated the grassland have disappeared;
   (b) the vegetation coverage is > 70%;
   (c) the average height of grasses per unit area of grassland is between 11 and 30cm; and
   (d) the average dry matter production is > 100kg per mu per annum.

Non-degraded grassland   A grassland is said not to be degraded if it has the following characteristics:
   (a) a vegetation coverage per unit area which is > 90%;
   (b) an average grass height per unit area which is > 50cm; and
   (c) the average dry matter production is around 250kg.

It is estimated that 40% of the degraded grasslands are desert pastures which have turned to desert.

The build-up in livestock numbers is the obvious direct reason for the rangeland degradation problem. In 1949, there were only 180,000 grazing animals (or about 0.5 million sheep equivalents) in Alukeerqin. By 1989, there were 2.1 million sheep equivalents. However, as elsewhere in the pastoral region, the basic underlying factors which have contributed to the growth in herbivorous animal numbers and hence to the increasing degradation problem are much more subtle (Williamson and McIver, 1993).

11.1.3 Population

- Total population (1989)   266,363   (100%)
  - rural population   243,627   (91.46%)
- Ethnic composition (1989)
  - Han   169,971   (63.81%)
  - Mongolian   90,495   (33.97%)
  - Other   5,897   (2.21%)
- Population density (1989)   18.657 people/km^2

11.1.4 Administrative Structure

Of the 27 townships in Alukeerqin, 13 are defined as pastoral and another 12 are classified as semi-pastoral (Table 11.2). Less than a quarter of all rural households are classified as being agricultural. Alukeerqin is clearly very different to Aohan in this respect.

Table 11.2 Administrative Units in Agricultural, Pastoral and Semi-Pastoral Areas of Alukeerqin County, 1989

Classification	No. of townships	No. of villages	No. of households
Agricultural	2	47	15,154
Pastoral	13	139	15,983
Semi-pastoral	12	169	34,286
Total	27	355	65,423

Source: Alukeerqin County Animal Husbandry Bureau

11.1.5 Livestock Population

Alukeerqin has the largest population of herbivorous animals of the five counties surveyed in the IMAR. In fact, it has more sheep, goats and cattle than any of the other counties. As in Balinyou, goats outnumber sheep although, as shown in Fig. 11.1, this is only a recent phenomenon in Alukeerqin.

Approximate mid-year numbers in 1990 were:
- cattle   218,000
- goats   546,000
- sheep   500,000
- total sheep equivalents   2,500,000

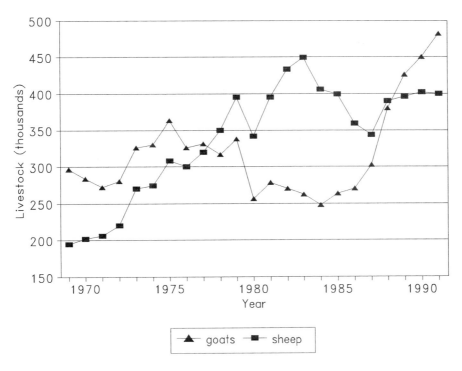

Fig. 11.1 End of Year Sheep and Goat Numbers in Alukeerqin County, 1970 to 1991

The long-term trends in total sheep equivalents and sheep numbers are depicted graphically in Fig. 5.1 of Chapter 5. As in Balinyou and Wongniute, total livestock numbers (as measured in sheep equivalents) dropped sharply at the end of the 1960s owing to the worst excesses of the Cultural Revolution. However, there has been a stronger upward trend in total sheep equivalents since 1970 in Alukeerqin than in Balinyou. In this regard, the contrast is even greater between Alukeerqin and Wongniute and Aohan. Grazing animal numbers in the latter two counties have actually declined since the early 1970s.

In general terms, sheep numbers in all four counties surveyed in Chifeng City Prefecture declined after about 1983 with the introduction of the household production responsibility system. In Alukeerqin, as can be seen from Fig. 11.1, goats outnumbered sheep until the late 1970s but there had been a steady upward trend in the sheep population for about 15 years prior to 1983. Sheep became more popular again in the late 1980s when wool prices rose to record levels. Nevertheless, Fig. 11.1 demonstrates that in Alukeerqin goat numbers almost doubled between 1983 and 1991 while the sheep population declined by more than 10%. The swing back to goats since households have been free to choose has been motivated by higher prices for cashmere relative to wool (see Section 11.3.2) and the easier care factor (lower costs) associated with goats compared with sheep, especially fine wool sheep.

11.2 Improvements to Production Conditions in the Pastoral Systems

11.2.1 Improved Pastures

In 1983, the County government introduced the following three laws which are still in force to protect the rangelands:
   (a) the collecting of grass and/or grass roots for cooking is forbidden;
   (b) the digging of medicine grass for sale is not permitted; and
   (c) land designated as pasture land is not permitted to be used for cropping.

More recently, Alukeerqin County has implemented four other policies aimed at overcoming the problem of deteriorating pastures.

(1)   The county government is encouraging the planting of artificial and semi-artificial pastures.

Officials report the use of semi-artificial pastures (pastures planted using aerial sowing) to be very successful, especially in desert areas. The total cost of planting semi-artificial pastures is shared by the Central government, local government and the individual households. Sixty percent of the cost is borne by the local government. Every year, the Central government specifies the area of semi-artificial pasture Alukeerqin County is expected to plant. In 1988, this quota amounted to 30,000mu. The actual area planted, however, was 73,000mu.

The county has conducted studies on the effectiveness of aerial seeding. Prior to replanting by aeroplane, the grass yield from desert pasture was around 110kg per mu. After replanting, this yield increased to between 250 and 300kg per mu. The cost of aerial seeding is around ¥15 per mu, which is about the same as the cost of artificial pasture planting without the use of aeroplanes. Aeroplane planting of pasture is preferred to manual or tractor planting for two reasons: aerial pasture improvement is less labour intensive, and it enables the planting of large areas of pasture in short periods of time. This is important because large areas of pasture can be planted just prior to forecast rains.

In Alukeerqin County, pasture is planted without fertiliser or irrigation.

(2)   The AHB is attempting to increase the area of fenced pasture land by encouraging collectives and individual farmers to invest in fencing.

The County government has introduced a policy in relation to fencing which provides the following rewards:
   (a) households which choose to fence their land are granted use of the land for a much longer period of time;
   (b) the contract rights to fenced land are now inheritable;
   (c) households are given access to credit at very low interest rates to finance fencing; and
   (d) households unable to borrow money from the Agricultural Bank or Credit Co-operative are given access to local government money loaned at a zero interest rate.

These measures appear to have been particularly effective. The total area of fenced pasture land has expanded remarkably during the 1980s, with increasing areas being fenced each year. In 1990, for example, an additional 278,000mu was fenced compared with only 84,000mu in 1985.

(3)   The county government has implemented a "user pays" system whereby a

pasture management fee is charged for the use of pasture.

Temporary Regulations Regarding Pasture Land Use with Payment were issued by the Alukeerqin Banner government in 1988. [Appendix 11A presents a translation of these Regulations.] These Regulations were intended to provide a local level interpretation of the provincial Rangeland Management Regulation (discussed in Section 5.2.4) which in turn was an interpretation of the National Rangeland Law.

The pasture management fee in 1989 was ¥0.2 to ¥0.3 per sheep unit. The penalty for overgrazing the pasture by 10% of the "allowed stock number" is an increase in the pasture management fee to ¥3 to ¥5 per excess sheep unit. Overgrazing of the pastures by 20% of the allowed stocking rate results in an increase in the pasture management fee to ¥6 to ¥8 per excess sheep unit.

These penalties, which are spelt out in the 1990 version of the Regulations (see Appendix 11B, Section 4), are extremely heavy compared with the base fee. It seems unlikely that local officials would impose such heavy fines. One consequence of writing such punitive penalties into the Regulations, therefore, is to discourage local officials from monitoring actual stocking rates too closely.

The County government has introduced the following means by which farmers can increase their allowed stock numbers:
(a) farmers are permitted to carry one extra sheep equivalent for each additional 7mu of *fenced* pasture land;
(b) for every 1mu increase in artificial pasture area the allowed stock limit is increased by one stock equivalent; and
(c) for every mu of pasture used for silage, a farmer is allowed to increase the allowed stock limits by one sheep equivalent.

(4) "Utilisation Certificates" were issued to all households in Alukeerqin county in 1990. As can be seen in Appendix 11B, these Certificates incorporate a revised version of the Temporary Regulations issued in 1988. The 1990 Regulations, which have been incorporated in the Pasture Utilisation Certificate, spell out in considerable detail the rights and obligations of the households to whom the pasture land has been contracted.

11.2.2 Improved Sheep and Wool Production

- *Use of artificial insemination*

The use of artificial insemination as a means of upgrading sheep is well established in Alukeerqin County. There are 420 AI stations in the county. Approximately 83% of all mature breeding ewes in the county are artificially inseminated. This percentage varies between villages, with some villages (in particular the pastoral villages) achieving 100% artificial insemination whilst other villages (primarily agricultural villages) achieve little or no artificial insemination. The reasons for these different levels of adoption of AI are essentially the same as discussed in relation to Aohan County.

- *Sheep breeds*

It was claimed that 90% of the sheep in Alukeerqin County are fine wool sheep or improved fine wool sheep. On the other hand, it was frequently stated that farmers in Alukeerqin prefer dual-purpose sheep since mutton is an important part of the diet for sheep-raising households and there is always a good market for mutton. Perhaps this is why very few sheep have been completely upgraded to fine wools. The farmers prefer to retain crossbreds which are classed as improved sheep and which are better mutton sheep than real fine wools. (See Table 5.10 and the discussion in Section 5.3.2.)

- *Specific sheep breeding programs*

Originally there were only Mongolian sheep in Alukeerqin County. In the 1950s and 1960s, Soviet and East German merino type rams were imported and crossed with the local Mongolian breeds.

Between 1975 and 1979, the Aohan Fine Wool was introduced into Alukeerqin County. In 1986, rams which were the progeny of crossing Australian merinos with Northeast Fine Wools were introduced into the county. By 1991, these crosses had become the most popular type of ram in Alukeerqin. The widespread use of AI has enabled the rapid introduction of this popular new type of fine wool sheep. The Australian-Northeast Fine Wool is said to produce good quality, long stapled 60/64s count wool. It is a large animal, good for mutton. The average production from these sheep is claimed to be 8kg of greasy wool per head.

In 1989, plans were being made to introduce more pure-bred Australian merinos into the county. There are problems, however, with pure-bred Australian merinos in the harsh Alukeerqin environment. In 1986, four Australian merinos were introduced into Alukeerqin County and of these four rams, one died, two had no sexual libido, and the other contracted a venereal disease.

In a comparison between the Australian-Northeast Fine Wool and the Aohan Fine Wool, officials reported very little difference in terms of strength of wool, fibre diameter and other characteristics. However, it was claimed that the County government can obtain special funds from the Ministry of Agriculture for the importation of Australian merinos. On the other hand, there are no special funds available for the importation of Aohan Fine Wools into the county.

The sheep improvement program operating in Alukeerqin County is very similar to programs operating in other major pastoral counties in Chifeng City Prefecture.

At the heart of the sheep improvement program in Alukeerqin County, there are two ram breeding stations under the control of the Animal Improvement Station of the AHB. These ram breeding stations are an integral part of an artificial insemination program aimed at improving sheep breeds in the county.

There are presently 280 breeding ewes on the two ram breeding stations. The usual lambing percentage obtained is in the order of 98%. These two stations are the first major source of high quality rams for the village AI centres. In 1989, the two ram breeding stations combined were producing up to 100 young rams annually for sale to village-level AI centres. The selling price for rams bred on the two stations is said to be between ¥150 and ¥300 each. The variation in sale price reflects variation in the quality of the rams. The rams and ewes used at the two ram breeding stations are obtained from the Aohan Stud Farm (Aohan Fine Wool) and the Chaganhua Stud in Jilin Province (Australian-Northeast Fine Wool). Chaganhua Stud is one of the four national Fine Wool Sheep Studs in China. The other three are Gadasu in Zhelimu Prefecture of IMAR, Ziniquan located near Shihezi City in Xinjiang, and Gongnaisi in Yili Prefecture of Xinjiang. It is highly likely that Alukeerqin County also obtains breeding stock from Gadasu Stud since this stud is closer to Alukeerqin and, although located in a different prefecture, it is nonetheless in the same province. For more information on fine wool sheep studs in China, see Longworth and Williamson (1993).

The procedure used by the two ram breeding stations to select rams suitable for the AI centres is as follows:

(i) The first selection takes place immediately after birth according to birth weight and the production performance of the ram lambs' parents. Ram lambs below 4kg at day 1 are regarded as rejects and as such are placed into a separate flock. These reject ram lambs are not castrated because, depending on their growth performance, they may later be returned to the non-reject ram lamb flock if their performance is considered to be outstanding. It was said that around 10% of ram lambs are normally culled at this stage in the selection process.

(ii) The second selection takes place immediately after weaning (approximately 6 months after birth) according to body conformation and development (i.e. the body weight must exceed 27kg for the ram lamb to be retained in the selected ram lamb flock). The normal number of ram lambs rejected at this stage is said to be something in the order of 10%.

(iii) The third selection is undertaken one year after birth and is based on the following characteristics:
   (a) the length of the wool must be > 8cm;
   (b) the wool fibre diameter must be equivalent to a > 60s spinning count;
   (c) the greasy level must extend at least two-thirds of the way up the length of the wool staple from the skin; and
   (d) the density of the wool must be fairly substantial (density is measured subjectively by feeling the wool in a manner similar to that used in Australia).

The normal number of ram lambs rejected at this stage is said to be something in the order of 10% of the original number of ram lambs born.

The second source of quality rams used by the village-level AI centres is the State-run sheep studs within and outside Chifeng City Prefecture. Each year, a total of 500 rams are purchased from State-run sheep studs at an average price of ¥500 each. While a number of these rams are utilised by the two county ram breeding stations, the vast majority are sold directly to village-level AI centres. The bulk of the rams purchased from the State-run sheep studs in Chifeng City Prefecture are from the Aohan Sheep Stud in Aohan County and the Haoluku Sheep Stud in Keshiketeng County.

In 1989, there were around 4,500 fine wool rams in Alukeerqin County. Of this total number of fine wool rams, 35% were of Special Grade, 53% were of Grade I, and the remaining 12% were of Grade II. According to the AHB officials, there was no shortage of good quality rams in the county in 1989. It was also noted that only five years previously there were no Special Grade rams in the county.

11.3 Socio-Economic Development

11.3.1 Income Growth

As with other pastoral counties (and China generally), nominal incomes increased sharply in the 1980s. However, real incomes have tended to decline in recent years. Table 11.3 provides some data on the average per capita income in rural areas (pastoral, semi-pastoral and agricultural) and for the whole county since 1978.

The rural income data have been plotted in Fig. 5.2 of Chapter 5. Income levels in Alukeerqin are similar to that in Balinyou but, while markedly higher than in Aohan, they are still well below both the average level of rural incomes in IMAR and the national average.

Table 11.3 Average Net Income Per Capita for Alukeerqin County, 1978 to 1990

Year	Rural net income per capita		County-wide net income per capita	
	Nominal	Real*	Nominal	Real*
	-------- (¥/year) --------		------ (¥/year) ------	
1978	49	49.0		
1979	70	68.6		
1980	69	63.8		
1981	81	73.2		
1982	292	258.9	281	249.1
1983	400	349.3	377	329.3
1984	319	271.0	319	271.0
1985	336	262.3	311	242.8
1986	314	231.2	282	207.7
1987	418	286.9	415	284.8
1988	545	315.8	545	315.8
1989	359	176.5	582	286.1
1990	479	230.6	556	267.7

*Deflated by "Overall Retail Price Index" in State Statistical Bureau of the PRC, *China Statistical Yearbook 1991*, p.199.
The base year is 1978.
Note: The method of data collection changed after 1981. It would seem that the 1978 to 1981 nominal income data is not comparable with the post 1981 data.
Source: Animal Husbandry Bureau, Alukeerqin County

11.3.2 Commodity Prices

The rapid increase in the prices paid to farmers for wool, mutton, cashmere, goat meat and beef during the 1980s was a major factor explaining the rise in nominal per capita incomes in pastoral counties such as Alukeerqin.

- As described in Section 5.6.2, the IMAR government delegated the authority to declare the wool market "open" to the prefectural-level authorities in 1986, 1987 and 1988. At the same time, the provincial Price Bureau set indicative prices which were intended to guide pricing in those prefectures which did not open their wool markets. Chifeng City was one prefecture which did not adopt an open market for wool during the 1986 to 1988 period. The prefectural Price Bureau, therefore, was responsible for establishing the prices which the SMC was supposed to pay farmers for their wool. These prices were not officially subject to variation at the county level although, as discussed in the Aohan case study, market realities often forced the county SMCs to pay more competitive purchasing prices.

To provide some official flexibility for the county level SMCs, the prefectural

Price Bureau in Chifeng City developed maximum/minimum pricing schedules for 1988 and 1989. The Alukeerqin SMC was able to more or less apply this approach to setting the purchase prices in Alukeerqin because the county is more isolated than Aohan and has no common borders with any province in which the wool market was open. The range of purchase prices which applied in Alukeerqin in 1988 and 1989 is set out in Table 11.4.

Table 11.4 Maximum and Minimum Wool Prices Offered by the SMC in Alukeerqin County, 1988 and 1989

Wool type	Grade	Maximum price		Minimum price	
		1988	1989	1988	1989
		(¥/kg greasy)		(¥/kg greasy)	
Fine wool	– special grade	9.52	10.88	8.72	9.80
	– grade I	8.74	10.00	8.04	9.00
	– grade II	7.52	9.40	6.91	8.50
Improved wool	– grade I	6.49	8.78	5.94	7.98
	– grade II	5.56	7.98	5.11	7.18
Coarse wool		5.56	7.98	5.11	7.18

- Under the administered pricing system for wool which had traditionally operated in the IMAR, the SMC official at the SMC purchase station had the responsibility to determine the grade of each lot of wool to be purchased and then to set the price for that particular lot. Once the grade had been established, the maximum possible purchase price for the lot in question became the officially determined price for that grade. The actual price for any particular lot of wool of that grade was then determined by a simple "points" system.

The highest possible number of points which could be awarded for any lot of wool was 100. In this case, the actual price was the maximum allowable price for that grade of wool. It was said that only very high quality wool received 100 points. The buyer would weigh the wool, then shake it vigorously for some time and weigh it again. Points were then deducted from the maximum possible 100 according to the percentage weight loss due to shaking out the sand, dust, grass seed, animal waste, twigs, etc. For example, a lot of wool which originally weighed 100kg but lost 5kg in weight due to shaking would receive 95 points. The price for this lot of wool would then be determined as 95% of the maximum possible price.

In Alukeerqin (and elsewhere in the IMAR), there was a great deal of dissatisfaction with the points system. It was said to be too easy to exploit and it left too much discretion in the hands of the SMC buyer at the purchase station.

- Another aspect of pastoral commodity pricing which has already been mentioned in relation to other counties and which was well illustrated in Alukeerqin, was the impact of the different pricing policies for wool and cashmere. Until 1992, the Chifeng City prefectural administration had retained control over wool prices but had allowed a free market to develop for cashmere. In remote pastoral counties such as Alukeerqin, the prices at which wool was purchased from farmers by the SMC on behalf of the State, closely approximated the official prices set by the prefectural Price Bureau in Chifeng City and shown in Fig. 11.2. The SMC, on the other hand, had to pay free-market prices for cashmere. As shown in Fig. 11.2, once the market for this animal fibre was "opened" in 1985, official prices paid by the SMC for white cashmere rose more than eightfold from ¥20 per kg in 1984 to a peak of over ¥170 per kg in 1989. Over the same period, the administered price paid for fine wool Grade I little more than doubled from ¥4.86 per kg to just over ¥11 per kg. Consequently, as indicated in Section 11.1.5, farmers increased their goat herds much faster than their sheep numbers. Officials of the County AHB considered goats to be more severe on pastures than sheep. The increasing goat population, therefore, is a serious problem since it intensifies the over-grazing pressure on the rangelands.

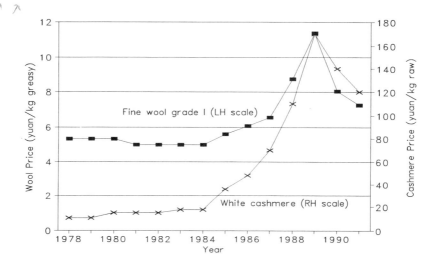

Fig. 11.2 State Purchase Prices for Wool and Cashmere in Chifeng City Prefecture, 1978 to 1991

11.3.3 Timing of Introduction of the Household Production Responsibility System

The first stage of the household production responsibility system was introduced in Alukeerqin County in 1980 when the livestock were contracted out to the households. Subsequently, the sheep were sold to the households. The contracting out of the pastures began in the mid-1980s. However, large areas of unfenced pasture land are still grazed in common.

11.3.4 Level of State Assistance

Alukeerqin was recognised by the Central government as being a "minority" poverty county in 1986. To qualify for assistance as a "minority" poverty county, a county must have a significant minority population and have recorded an average net income in 1985 of less than ¥200 per capita.

There are essentially two channels by which poverty alleviation funds are provided to the county. The first channel is concerned with funds which are known as agricultural and development aid funds. These moneys are channelled down through the various finance departments at each administrative level of government. In 1986, the first year the poverty alleviation scheme applied to Alukeerqin, a total of ¥0.72 million was made available through the government finance departments as agricultural and development aid funds. In 1990, ¥0.85 million flowed to Alukeerqin through these channels.

The second way in which poverty alleviation funds are made available is through the Agricultural Bank of China (ABC) as subsidised credit. Loans from these sources may attract an annual interest of less than 3% for a three-year term. In 1986, ¥1.5 million became available in Alukeerqin as loan funds for poverty alleviation. About one-third of these funds were loaned out for development of animal husbandry. By 1990, the annual allocation of loan funds for poverty alleviation had increased to ¥1.7 million with about ¥0.6 million being provided for animal husbandry related investments. All funds made available through the ABC under the poverty alleviation program are lent to households. The majority of the loans classified as being for animal husbandry related investments are for pasture improvement purposes such as pasture planting and fencing.

The ABC is the principal source of credit for the agricultural sector in Alukeerqin County. In 1991, commercial loans to households or groups of households were usually on a 3-year term at 9.35% annual interest. Over ¥2 million was lent for animal husbandry purposes in 1991. The comparable figure for 1980 was ¥794,000 but in that year almost all of the loans would have been to collectives not households. In real terms, 1984 was a record year for commercial loans, with the ABC lending around ¥1.88 million both to collectives and private households. Most animal husbandry related loans in 1984 were to households or groups of households since the county government was vigorously encouraging these units to invest in pasture construction. As indicated in Section 11.2.1 above, the major thrust of county government policy is still in this direction, with growing emphasis being placed on fencing.

Some farmers in the pastoral areas are reported to have over-extended themselves in regard to their borrowings. The County ABC has a mechanism for aiding farmers in the pastoral areas in genuine need of debt relief. This aid is not in the form of debt relief *per se*, but is provided in the form of advice and/or extra loans to develop their animal production capacity and thereby improve the ability of the household to repay the loan. The Agricultural Bank provides these advisory services in such areas as animal production, pastures, marketing, and farm improvement. The Bank has staff that are trained in providing this advice. In addition, the Bank sometimes requests the Animal Husbandry Bureau to provide advice on their behalf.

11.3.5 Development of Pastoral Product Processing

Construction of a wool scouring plant began in 1986 and was completed in 1987. The plant has the capacity to produce 600 tonne of clean wool per year. The main reasons for building the plant were the long distance to either Chifeng City or to Tongliao (and hence transport cost savings were likely to be considerable); the possibility of extra revenue being generated for the County government; and to provide a better means of monitoring raw wool quality and quantity in the county.

Total investment in the plant was ¥2.4 million and it came from three sources:
- a 3-year loan from the IFAD North China Pasture Development Project of ¥500,000;
- an Agricultural Bank loan of ¥1 million (with different interest rates applying according to the length of loan, namely 8.4% for the 1-year loan and 11.3% for the 2-year loan); and

- County SMC equity of ¥900,000.

The plant initially generated considerable tax revenues for the county. In 1987, it contributed 7% of the county's revenue and in 1988 almost 11%. The results for 1989 and subsequent years have been much less favourable. The main taxes to be paid by the plant to the County government are:
- a value-added tax of 14%;
- a progressive profits tax with eight different rates; and
- a scouring tax of 3% of the fee charged for scouring the wool of other counties. (As of 1992, no wool has been scoured for other counties.)

The purchase price and amount of wool scoured varies from year to year. In 1987 and 1988, the selling price of scoured wool was determined by the Price Bureau of Chifeng City. But in 1989, the price for clean wool was decided by negotiation between the scouring plant and the textile mills. In 1987 and in 1988, a profit was made from scouring wool. On the other hand, 1989 was a bad year for the plant. The total cost, including the purchase price paid for the raw wool, was ¥49,000 per tonne of clean wool. However, the average price received for scoured wool in 1989 was only ¥42,000 per tonne, resulting in an average loss of ¥7,000 per tonne of clean scoured wool produced.

Scouring plant - losses

The selling price averaged over the industrial grades for the three years 1987 to 1989 was ¥31,000, ¥46,250 and ¥42,000 per tonne of clean scoured wool. These prices can be compared with the average scouring costs shown in the last column of Table 11.5.

Table 11.5 Throughput and Scouring Costs in Alukeerqin County, 1987 to 1990

Year	Raw wool	Scoured wool	Clean yield	Scouring costs*
	(tonne)	(tonne)	(%)	(¥/tonne clean)
1987	420	151	36.0	24,930 (21,000)
1988	1,043	366	35.1	35,373 (31,000)
1989	754	245	32.5	49,160 (42,100)
1990 (prov.)	1,000	360	36.0	n/a

*The figure in brackets is the average purchase cost of the raw wool. This cost is included in the scouring costs shown.

11.3.6 Growth in Output of Major Pastoral Commodities

Production data for wool and cashmere were only available from 1983. This information, which is collected by the County Animal Husbandry Bureau, is presented in Table 11.6.

A longer data series relating to the amount of wool purchased by the County Supply and Marketing Cooperative (SMC) is presented in Table 11.7. This information must be interpreted carefully because not all the wool produced in the county is necessarily purchased by the SMC and it is likely that the SMC may purchase some wool not produced in the county. Nevertheless, the information in both Tables 11.6 and 11.7 suggests that wool output has plateaued since 1983. Table 11.7 indicates that wool output increased rapidly between 1978 and 1983 as the sheep flock expanded (Fig. 11.1). Prior to 1979, there had been a big swing towards fine and improved fine wool production. In recent years, virtually all wool grown in Alukeerqin has been classified as fine or improved fine wool.

Table 11.6 Wool and Cashmere Production in Alukeerqin County, 1983 to 1990

Year	Wool	Cashmere
	----------- (tonne) -----------	
1983	1,537.0	57.0
1984	1,291.5	58.5
1985	1,552.5	68.5
1986	1,450.0	70.0
1987	1,368.0	78.0
1988	1,409.0	105.0
1989	1,484.0	105.0
1990	1,440.0	109.0

Source: Alukeerqin County Animal Husbandry Bureau

Table 11.7 Amount of Wool Purchased by the Alukeerqin County SMC, 1973 to 1988

Year	Wool		
	Total wool purchased	Fine wool and improved fine wool purchased	
		Amount	Proportion of total wool
	(tonne)	(tonne)	(%)
1973	466.7	330.1	70.73
1974	680.5	470.4	69.13
1975	664.6	420.5	63.27
1976	705.0	450.1	63.84
1977	766.9	651.5	84.95
1978	830.7	626.4	75.41
1979	1,007.8	956.8	94.94
1980	1,092.6	957.5	87.63
1981	1,004.1	941.6	93.78
1982	1,228.6	1,167.0	94.99
1983	1,526.9	1,439.2	94.26
1984	1,291.0	1,243.7	96.34
1985	1,339.9	1,263.1	94.27
1986	1,289.2	1,272.6	98.71
1987	1,254.6	1,228.6	97.93
1988	1,436.3	1,424.1	99.15

Source: Alukeerqin County Supply and Marketing Cooperative

The officials of the County SMC estimated that the amounts of wool purchased by grades in 1989 were as follows:

Fine wool	– Special Grade	100 tonne
	– Grade I	450 tonne
	– Grade II	750 tonne
Improved fine wool	– Grade I	200 tonne
	– Grade II	70 tonne

In contrast to wool production which has plateaued since 1983, Table 11.6 shows that there has been a large increase in cashmere output since the mid-1980s. This is to be expected given that the size of the goat herd almost doubled between 1984 and 1991 (Fig. 11.1). As discussed above, the switch from wool to cashmere production reflects the big change in relative prices for wool and cashmere in favour of cashmere. Meat production in Alukeerqin also increased dramatically in the late 1980s (Table 11.8). Pastoral households in Alukeerqin clearly have the capacity to adjust their output mix in response to changing price relativities and other factors. Under these circumstances, wool production may increase in the future at the expense of the other products if wool becomes significantly more profitable than meat (and cashmere).

Table 11.8 Meat Production in Alukeerqin County, 1987 to 1990

Year	Mutton and goat meat production	Beef production
	(tonne)	(tonne)
1987	1,867	3,399
1988	2,133	2,796
1989	3,589	5,108
1990	3,065	4,222

Source: Alukeerqin County Animal Husbandry Bureau

Appendix 11A
The Temporary Regulations Regarding Pasture Land Use With Payment

Issued by
The People's Government of Alukeerqin Banner on the
16 August, 1988

Grassland is an important resource of the state. It is the basic means of production for animal husbandry, and the material basis for the development of the animal economy, and it is also an important protective screen for the maintenance of ecological environment.

The following temporary regulations have been set up for pasture land use with payment according to "The Constitution of the People's Republic of China", "The National Rangeland Law of the People's Republic of China", and "the Rangeland Management Regulation of Inner-Mongolia Autonomous Region", together with the consideration of the actual conditions in the Banner. These regulations have been introduced in order to strengthen pasture management, pasture conservation and pasture construction; to alleviate the intense conflict between animals and pastures; to insure the continuous use of pasture resources; to make the grassland resource well-performed in terms of economic efficiency and ecological efficiency; to promote the stable development of animal husbandry in the banner.

A. Pasture Land Use with Payment
1. All grassland within the banner, including hill and mountain pastures are owned by the socialist public of the whole people or the socialist working people of the collectives.
2. The developed, utilised and constructed grassland is a natural resource with value. In order to protect and rationally use the grassland resources in our banner and to carry out the responsibility and obligation of pasture construction, the system of pasture land use with payment must be implemented.
3. The system of pasture land use with payment is carried out in two forms. In one form the payment of pasture land use is collected by animal numbers, in the other by the pasture land area in use.
4. The collection of the payment by animal numbers
   (a) The payment of pasture land use must be collected from all the animals which graze the pastures of our banner according to the regulation.
   (b) If the animal number remains within the tested rational stocking rate, a proper payment will be collected by animal numbers according to the regulation.
   (c) If the animal number breaches the tested rational stocking rate, the payment will be doubled for the over-grazed numbers.
5. The Collection of the Payment by the Pasture Land area in use
   (a) In a certain administrative area where the ownership of the pasture land and the right to use the pasture land are clear enough, the grassland will be categorised into several grades and animal farming households are permitted to lease it.
   (b) The individually contracted pastures may also be leased after fencing and constructing.
   (c) A clear lease must be worked out and a proper rent must be determined. The rights and obligations between the lessors and lessees as well as the articles of the agreement for both sides to follow must be made clear in the lease and all these must be notarised legally. The rental should be determined according to the leased pasture land area, grades and facilities attached as well as the term of the lease.
6. Every sumu/township government can choose one of two forms of payment collection mentioned above or can adopt the two forms according to its own local conditions.
7. The proper payment collection and management method should be ruled out over the collectively fenced pastures, the established pasture stations at different levels and the fenced pastures for animal improvement and disease prevention; if these pastures are already contracted out to individuals. The lessors (leasing units) should collect the pasture-land-use

payment from the lessees in the light of fenced area; the lessees are permitted to rent the contracted fenced pastures to animal farming households and collect rental from them according to this regulation.

8. The individually fenced pastures can be either continuously used by the individual investors or rented out, and the land owners should collect the pasture use payment by the fenced area (exclude those who construct the desertified, heavily degraded pastures in accordance with this regulation).

9. The pasture use payment from poor households recognised by sumu/township governments can be reduced or even free in a certain period of time.

10. Animal farming households should be encouraged to develop the pastures without water supply and construct the desertified pastures. To those who develop the pastures without water supply and the desertified pastures, the pasture use payment is free in a certain period of time.

11. When the system of pasture use with payment is carried out by every sumu/township government, the payment can be reduced over breeding animals and draught animals.

12. The system of pasture used with payment is going to be carried out in three stages in the banner and completed within a period of three years.

In 1988, every sumu/township in the pastoral areas can select two or three gacas to begin the trial;

In 1989, every sumu/township should strive to extend the trial up to 60% of gacas under it.

In 1990, the system will be carried out throughout the banner.

B. Payment Collection Method

1. The Payment over all species of animals needs to be collected by sheep equivalents.

2. Gacas, groups or physical villages are considered a unit. The pasture monitoring stations or animal improvement/veterinary stations at sumu/township level are responsible for the examination of pasture stocking rate. The payment is collected from animal farming households on the basis of animal stock at year-end. The unit is responsible for collecting the payment from animal farming households with the aid of pasture monitoring stations and animal improvement/veterinary stations at sumu/township levels. The money collected as separate savings is subject to the unified control by finance sectors at sumu/township levels on behalf of each unit.

3. The payment can be collected once a year by year end or twice a year, depending upon the local conditions.

4. The concerned sectors (organisations) at banner and sumu/township levels need to make a regular examination of pastures so that the annual stocking rate and year-end animal stock can be figured out in the light of actual pasture production. In this way, the animal population can be determined by pasture production and the goal of pasture-animal balance can be achieved.

C. The Criterion of Payment Collection

1. If the animal number is under a rational stocking rate, 0.1-0.5 yuan of the payment needs to be collected per sheep equivalent.

2. If the pasture is generally overgrazed, the payment which needs to be collected should not be less than 50% of the value of the annual pasture intake grazed by the overgrazed animals in principle. At present, 5-10 yuan of the payment can be collected per sheep equivalent.

3. If pastures are seriously degraded and desertified and heavily overgrazed, the payment will be doubled on the overgrazed animal numbers, and can be even increased up to the value of the overgrazed animals to alleviate the conflict between pastures and animals.

4. At present, sumu governments can work out the concrete amount of the payment in accordance with the regulation, considering local conditions.

5. The payment can be reduced or free for one year if there is a serious natural disaster. But the gacas/villages must get an approval from sumu/township government, and sumu/township must get an approval from the banner government.

6. If one sumu temporarily grazes its animals on the other sumu's pastures, the amount of payment can be collected upon negotiations of both sides in accordance with the regulation.

7. If other banners temporarily use our banner's pastures, the payment needs to be collected in accordance with the regulation.

D. Final Management and Utilisation
1. The payment of pasture land use collated by gacas/villages from the pasture land must be returned to the pasture land. The concerned sectors can take 5% of the fund as management fees and the rest will be saved in a special account by sumu governments as "a pasture development fund" for each gaca/village.
2. The procedure to use the fund is that gacas/villages should make an application to sumu governments. After getting the approval from sumu governments gacas can use the fund concentrating on pasture constructions.
3. The fund can only be used for pasture constructions, because it is a special fund for a special purpose. A strict system must be established for the fund collation, management, utilisation and examination and ratification.
4. The fund must be subject to a regular auditing system.
5. Each sumu/township government needs to make regular reports to the people's congress of the banner about the collection, management and utilisation of the fund.

E. Encouragement and Punishment
1. The implementation of the system of pasture land use with payment needs to be considered an important part of the economic and comprehensive responsibilities of all level governments. Encouragement and punishment must be formalised.
2. The people's governments and pasture management institutions should encourage spiritually and materially the persons and organisations who had made a great achievement in the process of promoting the system of pasture land use with payment.
3. If any animal farming households do not pay the pasture land use fees on time, 10% at least of the total fees must be collected by the fee collection units from them for the delay of payment.
4. If anyone commits any of the following behaviours which are against the regulation, he will be criticised and warned; if very serious, the people's governments or pasture management sectors at all different levels must punish the leaders and the persons who are directly responsible and confiscate his illegal income; if anyone commits a crime, the criminal responsibility must be found out by laws.
- Does not pay the pasture land use fees on time against the regulation.
- Does not honestly disclose the animal numbers in his flock.
- Unjustifiably resist or obstruct the working staff from pasture monitoring institutions and other sectors concerned to conduct their duties.
- To convert the collected fees to private use, to misappropriate or corrupt the collect fees.

5. The people's government of the banner will organise some working staff from the concerned sectors to form a leading group and to set up a working institution responsible for the implementation of the regulation. The people's governments at sumu level should also set up the corresponding working institutions responsible for the implementation of the system of pasture land use with payment.
6. The concerned sectors under direct control of the banner government need to formulate a detailed and concrete regulation based on the regulation. The people's governments at sumu level also need to work out the concrete methods to carry out the regulation while considering the actual local conditions. However, this needs to be put on file by the banner's government.
7. If there are any special cases or problems in carrying out the regulation, the banner government will coordinate and handle them.
8. This regulation will come into use on the date issued.

Appendix 11B

Certificate for Pasture Land Contract with Payment

Issued by the People's Government of Alukeerqin Banner
1990

Name of pasture land user: ....
                                                                                                       date   month   year

Land owners: .... Gaca Committee (seal)

                                                                                                                                                  date   month   year

Issued by: .... Sumu Government (seal)

                                                                                                                                                  date   month   year

Examination of Credential from a Lawyer

This certificate has been examined and conforms to "The Rangeland Law of the People's Republic of China" and stipulations concerned in "The Pasture Management Regulations of Inner Mongolian Autonomous Region". The certificate holder has legal right to use the land. The certificate has come into effect.

Observer: Legal consultant to the People's government of Alukeerqin Banner

Lawyer: ....
                                                                                                          date   month   year

Used Area of Pasture Land with Payment

Pasture Land Type I	mu
Pasture Land Type II	mu
Pasture Land Type III	mu
Pasture Land Type IV	mu
Total Annual Average Pasture Production	kg
Total Stocking Numbers in a Normal Year	sheep equivalents

Regulations on Pasture Land Utilisation with Payment

1. The Ownership of Pasture Land

All the pasture land within the banner (county) belongs to gacas (villages) collectives of working people except for the pasture land which has been recognised by the People's Government of the banner as State-owned and sumu (township) owned according to law.

The land entitlement for basic cutting land is given to households. The land entitlement for grazing land is given to duguilongs or animal flock production group or joint households, even to individual households if possible.

The enforcement of pasture land entitlement needs to be registered and forms filled in at different levels of governments and put into files. The household's certificates need to be clarified with the basic cutting land area, grazing land area, pasture production, permitted animal numbers and real animal numbers, which can be used as the basis to collect fees from land users.

For the convenience of public constructions of a collective, sumu (township) governments can reserve some flexible area of land as sumu (township) collective owned pasture land according to local situations and this land can be contracted out and used with payment. Also each gaca (village) collective should reserve some flexible area of land.

2. Pasture Land Conservation and Construction

Institutions and individuals, whoever uses the pasture land which belongs to the Banner, are responsible for pasture conservation, rational utilisation and constructions.

The combination of economic efficiency and ecological efficiency must be taken into account in pasture construction. Pasture desertisation and degradation must be under control in the light of local conditions. Efforts must be made to fence pastures, to plant pastures and to conduct aerial seeding, to build small pasture enclosures for family use with an irrigation system and to engage in crop farming for animal feeding. Anyone who constructs and uses pasture land is entitled to enjoy the benefits if he/she is permitted to conduct pasture construction by gaca and sumu government in the light of pasture construction plans. The additional increase in stocking numbers is allowed if pasture production is increased correspondingly due to pasture construction. The criterion of the additional increase in stocking numbers is usually restricted to 730kg of grass in dry matter or 1300kg of silage per sheep equivalent per year.

The legal rights and benefits are protected by law. Farmers who have contracted in collective pasture land and equipment will discharge the original contract. If a household has a fenced pasture area which is more than the permitted, the additional fee should be collected from the household for the over permitted fenced pasture area, or the over permitted area can be arranged by collectives.

3. Standard Ratio between Pasture Land and Stocking Numbers

According to the figures collected during pasture census survey in 1984, all the pasture land is classified into:

(I) mountain plain pastures, 130kg of grass in dry matter was produced per mu;
(II) hill pastures with dry, 61.5kg of grass in dry matter was produced per mu;
(III) desert pastures with sand dunes, 82.5kg of grass in dry matter was produced per mu;
(IV) river valley and low land pastures, 88.5kg of grass in dry matter was produced per mu.

The grass production figure collected in 1984 is regarded as a normal year production. If it is a good year, the pasture production per mu can be increased by 10% on the basis of normal year figure. If it is a bad year, the pasture production can be decreased by 10% on the basis of the figure in a normal year. If there is a serious disaster, the pasture production can be decreased by 20%. However, all this needs to be approved by pasture management sectors of the Banner.

It is assumed that grass intake per sheep equivalent is 730kg in dry matter per year. The conversion ratio of different animals into sheep equivalents is as follows: 1:7 for camels, 1:6

for horses, 1:5 for cattle and mules, 1:3 for donkeys. Two young animals under one year old are converted into one adult animal.

The examination of stock numbers for an individual household is conducted twice a year. The first one is conducted by the end of June together with animal population census. The stocking numbers and destocking numbers are written down in the forms as a guideline of destocking rate. The second one is done by the end of the year and the fees are collected at once.

4. Fee Collection Standard

The payment of pasture land utilisation can be increased or decreased depending upon real situations in a year. The fees are collected according to stocking numbers by the end of the previous year. The payments of pasture utilisation by stocking numbers are shown below: 0.20–0.30 yuan per sheep equivalent within permitted stocking numbers; 3.00–5.00 yuan per sheep equivalent for general overstocking; 6.00–8.00 yuan per sheep equivalent for serious overstocking and it will be doubled next year if overstocking continues. The payment standards for permitted stocking, general overstocking, serious overstocking are determined by sumu (township) government. 0.40–0.50 yuan is collected per goat due to their damage to pastures relative to sheep.

If a very serious natural disaster occurs, the payment can be decreased or free in that particular year. However, this needs to be approved by up-level governments (gaca from sumu, sumu from banner).

5. Standards for General Overstocking and Serious Overstocking in Different Types of Pasture Land Areas

Pasture type	Name of sumu (township)	General overstocking	Serious overstocking
Mountain plain pastures in the northern part	Bayanbaolige, Bayanwundu, Han sumu, Kundu, Saihantala	less than 20%	more than 20%
Hill dry pastures in the middle part	Bayanhua, eastern Shabutai, Xianfeng, Xinmin, Wulanhada, Qiaomaitala, Tianshan, Shuangsheng, Gangtai, Bayanbaote, Tianshankou, Pingandi, Baichengzi, Daolunbaixing	less than 15%	more than 15%
Desert pastures with sand dunes in the southern part	Daode, Chaidamu, Shaogen, Baolizhao, Huhegeri, Bayannuoer, Balaqirude, Zagashitai	less than 15%	more than 15%

6. Pasture Land Subcontract and the Payment Standard

Pasture land subcontract should be signed and notarised through negotiations under the principle of volunteer, mutual benefit and mutual development. In the subcontract, the pasture land area and the term to use should be clarified; stocking numbers and grazing period should be restricted; payment should be made according to either the land area used or stocking numbers.

Pasture land subcontract between sumus/townships can be made by sumu township governments through bilateral discussions. Sumu/township governments are responsible for collecting the payment and the matters concerned.

Pasture land subcontract between gacas/villages in a sumu can be made through bilateral discussions of gaca/village committees.

The basic cutting land which is contracted to those households who have no animals or a few number of animals is allowed to subcontract out their land to others.

Payment standard of pasture land subcontract is shown as follows: 2.00-4.00 yuan per sheep equivalent. If the payment is collected by land area, payment standard can be determined by bilateral discussions in the reference of payment by sheep equivalent. The payment standard can be higher than usual if the pasture land is constructed.

7. The Capital Utilisation and its Management

The capital can only be used with interest in the principle of special capital for the special purpose.
(1) the capital can only be used for pasture construction, nor for other purposes
(2) the capital must be used in the form of working capital
(3) in principle, the capital can be used by a gaca where the capital is collected. But sumu/township governments can use the capital collected from those gacas/villages which are going to be benefited for public pasture constructions
(4) institutions and individuals are entitled to apply to gaca and sumu governments for the capital on the basis of pasture construction plan and pasture construction project; along with the monitoring of sumu/township pasture management station and the approval of sumu/township governments.

8. Reward and Penalty

Spiritual encouragement and material reward can be given by the people's government or pasture management station to those who have made a great achievement in the process of enforcing "user pay" system.

The institutions and individuals who have made a significant achievement in the process of pasture protection, construction and rational utilisation should be highly praised and encouraged. Whoever treat the movement or semi-movement sand dunes and develop or contract new pasture land, he is entitled to use the land and given a loan and material on liberal terms. The payment for the pasture land use is not collected until the pasture construction investment is paid off.

Those who break the regulations regarding the pasture land use with payment should be punished according to how seriously they break the regulations, especially those who conceal their stocking numbers should be heavily punished by additional taxation, apart from collecting overgrazing fees from them.

9. Others

Sumu/township governments can work out the concrete regulations to enforce the "user pay" system of pasture land use according to local situations in order to promote animal improvement, to increase animal destocking rate, to move to different seasonal grazing land in time, to improve management and to speed up the development of commercial animal husbandry.

Table 1 Stocking Numbers and Stocking Rate by the end of June

	Year				
	1991	1992	1993	1994	1995
Cattle:					
sheep equivalents					
young cattle					
adult cattle					
Horses:					
sheep equivalents					
young horses					
adult horses					
Donkeys:					
sheep equivalents					
young donkeys					
adult donkeys					
Mules:					
sheep equivalents					
young mules					
adult mules					
Sheep:					
sheep equivalents					
young sheep					
adult sheep					
Goats:					
sheep equivalents					
young goats					
adult goats					
Annual permitted stocking rate:					
Destocked sheep equivalents:					
Permitted increase in sheep equivalents:					

Table 2 Changes in Pasture Land Productivity and Stocking Rate

	Year				
	1991	1992	1993	1994	1995
Total pasture production					
Rational stocking rate					
Annual rational stocking rate:					
normal annual pasture production					
weather change:					
increased by %					
decreased by %					
annual pasture production					
annual rational stocking rate					
Permitted increase in stocking rate after pasture constructions:					
fenced pastures					
area					
production					
artificial pastures					
area					
production					
silage					
area					
production					
crop planted					
area					
production					
Increased pasture production					
Permitted increase in stocking rate:					

Table 3 Total Destocked Animals and the Animals Bought in

	Year				
	1991	1992	1993	1994	1995
Destocked animals:					
Destocked sheep equivalent more than the permitted quota					
Destocked animals (sheep equivalents) less than permitted quota					
Total destocked animals:					
Animals sold:					
cattle					
horses					
donkeys					
sheep					
Animals killed:					
cattle					
horses					
donkeys					
sheep					
Animals died:					
cattle					
horses					
donkeys					
sheep					
Total:					
sheep equivalents					
animal numbers					
Total animals bought in:					
Cattle:					
adult					
young					
Sheep/Goats:					
adult					
young					
Horses:					
adult					
young					
Mules:					
adult					
young					
Donkeys:					
adult					
young					
Total					
adult					
young					

Table 4 Year End Stocking Numbers

Year end stocking numbers	Year				
	1991	1992	1993	1994	1995
Cattle:					
adult					
young					
sheep equivalents					
Horses:					
adult					
young					
sheep equivalents					
Donkeys:					
adult					
young					
sheep equivalents					
Mules:					
adult					
young					
sheep equivalents					
Sheep:					
adult					
young					
sheep equivalents					
Goats:					
adult					
young					
sheep equivalents					
Total:					
stocking numbers					
sheep equivalents					

Table 5 Year End Stocking Rate and the Total Payment

	Year				
	1991	1992	1993	1994	1995
Total pasture production					
Rational stocking rate					
Year end sheep equivalents					
Of the year end stock:					
Overcarried sheep equivalents					
Permitted increase in sheep equivalents					
Payment needs to be collected:					
Normally grazed					
Generally overgrazed					
Heavily overgrazed					
Concessions:					
Disasters					
sheep equivalents					
money paid					
Others					
sheep equivalents					
money paid					
Total payment:					
Total					
sheep equivalents					
total payment					
Additional payment for goats					
goat numbers					
payment standard					
money paid					

Chapter Twelve

Wushen County (IMAR)

Wushen County is located in what was once the famous Eerduosi Grasslands in Yikezhao Prefecture of the Inner Mongolia Autonomous Region (IMAR). These days almost half of Wushen is covered by large sand dunes which dominate the landscape. Yikezhao Prefecture lies south of the Yellow River and south west of Huhehot, the capital of the IMAR (Map 12.1). Wushen County has a long eastern border with Shaanxi Province.

While about one-third of the total population of Wushen are Mongolian, around 90% of the pastoral households in the county belong to this nationality. Sheep raising is the dominant pastoral activity. Wushen produces more wool than any

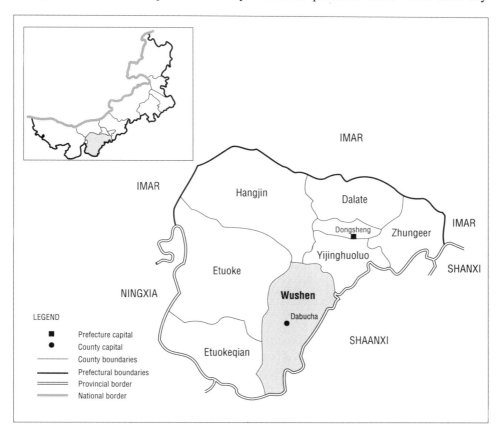

Map 12.1 Wushen County in Yikezhao Prefecture of IMAR

other county in the IMAR and it is well known as the "home" of the Erdos Fine Wool breed of sheep. As Longworth and Williamson (1993) point out, this provincially recognised breed has been developed by mass selection techniques and is not based on a stud or studs.

12.1 Resource Endowment

Wushen is regarded by all levels of government as a pastoral county. The Central government also recognises Wushen as a minority poverty county. It is a large remote desert county just north of the Great Wall.

12.1.1 General Features

The history and geographical features of Wushen County (and the surrounding areas) contrast starkly with all the other counties surveyed.

Originally, as already mentioned, Wushen County was part of the famous Eerduosi Grasslands on which nomadic Mongolian tribes had lived for millennia. In fact a spectacular pseudo-Mongolian style memorial to Genghis Khan has been constructed by the Central government just north of Wushen in Yijinghuoluo County (Map 12.1). This major tourist attraction is about 330km from Huhehot by sealed road. Genghis Khan (1167–1227) died in what is now eastern Gansu Province several hundred kilometres to the south west of the memorial site.

After the first Opium War (1839–1842), the European Powers imposed heavy reparations on the Qing Dynasty. The first Opium War occurred because the Qing Government placed trade restrictions on the illegal importation of opium by Western traders (mainly British). Under the Treaty of Nanking (1842), the Chinese government was required: to pay an indemnity to cover the cost of the war; to cede Hong Kong to the British; to open five ports for British trade; and to allow British citizens in China to be tried in British courts. The Treaty of Nanking was the first of the so-called unequal treaties forced on China by the Western powers. One approach adopted by the Qing Government to enlarge the tax base so that they could meet the reparation obligations was to increase the number of Chinese farmers by settling new areas.

The Eerduosi Grasslands were one such area settled. The Han farmers sent to the Grasslands farmed the best pasture land and cut down all the trees for fuel. To the north west of Wushen (in the centre of Yikezhao Prefecture), there was a small desert known as the Mu Us Desert (or "bad water" desert). The light soils and the dry climate were unsuited to farming and the Mu Us Desert grew bigger and bigger until it joined the Kuboui Desert.

Before this first wave of Han settlement, the Eerduosi Grasslands were middle-grade natural grasslands (i.e. typically the native grass species grew to 40 or 50cm in good seasons). There were patches of forest and the better areas had dark brown calcium-rich clay soils. Over the last 140 years the Grasslands have almost all disappeared, the forests no longer exist, and the soils have deteriorated to sandy loams at best but mostly they are now sandhills and desert.

After Liberation in 1949, another wave of new settlers were sent to Wushen and the surrounding areas. Of course, the resulting increase in the population put even more pressure on the fragile ecosystem.

Since the mid-1980s Wushen County and the surrounding pastoral counties in the old Eerduosi Grassland area have greatly improved. Both households and the local governments have planted huge numbers of trees (willow and poplar species) and shrubs to stabilise the sandhills.

Remarkably, many of the sandhills contain moist sand. Even on the highest plateau in Wushen (just north of the county capital), the sandhills contain moist sand within 30cm of the surface. This explains why the "clothes-line poles" which are stuck into the sand dunes everywhere in Wushen, suddenly sprout leaves and grow into poplar or willow trees.

In the flat areas between the dunes, the water-table is often close to the surface, sometimes causing salting problems, but generally edible shrubs and grasses seem to grow well provided they are protected from grazing animals. To control their livestock, the households in Wushen have erected an amazing amount of fencing.

The sheep-herding families live on their land as farmers do in Australia. Their homes are isolated and scattered across the countryside. In all other parts of China surveyed, the sheep-raising families lived in small villages.

Rainfall in Wushen is said to average 350mm per annum, with most of the rain falling in July/August and September. Temperatures range from 36°C in summer to -32°C in winter.

There are no mountains but the countryside is scoured with dry creeks and rivers which become substantial streams for short periods, at least in some years. Water erosion is extensive in some parts of Wushen, presumably because the intensity of the rain (when it comes) is too much for the unprotected light sandy soils.

12.1.2 Land Use

Very little land in Wushen County is suitable for cropping and only just over half the total area is described as rangeland. As mentioned earlier, almost half the county is covered by sand dunes.

- Total land area   17,460,000mu   (or 11,640 km²)
- Total arable land   166,000mu
- Total rangeland   9,090,000mu
  - usable pasture   7,840,000mu
  - artificial pasture   40,000mu
  - semi-artificial pasture   1,200,000mu

Most of the pasture can be described as desert pasture.

As outlined above, land degradation in Wushen has a long history. Virtually all of the land in the county has been degraded to some extent although there are large areas where substantial progress has been made towards rehabilitation.

12.1.3 Population

Of the five counties surveyed in the IMAR, Wushen is easily the least densely populated.

- Total population (1990)   90,000
  - rural   75,000
  [pastoral   30,000]
  - urban   15,000

- Ethnic composition (1990)
  - Mongolian — 32% of total population
    [90% of pastoral population]
  - Han and others — 68% of total population
- Population density (1990) 7.73 people/km^2

12.1.4 Administrative Structure

The sparse population is organised into the county capital and 13 administrative townships of which only two are classified as being agricultural.
- one town (the county capital)
- six pastoral townships (sumus)
- five semi-pastoral townships
- two agricultural townships

Many of the pastoral households live on their land rather than in natural villages or hamlets.

12.1.5 Livestock

Sheep are the dominant grazing animals in Wushen but there are also significant numbers of goats and large herbivores.
- Livestock numbers (end of 1990)
  - Sheep — 480,000
  - Goats — 63,000
  - Large livestock — 33,000
    (horses, mules, donkeys and cattle)

12.2 Improvements to Production Conditions in the Pastoral Systems

12.2.1 Improved Pastures

- *Area of improved pastures*

In 1991, Wushen County had 40,000mu of artificial pasture and 1,200,000mu of semi-artificial pasture. Widescale fencing has been undertaken since 1984 (when the land contracts were introduced); however, no actual figures were available.

- *Specific pasture improvement programs*

When the pasture land was contracted out in 1984, the contract included a requirement that the household take steps to treat their land to arrest the desertification process. After five years, the State could take back the land and reallocate it if the household did not treat its land as required in the contract.

To assist the households, the county government provides "sticks" to plant for growing trees. These are free of charge but the farmer must take care of his "sticks" to ensure that at least 70% survive to become trees.

The Central government is operating a large project to treat the desert in Wushen County. This project, which involves an ¥8 million loan, follows the success of the county in planting trees and artificial pastures. County officials would like to have an internationally-sponsored desert reclamation project established in the county.

There was very little fencing of pasture land before the land contracts were introduced in 1984. Since then, some farmers have invested more than ¥10,000 in their land (including fencing). By 1991, the average amount invested was ¥1,500 per household.

Herdsmen households understand they must treat their pasture land carefully. Most households now try to practise rotational grazing – this is the major reason for the substantial investment in fencing.

12.2.2 Improved Sheep and Wool Production

- *Use of artificial insemination*

Of the 280,000 adult ewes in the county, about 180,000 (or 64%) are mated using artificial insemination. The rest, and those of the artificially inseminated ewes which do not conceive after two cycles, are mated naturally.

AI started in 1953. By 1966 (i.e. before the Cultural Revolution), the percentage of ewes in the county mated by AI had reached almost 100%. This proportion remained more or less at this high level until the late 1970s. With the introduction of the first stage of the household production responsibility system in 1980 (i.e. the contracting of the sheep to the households), the percentage mated by AI dropped from around 90% in 1978 to about 50% in 1985. Since 1986, the percentage mated by AI has increased and in 1992, as mentioned above, it was about 64%. Of the ewes which have met the Erdos Fine Wool Breed Standard, about 90% are mated by AI.

There are 640 artificial insemination stations in the county.

The warm semen AI process is standard throughout China. It involves two inseminations (one in the morning and one in the evening) for each of two cycles if necessary. Normally, about 75% are successfully mated after the first cycle and up to 95% after two cycles. This is the system used in Wushen, and the success rate here is over 90%.

The charge for two AI treatments varies from ¥0.80 to ¥1.5 depending on the quality of the rams.

- *Sheep breeds*

Of the 480,000 sheep in Wushen County, about half are classified as having reached the Erdos Fine Wool Sheep Standard. Another 30% are classified as "improved fine wool sheep" because they have not formally been assessed as having reached the Erdos Fine Wool Breed Standard. However, their wool is almost homogeneous and for wool grading purposes, most of it is classified as "fine wool", not as "improved fine wool".

The remaining 20% of the Wushen sheep flock are almost of sufficient quality to be called "improved fine wool" sheep.

- *Animal productivity*

Greasy wool cut per head and clean yield have increased in recent years with a consequent big increase in clean wool produced per head. While the changes in the general flocks would be somewhat less dramatic, the improvement in wool cut for the top ewes in the 300 nucleus flocks is shown in Table 12.1.

There has been a marked improvement in the clean wool yield, especially since 1987 when Australian rams were introduced. Between 1985 and 1991, the clean yield of rams' fleeces increased from 36% to 45% with most of this improvement occurring in the last two years of that period owing to the impact of Australian merino blood. In the case of the ewes referred to in Table 12.1, clean yields jumped from 34% to 48% in the 1985 to 1991 period. Again most of the change occurred in the last two years of that period owing to the infusion of Australian merino blood. The data in Table 12.1 and comments about clean yields refer only to the nucleus flocks kept by the 300 specialist ram breeding households (see below).

Table 12.1 Fleece Weights for the Best Erdos Fine Wool Sheep, 1985 to 1991

Year	Special and Grade I Ewes in the nucleus flocks	Grade I Rams at the County Ram Stations
	kg/head (greasy)	kg/head (greasy)
1985	3.5	10.05
1986	3.8	Not available but there has been no significant change during this period.
1987	4.0	
1988	4.1	
1989	4.3	
1990	n.a.	
1991	5.0	

- *Specific sheep breeding programs*

Wushen County began a sheep improvement program in 1957 by importing fine wool rams from Xinjiang (Longworth and Williamson, 1993). In 1964, about 300 rams were imported from Russia and East Germany.

By 1974, the general quality of the sheep had improved substantially and the decision was taken to establish a local breed. In 1985, the IMAR government accepted the new breed and it was named the Erdos Fine Wool Sheep. Since 1987, there has been a major infusion of Australian merino blood so that in 1992 all the ewes in the nucleus flock have at least 50% Australian merino blood.

There are no sheep studs or State ram breeding stations in Wushen. That is, the Erdos Fine Wool Breed is not based on a stud or studs. Instead, the best 20,000 Erdos Fine Wool ewes are owned by about 300 selected households. These 20,000 ewes constitute the "Nucleus Flock". Each of the 300 specialist households runs its nucleus flock ewes (usually about 60 to 70 per household) with its commercial sheep. The average total flock size in Wushen for both the specialist households and general sheep-raising households is about 150 sheep.

The 300 households keeping the nucleus flock ewes are called ram breeding specialist households. Each year, the two county-level "ram care" stations select the best available rams to mate with the 20,000 nucleus flock ewes. The nucleus flock consists of about 50% Special and Grade I ewes, 30% Grade II and III and 20% of ewes which are culled each year.

At weaning time (i.e. in the month of October), the young ram lambs produced by the nuclear flock ewes are assessed and about 450 ram lambs are selected and taken to the two county-level ram care stations.

The ram lambs are selected on two sets of criteria (not according to their pedigrees which are not recorded).
(i) The ram lamb's parents must satisfy the following:
- the sire must be one of the Grade I rams from the county-level ram care stations (these are the only rams mated by AI, or by natural methods when AI fails, to the nuclear flock ewes)
- the ewe must have qualified for the nuclear flock and also must meet the following criteria:
    - staple length > 9.5cm
    - wool must have white colour (grease)
    - crimp well defined
    - count 64s to 66s
    - greasy fleece weight > 5.5kg
    - body weight after shearing > 42kg.

(ii) The ram lamb itself must meet the following criteria:
- staple length > 6cm
- count 64s to 70s
- body weight at weaning >35kg.

The ram lambs selected are purchased from the specialist ram breeding households at prices determined by market conditions for wethers, wool, etc. There is a quota of 10% which each ram-raising specialist household is supposed to meet. However, the selection team does not select equal numbers of ram lambs from each household. With 20,000 nucleus flock ewes, 90% lambing and 95% survival rates, there are usually about 8,550 ram lambs from which to select the best 450. That is, only between 5 and 6% of the available rams are selected at weaning as potential replacement sires.

Of the 450 rams selected at weaning and raised until one year old at the two county stations, about 50% are assessed as Grade I rams and retained for mating with the nucleus flock ewes. About 20% are culled. The remaining 30% are classed as Grade II or III rams and most of these are sold to township (sumu) level ram care stations in Wushen or to stations in neighbouring counties. There are 16 township level ram care stations in Wushen, 10 in Yijinghuoluo Banner and six in each of Etuoke and Etuokeqian Banners (see Map 12.1). All of these stations keep only Erdos Fine Wool rams.

The selection pressure to obtain the Grade I replacement rams is high (roughly 200 are selected from 20,000 ewes with 90% lambing and 95% survival rates – that is, less than 1% of male offspring are used for breeding in the nuclear flock). However, since the primary selection pressure is applied at weaning, this process may not achieve the rapid rates of genetic improvement which should be possible.

Progeny testing is not used because of the cost of having the measurements determined and the lack of (and cost of) skilled labour to keep the necessary records.

The two county-level ram care stations raise the weaner ram lambs to one year of age, then maintain the 200 selected Grade I replacement rams until they complete their first mating season at almost two years of age. These still-young rams are then sold on to the 16 township (sumu) ram care stations for use as AI sires for the general flocks.

All rams used for breeding in Wushen should come from the 18 ram care stations (two county-level and 16 township-level). These stations receive all their ram replacements from the nuclear flock.

The sheep-breeding system in Wushen is a cheap and effective means of obtaining good rams for breeding purposes. The officials of the Wushen Animal Husbandry Bureau consider that it is too expensive to operate a State stud farm and that the rate of genetic improvement could be faster with the above system than on a stud with many fewer ewes from which to select replacement rams. The success of the system depends on the effectiveness of the selection procedure.

12.3 Socio-Economic Development

12.3.1 Income Growth

As with the rest of China, nominal incomes increased sharply in Wushen during the 1980s. Improvements in both the quantity and the quality of the fine wool clip have combined with higher prices to substantially raise nominal pastoral incomes. Many of the sheep-raising families have built (or are building) new and bigger houses on their land. As the average income data suggests, it would seem that the Mongolian pastoral households in Wushen are considerably better-off than comparable households in the counties in the Eastern Grasslands of Chifeng City Prefecture.

Nominal net per capita income (1991)
- pastoral households   ¥688
- agricultural households   ¥412

12.3.2 Commodity Prices

Wushen and the three surrounding counties in Yikezhao Prefecture in which Erdos Fine Wool Sheep are raised have developed a reputation in China for the good quality of their wool. The Wushen SMC officials seemed to be most enlightened about the advantages of strengthening their "brand name".

Even though the wool market was supposed to be "open" in the IMAR in 1992, the Wushen SMC was expecting to buy wool from households strictly according to the prices set by the IMAR Price Bureau. These prices were as set out in Table 12.2. It is interesting that the premiums and discounts do not correspond to the quality differentials in the National Wool Grading Standard. (See Section 4.5.2.) Furthermore, the premiums for the higher grades were expected to increase in 1992 relative to 1991.

In early June 1992, before shearing and purchasing had really begun, contracts had been signed with Nanjing and Lanzhou wool topmaking mills, and with the Animal By-Products Company in Huhehot (provincial level), Animal By-Products Company in Yikezhao Prefecture (prefectural level), and the Animal By-Products Company in Wuhai City (county/city level). The total expected Wushen wool production was covered by these contracts.

As an example, the Manager of the Wushen County SMC quoted the following prices per kg (greasy) in the contract signed with Nanjing Topmaking Mill:

    Special Grade Fine Wool   ¥11.18
    Fine wool   –   Grade I   ¥10.2
                      –   Grade II   ¥8.4

Table 12.2 Average Prices Paid for Raw Wool (by grades) in Wushen County, 1991 and 1992

Grade		Grade	1991 (actual)		1992 (expected)	
			(¥/kg greasy)	(%)	(¥/kg greasy)	(%)
Fine wool	–	Special Grade	8.72	122	8.8	133
	–	Grade I	8.02	112	8.2	124
	–	Grade II	7.22	101	6.6	100
Improved fine wool	–	Grade I	7.16	100	6.6	100
	–	Grade II	(none produced in Wushen County)			
Belly/head/leg wool			2.40	–	2.0	–

The Manager said that the wool sold under contract to other places had been sold at lower prices, presumably because the "best" wool was going to Nanjing and the freight costs would be a little higher. The SMC officials emphasised that they must "police" the buying from farmers in 1992 to make sure they satisfied the contracts as regards quality. They would probably only buy the best quality wool from the neighbouring counties so that they could maintain the Wushen County "image" for quality.

The Wushen County SMC buys Erdos fine wool from the other three pastoral counties in Yikezhao Prefecture and markets it as if it came from Wushen County. This is because Wushen County has developed a reputation for wool quality. Presumably this is why the management of the Wushen County SMC said that they had already sold all this year's production under contracts, although earlier they had indicated they would sell some wool through the new auction system in 1992. This also helps to explain why the Wushen County SMC has built a scouring plant with a maximum capacity more than double the quantity of greasy wool grown in Wushen County (see below).

12.3.3 Timing of Introduction of Household Production Responsibility System

Livestock were distributed to the households under contractual arrangements in 1980 in Wushen County. The land was contracted out in 1984 and the livestock were sold to the households at the same time.

The land contracts stipulated that the households must take positive action to preserve and improve the pastures. On average, each farmer was also required to plant 3mu of trees.

12.3.4 Level of State Assistance

Wushen County is one of the pastoral counties participating in the Poverty Alleviation Scheme financed by the Central government. [See the discussion of these schemes in the Aohan and Alukeerqin County case studies (Chapters 10 and 11).]

12.3.5 Development of Pastoral Product Processing

A scouring plant with a 5,000 tonne of greasy wool per year maximum capacity has been built and equipped with six brand new scours. There is virtually no industry in the county and the scouring plant is an obvious start to industrialisation. On the other hand, unless the wool is properly sorted before scouring and the wool washing process is carefully managed, irreversible damage may be done to the quality image of Erdos wool (Brown and Longworth, 1992b).

The annual production of wool in Wushen County is only about 2,250 tonne. Thus the scouring plant has the capacity to process twice the amount of wool grown in the county. From 1993, it is expected that the County government will require all wool grown in the county to be scoured at the new plant. The SMC owns the scouring plant. They plan to continue buying-in wool from neighbouring counties. There are a lot of Erdos Fine Wool Sheep in the adjoining three counties (Etuokeqian, Etuoke, and Yijinghuoluo). Etuokeqian and Etuoke counties/banners (along with Wushen) are designated pastoral counties by the Central government while Yijinghuoluo Banner is called a semi-pastoral county. All four of these counties are included in the 266 counties defined as the pastoral region by the Central government (see Table 5.2 in Chapter 5 for more details). The farmers in these neighbouring areas want to obtain the good prices paid for Wushen SMC wool. Hence, as discussed earlier, it seems that much of this wool will be marketed by the Wushen SMC.

The managers of the SMC seemed to be well aware of the need to regrade the wool properly according to the industrial grading system *before* scouring. They seemed less concerned about the difficulties of carrying out the scouring process properly. They said they did not have the grading technicians yet. They planned to send suitable people away for training – perhaps to the new Wuxi Training Centre for the Modern Management of Wool in Jiangsu Province. They also planned to invite grading technicians to come to the county to train local graders. These plans to develop grading skills seemed very vague and indefinite considering the six brand new scours were already in place and were being tested. The need for skilled graders may be recognised but has not been given the necessary priority.

The SMC funded the building of the scour from three sources: Poverty Alleviation Funds; loans from the Textile Corporation in Dongsheng (the prefectural capital); and loans from the Agricultural Bank of China.

12.3.6 Growth in Output of Major Pastoral Commodities

Total greasy wool production in 1991 was 2,250 tonne. The proportion which is of Special Grade and Grade I has increased steadily. The approximate percentage distribution by grade of the Wushen clip in 1992 was:

Special Grade			30%
Fine wool	–	Grade I	40%
	–	Grade II	20%
The "rest"	(including improved fine wool, coloured wool, etc.)		<10%

Wushen was the only county surveyed where both officials and farmers seemed genuinely keen to continue upgrading their sheep by using good quality fine wool rams. Consequently, there should be a steady increase in the quantity of fine wool grown in Wushen and the surrounding counties during the 1990s. This increased output will result from improving the wool cut per head by upgrading rather than from an increase in sheep numbers.

Chapter Thirteen

Sunan County (Gansu)

Sunan is a Yugur Minority Autonomous County in Zhangye Prefecture of Gansu Province. As can be seen from Map 13.1, this county consists of four separate geographic areas all of which have a significant population of the Yugur Nationality. People of this nationality make up about one-quarter of the county population but they are rare outside Sunan County. Tibetans represent another quarter of the population. There are some Mongolian, Hui and Tu people as well so that around 53% of the total population of Sunan belong to minority nationalities.

With a population density of less than 1.5 people/km², Sunan is extremely sparsely settled. The raising of sheep, and to a lesser extent goats, is the dominant rural activity. Over 90% of the sheep are Gansu Alpine Fine Wools.

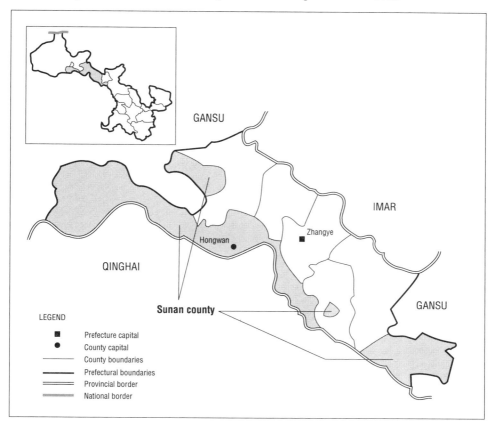

Map 13.1 Sunan County in Zhangye Prefecture of Gansu Province

13.1 Resource Endowment

Sunan is officially recognised at Central and Provincial government level as being a pastoral county. Three of the four areas which make up Sunan are in remote alpine locations. The fourth area is semi-desert on the northern side of the Hexi Corridor.

The biggest part of the county, which includes Hongwan the county capital, and a smaller area, in which Huang Cheng State Farm and Sheep Stud is located, were surveyed. Both of these parts of Sunan are remote alpine areas on the border between Gansu and Qinghai in the Qilian Mountains.

13.1.1 General Features

Most of Sunan is extremely mountainous. Altitudes range between 1,400 and 5,560 metres. The altitude of Hongwan is 2,300 metres. The relatively high altitude of most of the county results in fairly low temperatures in summer (27°C in July) and very low temperatures in winter (-18°C to -23°C in January). The predominant soil type in Sunan County is a dark brown soil with significant quantities of calcium. Soils are mainly deficient in nitrogen and phosphorus.

Precipitation varies a great deal from year to year and from place to place depending upon altitude etc. The annual average rainfall recorded by the Meteorological Station at the county capital for the years 1960 to 1989 is shown in Table 13.1. The lowest recording during this period was 175mm in 1968 while the highest was 411mm in 1979. Most of the rain falls in June–July–August although good falls can occur in May and September.

Table 13.1 Annual Rainfall at Hongwan the Capital of Sunan County, 1960 to 1989

Year	Annual rainfall* (mm)	Year	Annual rainfall* (mm)
1960	225.3	1975	253.3
1961	198.9	1976	239.4
1962	219.5	1977	233.4
1963	249.9	1978	244.1
1964	319.3	1979	410.7
1965	204.3	1980	245.7
1966	242.2	1981	302.8
1967	242.0	1982	250.5
1968	175.0	1983	234.3
1969	237.3	1984	230.8
1970	255.1	1985	192.0
1971	258.1	1986	255.7
1972	273.9	1987	279.3
1973	248.6	1988	325.9
1974	248.2	1989	258.8

*Recorded at the Sunan Meteorological Station
Source: Sunan County Animal Husbandry Bureau

13.1.2 Land Use

The mountainous terrain in Sunan means that there is little land available for cropping or for improved pastures. Almost all rural activities are based on the rangelands.

- Total land area   35,800,000mu   (or 23,887km^2)
- Total rangeland   25,640,000mu
  - usable pasture   21,030,000mu
  - "cutting" land   38,000mu
- Arable land   39,000mu
- Forestry land   1,293,000mu
- Total degraded land (1990)   10,682,000mu
  - heavily degraded   4,037,000mu
  - lightly degraded   6,645,000mu

Government officials in Sunan County recognised that degradation of these rangelands was a major problem. They attributed the pasture degradation problem to two factors: too many animals, and the recent run of poor seasonal conditions (drought). They commented that the pasture can only grow for a short period each year owing to the climate and altitude of the county. Under these circumstances, excessive numbers of animals put great pressure on the natural ecosystem.

According to the Pasture Management Station under the Sunan County AHB, the pastures most susceptible to degradation in the county are the alpine cold desert and semi-desert pastures. These pasture types are described in Section 6.2.4 of Chapter 6.

13.1.3 Population

Sunan is a sparsely settled county but the high altitudes and mountainous terrain make living conditions extremely harsh.

- Total urban and rural (1989)
  - total   35,259   (100%)
  - rural   25,533   (72.42%)
  - urban   9,726   (27.58%)
- Ethnic composition (1989)
  - Han   16,533   (47.2%)
  - Mongolian   291   (0.82%)
  - Tibetan   8,503   (24.1%)
  - Yugur   8,940   (25.3%)
  - Huizu   520   (1.4%)
  - Tu   472   (1.3%)
- Population density (1989)   1.47 people/km^2

13.1.4 Administrative Structure

Sunan County includes 1 town, 6 districts, 23 townships and 96 villages. The number of rural households is 4,327 which is 58 per cent of all households in the county.

13.1.5 Livestock Population

The most recent comprehensive inventory of livestock available was for the end of 1977:

–	cattle and yaks	43,829
–	sheep	436,750
–	goats	115,367
–	donkeys	4,073
–	mules	961
–	horses	5,957
–	camels	2,137

At the time of the survey in 1990, the Sunan County officials claimed that the only census data they had for livestock related to the end of 1977. There are many possible reasons why they did not wish to reveal more recent data. However, perhaps the most likely explanation is that herbivorous livestock numbers have increased markedly since 1977 with disastrous consequences for the rangelands. For example, as indicated in Section 13.2.4 below, the number of mature ewes in Sunan in 1989 was 466,700. This would suggest that total sheep numbers at the end of that year must have been over 750,000 or approaching double the 1977 figure.

13.2 Improvements to Production Conditions in the Pastoral Systems

13.2.1 Improved Pastures

- Total area of improved pastures by type (1990)

– improved pasture	571,000mu
– artificial pasture	35,000mu
– semi-artificial pasture	536,000mu
– fenced pasture	570,000mu

Prior to 1983 there was said to be no fenced or improved pastures in the county.

13.2.2 Pasture Improvement Programs

Since the breakup of the communes, pasture productivity has been increased through the planting of improved pastures and the fencing of pasture land. This has largely been undertaken by private households.

There has been no large-scale aerial seeding of pastoral land in this county by government authorities. One possible reason for this could be the very mountainous terrain of the county which fragments potential pasture improvement sites, as well as making aircraft travel hazardous.

There are no government subsidies available to farmers in Sunan County for planting improved pastures. Pasture seed for planting improved pastures can be obtained from the Pasture Management Station of the County AHB, and the cost is said to average ¥0.4 per kg. Total cost of planting improved pasture is ¥18 per mu, this figure being an average for both artificial and semi-artificial pastures. Since the overwhelming majority of the improved pasture is likely to be semi-artificial pasture, this cost would closely approximate that of semi-artificial pastures.

The "public accumulation fund" which is collected by the collectives at the village level is used mainly for pasture restoration in Sunan County.

Winter pastures cannot be improved as they are subjected to extreme grazing pressure and usually there is no irrigation water available nearby.

13.2.3 Pasture Management Programs

The County government policy to overcome the problem of pasture degradation has four basic strategies:
- control the growth of animal numbers;
- improve the quality of animals;
- increase the productivity per animal; and
- fence the pastures to enable the use of rotational grazing systems, etc.

The county government has been thinking about developing a compulsory stocking rate policy similar to that in use in the Eastern Grassland counties in IMAR. The merits of limiting stocking rates to levels which take into account the potential productivity of the pastures are well understood. However, the officials also pointed out how difficult it was to determine the carrying capacity of each particular area of pasture in Sunan because the weather is so variable and because each area had its own peculiarities. The major aim of the county government at present is simply to stabilise the animal numbers at their current levels.

The county government has also been considering a "user pays" system in connection with the management of pastures. According to county officials, the implementation of such a user pays system would involve working out the productivity of each pasture and estimating its carrying capacity before deciding upon an appropriate fee. As just mentioned, this would be very difficult in Sunan.

The possibility of levying the fee on a per sheep equivalent basis hence avoiding the need to estimate pasture carrying capabilities was raised. As discussed in relation to the counties surveyed in the IMAR, there are theoretical as well as practical advantages in setting the pasture management fee on a per sheep equivalent basis (see Section 5.5.2). The county officials being interviewed did not seem to be interested in this approach to implementing a user pays system.

Another major problem with the introduction of a pasture use fee in Sunan County is that this arrangement was not in the original pasture contracts. Hence, if the government implemented a pasture use fee, farmers would see this as a major change in their pasture contracts. The government does not wish to change the contracts and hence increase the tenure uncertainty in the minds of the farmers.

The introduction of "pasture utilisation certificates" by the County government was aimed at defining more precisely the responsibilities the farmers have towards maintaining and, if possible, increasing the productivity of the grassland for which they have contracted. By mid-1990, approximately 70% of rural households in the county had been issued with pasture utilisation certificates [see Appendix 13B]. These certificates are much less sophisticated than the corresponding documents in use in Balinyou and Alukeerqin counties. [See the Appendices to Chapters 8 and 11.]

13.2.4 Improved Sheep and Wool Production

In 1989, almost all of the adult fine and improved fine wool sheep population in

Sunan were assessed for quality. The results of this assessment were that the county flock was classified as follows: 1st class (73.9%); 2nd class 16.9%; 3rd class (6.83%); and 4th class (2.3%).

The sheep classified as 1st class were said to have the following characteristics:
- height > 80cm
- large body
- wool length > 9cm
- fibre density between 5,000-7,000 fibres per cm^2
- fibre thickness of between 60 and 64 spinning count
- strong constitution, a deep chest, and large rump.

Sheep classified as 1st class are regarded as being pure-bred Gansu Alpine Fine Wool sheep. The sheep classified as 2nd class were said to have the above characteristics of 1st class sheep, except for the appearance of coarse wool in an area between the knee and 12.5cm above the knee. Sheep classified as 3rd class are crossbred sheep with some fine wool characteristics.

Sheep classified as 4th class include all the coloured wool sheep and crossbreds not yet showing enough fine wool characteristics to be classified as 3rd class sheep.

On the basis of the above classification, roughly three-quarters of the sheep in Sunan are pure-bred Gansu Alpine Fine Wools while another 17% are "almost" of this quality. The remaining 8% are improved fine wools.

In addition to these fine and improved fine wool sheep, there are also a significant number of semi-fine wool sheep in Sunan County. The semi-fine wools are raised in the non-alpine and most northerly segment of the county (see Map 13.1).

- *Sheep production systems*

The sheep production system involves moving the sheep up to 100km to summer pastures in the high mountains (most flocks travel only 15 to 20km). Both on the way to and from the summer pastures (where they are shepherded for only a couple of months at the most), the sheep graze the so-called spring/autumn pastures in the hills and valleys of the middle mountains. Winter grazing is in the lower valleys and hills (around 1,500 to 2,000m above sea level) where the permanent settlements are located.

- *Use of artificial insemination*

The proportion of ewes artificially inseminated in Sunan is less than 50%. The precise figures for 1988 and 1989 were:

	1988	1989
Total number of mature ewes	467,300	466,700
Total number of ewes mated by AI	191,600	193,400
Percentage mated with AI	41%	41.4%

- *Animal productivity*

On average, sheep in Sunan County produce between 3 and 3.5kg of greasy wool per year.

Excluding the wool grown on Huang Cheng Stud Farm, the total output in 1988 was 1,540.7 tonne of which 1,369.5 tonne was fine or improved fine wool and 170.2 tonne was semi-fine wool.

- *Specific sheep breeding programs*

Animal improvement is the responsibility of the Animal and Veterinary Station which is one of the stations under the county Animal Husbandry Bureau.

Sheep improvement in Sunan County commenced in 1954 with the introduction of Xinjiang Fine Wool rams to cross with local Tibetan ewes. Between 1987 and 1989, rams from Huang Cheng Stud Farm were introduced. The primary reason for doing so was to introduce Australian merino blood, as Australian merinos were said to have bigger bodies and longer wool than the Gansu Alpine Fine Wool sheep. Australian merino blood was first introduced to Huang Cheng Stud Farm in 1982 via Gongnaisi Stud. For more detail on the history of the breeding program at Huang Cheng, see Longworth and Williamson (1993).

Large numbers of artificial insemination stations were established throughout the county at an early date in order to facilitate breed improvement. Currently there are around 222 AI stations in the county, with the vast majority of these stations having been established at the village level.

Replacement rams for these AI stations are either bred at the two State farms operated by the county AHB or selected from the best private flocks in the local area. The two AHB State farms obtain rams from Huang Cheng Stud Farm. Sometimes Huang Cheng rams are also distributed to some of the AI centres. However, there does not seem to be a free flow of the superior genetic material available at Huang Cheng into the Sunan commercial flocks. Huang Cheng Stud is the most important of the studs breeding Gansu Alpine Fine Wools.

Farmers are charged between ¥0.27 and ¥0.5 per sheep for artificial insemination. The variation in prices for AI services arises owing to a fixed cost structure being imposed upon villages having different flock sizes. The cost of feed grain used to feed rams during the winter is also added to the cost of artificial insemination services. The farmer is charged the grain costs on a per ewe basis.

The AI stations are said to be run as non-profit operations.

All of the 9,681 rams in Sunan County in 1990 were said to be fine wool rams with spinning counts between 60 and 64. All the rams are owned by collectives.

To facilitate animal improvement programs, the Animal Husbandry Bureau has established Animal Improvement Stations in every township in the county. Technicians at these stations are responsible for making sure that improvements in animal quality are facilitated and artificial insemination technology is disseminated.

- *Specific animal health programs*

Technicians at Animal Improvement Stations are charged with ensuring that disease prevention and disease treatment are undertaken.

The Division of Veterinary Services is responsible for:
- animal disease prevention; [This function relates mainly to the provision of technical advice and vaccines to the village level technicians. Any vaccines supplied to the village are paid for by the collective.] and
- a planned dipping and drenching program. [Sheep in the county are dipped and drenched twice a year (spring and autumn) regardless of the incidence of parasites. Again, advice is provided to farmers on how to dip and what chemicals should be used. The chemicals for dipping and drenching are available from the collective.]

13.3 Socio-Economic Development

13.3.1 Income Growth

Although they are living in a remote and hostile environment, the people of Sunan have relatively high net incomes per capita. While their nominal net incomes more than doubled during the 1980s, real incomes rose rapidly until 1985, plateaued until 1988 and then declined. Table 13.2 presents the available time series data on rural net per capita incomes in Sunan for the years 1978 to 1990. These data are presented as graphs in Fig. 6.2 of Chapter 6.

Table 13.2 Average Net Income Per Capita for Sunan County, 1978 to 1990

Year	Nominal net rural income	Real net rural income*
	(¥/capita)	(¥/capita)
1978	167.47	167.47
1979	n/a	n/a
1980	n/a	n/a
1981	n/a	n/a
1982	n/a	n/a
1983	385.60	336.8
1984	420.65	357.4
1985	629.55	491.5
1986	704.13	518.5
1987	778.77	534.5
1988	952.00	551.6
1989	972.00	477.9
1990	876.00	421.8

*Deflated by "Overall Retail Price Index" in State Statistical Bureau of the PRC, *China Statistical Yearbook 1991*, p.199.
The base year is 1978.

13.3.2 Changes to Commodity Prices

When county officials were asked whether the price differentials for various grades of wool were sufficient to encourage the production of fine wool as distinct from coarser types of wool, the answer was that the differentials were indeed sufficient for this purpose. However, later in the interview when the question was raised as to why the quality of wool had deteriorated since 1985, one of the answers given was that the price differentials between the grades were insufficient to encourage farmers to be more careful in the production of their wool and to improve their sheep.

In the past, the grass-roots SMCs in Sunan County have purchased wool from households on a mixed grade basis as described in Table 13.3. However, from 1990 they are planning to buy from farmers on a more specific grade-by-grade basis, separating fine wool into Grades I and II, and improved wool into Grades I and II.

The prices in Table 13.3 are the actual prices paid for quota wool in the years under consideration. The SMC was part of the Ministry of Commerce (now Ministry for Domestic Trade) up until 1984. From 1985 onwards, the SMC in Gansu Province separated from the Ministry of Commerce and now has no direct vertical responsibility to this Ministry. The county SMC officials interviewed could not provide details of prices prior to 1985.

Table 13.3 Prices Paid for Quota Wool in Sunan County, 1985 to 1989

Category of wool	1985	1986	1987	1988	1989
	\-\-\-\-\-\-\-\-\- (¥/kg greasy) \-\-\-\-\-\-\-\-\-				
Fine wool Grades I & II mixed*	6.4	6.6	7.39	10.4	9.4
Improved fine wool Grades I & II mixed*	6.0	6.2	6.94	9.0	8.1
Local wool	5.52	5.72	6.32	8.0	7.2
Coloured wool	3.96	4.1	4.58	6.0	5.5

*Purchased from the farmers as mixed grades.

Although the Gansu provincial government permits free trade in wool, the Sunan County government does not permit private traders to buy wool within Sunan. The only organisation officially permitted to buy wool in Sunan County is the SMC. Furthermore, the prices paid for wool in Sunan continue to be administratively determined.

Every year, the Price Bureau of the provincial Prices Commission and the provincial-level SMC officials meet and discuss the price for wool. They decide upon "instructive prices" for the whole province. In those prefectures and counties where the local authorities do not permit free trade in wool, the local governments then adjust these so-called "instructive prices" as they see fit. Normally the final decision is made by the county government but only after extensive discussions with the SMC and the Price Bureau at the county level. The provincial instructive price for fine wool was ¥12.54 per kg in 1989. However, the Sunan County government set the actual price for that county at ¥9.4 per kg for quota wool. In 1989, there was no separate price for over-quota wool. However in 1988, over-quota wool attracted a premium of ¥.6 per kg, and this premium was the same for all grades of wool. The premium in 1987 for over-quota wool was ¥0.4 per kg and in 1986, ¥0.2 per kg.

Cashmere prices, which are determined more or less in a free market, rose dramatically over the period 1985 to 1989 (Table 13.4). Obviously, the relative prices of wool and cashmere shifted significantly in favour of cashmere in the second half of the 1980s.

Table 13.4 Prices Paid for Cashmere in Sunan County, 1985 to 1989

Year	Cashmere price
	(¥/kg)
1985	34
1986	60
1987	70
1988	100
1989	190

*The price for cashmere in 1990 was expected to be ¥80 to ¥100 per kg.
Source: Sunan County Supply and Marketing Cooperative

Sheep and goat meat prices, which are set by the county Price Bureau, were constant in Sunan over the 1986 to 1989 period. The Food Company which is responsible for meat marketing under the Central government Ministry of Domestic Trade, has subsiduary companies at provincial, prefectural and county level. The county-level Food Company in Sunan compulsorily purchased all sheep and goats for slaughter at the liveweight prices shown in Table 13.5.

Table 13.5 Prices Paid for Livestock Destined for Slaughter, Sunan County, 1986 to 1989

Livestock type	Meat grade[1]	Price 1986-89
		(¥/kg liveweight)
Sheep	Grade I	3.74
	Grade II	3.40
	Grade III	3.06
	Outgrade	2.38
Goats	Grade I	3.29
	Grade II	2.99
	Grade II	2.69
	Grade IV	2.09
Cattle	No grade	2.64*

[1]Definitions of grades for sheep: Grade I > 17.5kg of estimated meat content; Grade II > 15–17.49 kg of estimated meat content; Grade III > 12.5–14.99kg of estimated meat content; and Outgrade < 12.49kg of estimated meat content.
*Free market price. All other prices are County government list prices.
Source: Sunan County Food Company

It was stated that from 1979 to 1986, the Food Company purchased all cattle as well as sheep and goats being sold for meat at a constant fixed price. [During these years, the price for sheep was ¥1.58/kg liveweight and the price for cattle was ¥1.9/kg liveweight.] After 1986, the cattle market was "opened" to other buyers and hence the price shown in Table 13.5 is a rough average of the free market prices paid for cattle over the 1986 to 1989 period.

13.3.3 Timing of Introduction of the Household Production Responsibility System

Government officials of Sunan County attribute the new initiative and enthusiasm of the farmers to the introduction of the household production responsibility system. The system was introduced in the early 1980s.

Initially, the village collectives contracted the sheep out to households. Within a year or so, the sheep were sold to the households and most of the collectively owned land was contracted out to households. The land contracts between the village collective and the private households were relatively simple documents. An example is presented in Appendix 13A. Later, a more detailed supplementary legal document called a "pasture utilisation certificate" was issued to most households with a contract for pasture land. This certificate was designed to clarify the rights and obligations of the household and the collective in regard to the use of the pastures. Appendix 13B contains an example of a Sunan County pasture utilisation certificate which, as mentioned above, is a much less detailed document than those in use in the Eastern Grasslands of IMAR.

13.3.4 Level of State Assistance

As indicated earlier, there are no government subsidies available to farmers in Sunan County for planting improved pastures. However, between 1986 and 1990, significant government subsidies were made available to farmers by all three levels of government (central, provincial and county) to offset the capital cost of fencing.

In 1986 and 1987, the total subsidy provided to the farmers was ¥15 per mu. For the two years 1988 and 1989, the subsidy was reduced to ¥10 per mu. In total, the cost of fencing is said to be around ¥20 per mu.

For fencing and pasture improvement from 1989 onwards, the arrangement is that the farmers can obtain an interest-free loan on a three-year term. When officials were questioned further about this interest-free loan approach to encouraging fencing and pasture improvement, it was said that there was only around ¥300,000 in total available for such loans in 1990 for the whole county.

Since Sunan is not considered to be a poverty-stricken county, no general poverty alleviation funds are provided by either the provincial or Central governments.

13.3.5 Development of Pastoral Product Processing

Given the location of Sunan County and its limited resource base, first-stage wool processing is one of the few feasible forms of industrial development. The Sunan County wool scouring plant was built in 1986 by the Economic Commission of the Sunan County government and commenced scouring in September of that year. The total cost of the plant was ¥1.25 million. Of this, ¥825,000 was borrowed from the Industrial and Commercial Bank of China and the remaining ¥400,000 was contributed by the 80 permanent staff of the factory. The plant employs a total of 150 staff, 70 of which are part-time seasonal workers.

The total capacity of the plant (working 24 hours per day 365 days per year) is 1,900 tonne of greasy wool per annum. Total wool production in Sunan County, excluding Huang Cheng Stud Farm, is around 1,550 tonne. Compared with many county wool scours, the Sunan plant was well managed. For example, all raw wool was graded into industrial grades prior to scouring which, as discussed by Brown and Longworth (1992b), is a pre-requisite for the production of good quality scoured wool. The gross value of output for the plant for the three years 1987, 1988 and 1989, was ¥1.2 million. [The gross value of output is determined using a National Standard output value for wool scouring plants of ¥.3 per kg of raw wool scoured.]

Raw wool scoured in the plant is obtained from the county-level SMC and the two county-level Animal Husbandry Bureau State farms. Both the SMC and the State farms are required by the County government to sell their wool to the wool scouring plant. Wool is purchased from the SMC at farm-gate price plus 21.7%. Wool purchased directly from the two county-level State farms is purchased at a farm-gate price plus 10%. Total production of wool from these two State farms in 1989 was said to be 30 tonne greasy.

Significantly, the wool scouring plant does not purchase any wool from Huang Cheng Stud Farm. This State farm is under the direct control of the provincial Animal Husbandry Bureau and not the Sunan county authorities. The issue of administrative control is extremely important because under the current arrangements the county authorities are powerless to force the Huang Cheng Stud Farm to supply wool to the Sunan County wool scouring plant. Huang Cheng Stud Farm presently produces about 65 tonne (greasy) of good quality fine wool each year. This wool attracts significant price premiums through direct sale to processors. Given the relatively low quality of the output from the county scouring plant, it would not be in the best interests of Huang Cheng to supply their wool to the county plant. Interestingly, the officials of the wool scouring plant indicated that they had no knowledge of how much wool is produced on Huang Cheng Stud Farm.

In 1989, total throughput for the wool scouring plant was 450 tonne clean, which is equivalent to about 1,000 tonne greasy. About 85% of the scoured wool was considered to be suitable for wool textile processing. The remaining 15% (cotted, stained, head, belly and tail wool) was too short for processing in a wool textile factory but was useful for felt-making.

The entire output of clean scoured wool is sold to the textile mills in Lanzhou, the capital of Gansu Province. Of the 450 tonne clean scoured production for 1989, 350 tonne was fine wool of 60s to 66s spinning count but the majority of this wool (about 280 tonne) was of 64s type. About 80% of the remaining 100 tonne of cleaned scoured output in 1989 was improved wool class I, with the remainder being improved wool class II, improved wool class III, and improved wool class IV. For detailed information about wool grading systems used in China, see Longworth and Brown (1994).

The Food Company in Sunan County has under its control the only meat-processing factory in the county. The factory is made up of five divisions:
– an abattoir division;
– a cake-making division;
– a soy sauce and vinegar manufacturing division;
– a bean curd manufacturing division; and
– a cold storage division.

The abattoir division operates seasonally between the months of early October and mid-December. All animals purchased by the Food Company are processed by the abattoir. In 1989, the County Food Company purchased a total of 15,000 sheep, 300 goats and 300 cattle from producers. Unlike sheep and goats, since 1986 cattle in Sunan County have not been subject to compulsory purchase quotas. If the County Food Company wishes to purchase cattle, it must do so at free market prices. The total numbers of sheep and goats purchased by the Food Company in 1989 were in line with the planned acquisition levels for that year. There were no sheep or goats purchased outside the planned amount.

Approximately two-thirds of the total animals purchased by the County Food Company are consumed within the county. All the animals acquired by the Food Company are slaughtered in the county abattoir and put into cold storage. Buyers from outside the county purchase meat from the Food Company and transport this meat to other destinations using cold storage vans. The Sunan County Food Company does not deliver meat to areas outside the county.

Hides taken from the slaughtered sheep, cattle and goats are one of the major by-products of the abattoir division. These hides are delivered to the Leather Processing Factory located in the county. The Leather Processing Factory, like the wool scouring plant, is owned by the Economic Commission of the County government.

Appendix 13A
Dual Management Land Contract

The Laohugou Village Cooperative, Xuequan Township
The People's Government of Sunan Yugur Autonomous Country, Gansu Province

Land Contractor: ....

The Land Contract is formulated on the basis of dual management operation (i.e. the land is managed according to both collectively unified operation and family operation). In order to establish a clear relationship between the cooperative and its members, the responsibility, rights and obligations to the land are listed as follows:

A. The Rights and Obligations of the Cooperative

1. The cooperative is the unit who contracts out the public pasture land. It owns and is entitled to manage the land and all kinds of public facilities and natural resources. It is responsible for the administration and coordination of the Land Contract as well.
2. The cooperative must undertake appropriate planning in order to further its economic and social development; ensure that members implement an annual production plan; provide households with all the necessary services prior to, in the middle of, and after the production season has ceased.
3. The cooperative must play a role in unified planning on the basis of stabilizing the family operation system; help the households to undertake activities which are neither easy nor possible for them to undertake themselves; paying particular attention to continuing "the five unified plannings" for sheep improvement (i.e. unified planning of public pasture land for sheep breeding, unified planning of the locations for sheep breeding stations, unified planning of the management over breeding rams, unified planning of mustering adult ewes to the sheep breeding stations, unified planning of training technicians responsible for sheep improvement). Economic punishment is to be imposed on Cooperative members who do not follow "the five unified plannings".
4. All capital accumulations generated from cooperative properties belong to the collective. The cooperative is entitled to collect, manage and use the cooperative levies and all capital accumulations in the light of the resolution made at the township level people's congress and the cooperative financial management regulations.

B. The Rights and Obligations of the Household Contractors

1. The Household Contractor is entitled to manage, construct and use the contracted pasture land. All the pasture facilities built with private investment belong to private members. The contracted pasture land is able to be inherited and subcontracted out to other individual units.
2. The Household Contractor is entitled to a self-decision making role in production, labour power control and profit distribution.
3. The Household Contractor is required to meet the animal product quota set by government every year.
4. The Household Contractor is requireed to pay taxes according to the taxation law and pay the cooperative levies. In addition, taxes and levies should be paid on time. The amount and coverage of the levies is discussed by the people's congress at township level and approved by the county government.
5. The Household Contractor is required to pay the collective for animals purchased according to the rules set by the collective when the animals were sold to each household. In addition, debt owed by the household contractor must be paid to the collective.
6. The Household Contractor is required to actively meet the quota of labour accumulation for both state and collective construction project.

C. Punishment for Those Who Break the Contract

1. The Agrarian Law must be strictly followed. The cooperative is entitled to withdraw from the contractors the land which is wasted for two successive years.
2. In accordance with the rules and regulations, the cooperative is entitled to punish anyone who does not discharge the contract and refuses to meet the animal product quota or who does not pay the taxes and cooperative levies, even to the extent of withdrawing the contracted land from him.
3. If one side of the contract does not discharge any clauses in the contract and subsequently causes economic losses to the other side, that side will be responsible for compensating for the said losses in accordance with the actual costs of the losses.
4. This contract comes into effect upon the notary verification.
5. The contract is legally protected. If one side breaks the contract and the notary organ fails to adjudicate, the other side can go to the court for final adjudication.

Contracting Institution:	Seal
Land Contractor:	Seal
Notary Organ:	Seal

Appendix 13B
Pasture Land Contract Utilization Certificate

The People's Government of Sunan Yugur Autonomous County, Gansu Province

This certificate is specially formulated in order to tighten up pasture land mangement, ensure rational use and construction of the pasture land, and to protect the contractor's right to use the pasture land.

The household's rights to use, manage and construct the pasture land are recognised over a long period of time. Whoever uses, manages and constructs the pasture land is entitled to enjoy the benefits from the operations. The pasture land is allowed to be inherited and subcontracted out, but is not able to be sold or rented out.

Pasture land reclamation is forbidden to protect the pastures. Pasture management organisations at different levels should be established and officially recognised. Pasture management committees should be organised at county and township levels, and pasture management groups should be formed in each village. These organisations are responsible for pasture conservation, utilisation and construction and adjudication of the dispute between households in order to guarantee no pasture land reclamation, no overgrazing, no damage to the vegetation and no pasture land degradation.

No government official or individual is entitled to revise the certificate without the permission of the township government.

The People's Government of Sunan
Yugur Autonomous County
(Seal)

Household	
Population	Male　　　　　　Female
Address	Village　　　Township　　　District

		Pasture area (mu)	Suitable stocking capacity (sheep unit)	Animal species and quantity			Pasture boundaries			
				Large animals	Small animals	Sheep units	East	West	South	North
Four season pastures	Spring pasture									
	Summer pasture									
	Autumn pasture									
	Winter pasture									

Chapter Fourteen

Dunhuang County (Gansu)

Dunhuang County is a very large desert area at the western end of the Hexi Corridor in Gansu Province. The county has an extensive common border with Xinjiang (Map 14.1). Dunhuang, the county capital, is famous for its ancient cave paintings (frescos) and because it lies at the western end of the Great Wall.

Although Dunhuang is geographically in a pastoral area, it is not included in the 266 counties classified as the pastoral region by the Central government. The inhabitants of Dunhuang are 99% Han although ten different minorities make up the remaining 1% of the population. Perhaps surprisingly since it is a desert county, the economy of Dunhuang is based on cropping. Almost all the sheep in the county are of the native Mongolian breed.

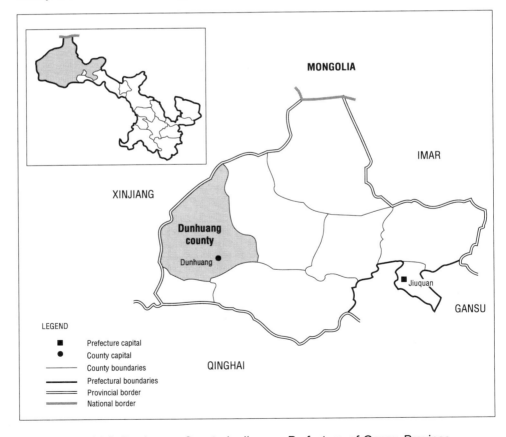

Map 14.1 Dunhuang County in Jiuquan Prefecture of Gansu Province

14.1 Resource Endowment

Dunhuang is classified for statistical purposes as an agricultural county. It is not classified as a poor county by either the Central government or the Gansu provincial government.

14.1.1 General Features

The densely settled parts of Dunhuang County are large oasis-like pockets of land with substantial irrigation development. However, vast areas of the county are flat to undulating stony (gibber) desert country. The western end of the Qilian Mountain range lies to the south of Dunhuang.

The average altitude of the county is 1,138.7m. The environment is a typical dry inland climate, with characteristically low rainfall (an average of 39mm per annum) and high evaporation rates (2,490.6mm per annum). The annual mean temperature for the year is 9.3°C. Summers are typically hot and occasionally wet, while winters are very cold and dry. Winds can sometimes be particularly bad during springtime. It was said that dust storms can constitute major disasters in Dunhuang County.

Irrigation water used for agricultural production is obtained from both underground and surface water sources. Both sources are recharged annually through snow waters funnelled from the Qilian Mountains down the 100km long Qilian River to Dunhuang County. In past geological times, the oases (or greenlands as they are called in Dunhuang County) were once terminal lakes for the Qilian River. Over time, these lakes have silted up providing rich alluvial soil for agricultural and animal husbandry production. Where water supplies are not salty, agricultural production takes precedence over animal husbandry.

Monthly and annual rainfall for the county capital is shown in Table 14.1. Obviously the county capital, Dunhuang, is located in the midst of a desert. Without the snow-fed Qilian River, settlement opportunities would be extremely limited.

14.1.2 Land Use

As the figures below demonstrate, very little of this vast county is usable pasture land or arable cropland.
- Total land area   46,965,000mu (or 31,200km²)
- Total pasture land   3,120,400mu
- Total usable pasture   1,215,000mu
    - marginal desert pasture land   972,000mu
    - semi-marginal desert pasture land   121,500mu
    - marshland   116,500mu
    - oasis pasture land   5,000mu
- Arable land   231,300mu
- Forestry land   2,260,000mu
    - planted forestry   70,000mu
    - natural forestry   2,190,000mu

Pasture degradation has become an increasingly serious problem. The reasons given for degradation were, first, the dry weather over recent years; and second, the increasing livestock numbers resulting in a decline in the quantity of grass available to animals.

Table 14.1 Monthly and Annual Rainfall in the Capital of Dunhuang County, 1960 to 1989

Year	J	F	M	A	M	J	J	A	S	O	N	D	Annual rainfall
							(mm)						
1960	0.0	1.5	0.0	0.0	0.9	0.0	2.9	0.7	0.5	0.0	1.3	0.0	5.4
1961	0.0	1.1	0.0	0.0	6.0	0.1	3.6	3.7	2.3	0.0	0.0	0.0	9.7
1962	0.0	0.0	1.7	0.0	0.0	0.2	3.5	12.9	11.5	0.0	2.7	0.0	32.5
1963	0.0	0.2	0.8	0.1	0.0	18.7	5.4	6.2	0.0	0.0	0.0	0.0	31.2
1964	2.5	1.1	0.0	7.4	5.3	3.3	3.0	0.0	0.1	4.8	0.0	0.1	18.7
1965	0.0	0.4	0.0	6.2	0.0	4.9	4.0	3.0	0.0	0.0	0.0	0.1	18.2
1966	0.2	0.3	0.0	0.3	2.6	1.3	7.4	1.6	1.4	1.6	0.0	0.6	14.2
1967	0.0	2.8	2.3	0.0	13.4	0.3	7.8	8.2	0.0	0.0	2.9	0.2	21.7
1968	0.0	0.0	2.3	0.0	0.4	5.9	10.0	0.9	0.0	0.0	0.0	0.0	19.1
1969	0.8	4.0	6.3	0.0	0.1	11.3	28.7	0.9	5.3	0.0	0.0	0.0	52.5
1970	0.0	0.5	1.8	8.9	4.4	1.3	6.1	7.8	0.5	2.0	0.0	0.0	28.4
1971	0.0	0.7	0.7	0.0	0.0	3.6	0.7	9.5	2.4	0.0	1.2	0.4	78.5
1972	0.5	2.2	0.0	0.3	0.0	13.0	18.9	8.6	0.0	0.0	4.1	1.6	46.5
1973	1.4	3.5	0.0	6.5	1.8	32.1	9.0	9.6	0.0	1.3	0.0	0.0	58.5
1974	0.1	0.7	11.0	0.0	0.0	3.9	2.3	4.7	0.0	0.0	0.4	1.3	23.6
1975	0.1	0.0	0.0	5.1	0.0	1.5	2.5	23.6	1.0	0.0	1.1	1.0	34.7
1976	0.0	0.0	0.0	16.2	6.0	4.0	26.5	3.0	0.0	4.1	1.1	0.0	53.8
1977	0.9	0.0	0.0	2.1	0.7	17.9	12.6	0.2	1.1	0.0	3.0	0.6	37.5
1978	1.2	3.6	3.3	0.4	0.1	13.8	1.3	1.4	9.1	3.0	0.1	0.0	32.4
1979	5.7	0.0	0.0	6.2	0.8	34.9	45.4	7.4	0.1	0.0	5.0	0.0	99.0
1980	0.7	2.0	0.0	0.5	0.5	0.4	15.6	1.7	0.0	0.0	0.0	0.4	18.6
1981	0.9	4.2	0.3	0.0	0.0	8.9	6.8	28.9	7.3	0.0	2.3	0.0	54.5
1982	0.0	0.0	0.0	1.8	9.8	16.4	4.9	7.1	0.0	0.0	0.7	0.0	30.9
1983	0.3	1.2	0.0	0.0	5.5	3.5	5.4	11.3	2.5	0.0	0.0	0.0	22.7
1984	0.3	0.7	1.2	0.0	0.1	11.0	18.9	6.7	0.0	0.0	0.0	1.2	39.0
1985	0.9	0.0	0.0	1.4	3.2	1.7	6.9	1.0	1.4	0.0	0.1	0.2	12.7
1986	0.0	0.0	0.0	0.6	4.2	3.0	6.1	0.9	0.0	0.0	1.2	1.3	13.1
1987	0.1	0.0	0.8	9.8	11.1	13.0	6.1	0.7	1.9	0.0	0.3	0.0	32.6
1988	1.2	0.6	0.0	0.1	13.1	3.7	9.8	5.5	0.0	0.0	0.1	2.8	22.0
1989	2.1	0.0	0.0	1.5	0.0	2.2	16.5	0.0	1.0	0.1	0.0	0.6	21.9

Source: Dunhuang County Animal Husbandry Bureau

14.1.3 Population

- Total, rural and urban (1990)
  - total   112,085   (100%)
  - rural   89,220   (79.6%)
  - urban   22,865   (20.4%)

- Ethnic composition

The Han are the overwhelmingly dominant ethnic group. There are a significant number of Huizu people but only a single household or two of each of the other nine minority nationalities which live in the county (Table 14.2).

- Population density   3.59 people/km^2

Table 14.2 Ethnic Composition of Dunhuang County Population, 1990

Nationality	No. of households	Population
Han	28,272	111,061
Huizu	211	924
Tibetan	3	19
Mongolian	9	51
Hasake	1	4
Yugur	1	2
Man	3	15
Korean	1	1
Tu	1	3
Dong	1	2
Miao	1	3
Total	28,504	112,085

14.1.4 Administrative Structure

Dunhuang County consists of two towns, 10 townships and 77 villages. Of the 28,504 households in the county, 21,451 live in the agricultural areas and 7,050 are classified as pastoral. There are four State farms in the county, none of which raise sheep.

14.1.5 Livestock Population

The livestock in Dunhuang County in 1989 were as follows:
- Large animals
  - cattle 2,900
  - horses 2,500
  - donkeys 13,000
  - mules 4,800
  - camels <u>1,300</u>
  - Total 24,500

- Small animals (excludes rabbits and poultry)
  - goats 32,900
  - sheep 93,600
  - pigs <u>60,200</u>
  - Total 186,700

14.2 Improvements to Production Conditions in the Pastoral Systems

14.2.1 Improved Pastures

In 1990, there was 4,000mu of improved pasture in and around the cultivated land areas of Dunhuang County. The majority of this pasture was lucerne (*Medicago sativa*).

The lucerne is used primarily as cutting land pasture to provide hay for large animals such as horses and cattle. Very little lucerne is fed to small animals such as sheep and goats. Lucerne is cut twice a year and yields between three and five tonne of dry matter per mu per year. Lucerne is irrigated three times a year and is fertilised in spring with a mixture of animal manure and artificial fertiliser. There are no direct restrictions on the use of artificial fertiliser on lucerne. An indirect restriction does exist, however, in that the total amount of lucerne that is allowed to be planted in the county is restricted. Lucerne is said to survive for between three and five years.

The Chinese bean (*Wando*) is another crop grown for animal feed, as well as to enrich the fertility of soil. This crop is planted in July and cut in October. Like lucerne hay, the cut forage is fed to large animals rather than sheep and goats.

The Pasture Management Station of Dunhuang County is responsible for ensuring that the National Rangeland Law is adhered to by households in the county. In this connection, staff attempt to ensure that pastures are not damaged through the unauthorised digging of holes, cultivation or digging for Chinese medicines.

Another role of the Station is to monitor the level of rodents and insects infesting pasture lands. The Station staff claim not to have any major problems with rodents or insects in the pasture lands of Dunhuang.

14.2.2 Supplementary Feeding

As mentioned above, lucerne hay and bean crop forage is fed to large animals such as horses and cattle. Very little lucerne hay is fed to small animals such as sheep and goats. Sheep are fed leaves, cotton stubble and occasionally grazed on the lucerne after it has been cut for hay.

14.2.3 Improved Sheep Breeds

- *Sheep production systems*

The main oasis in Dunhuang County, where the city of Dunhuang is located, is primarily devoted to agricultural production. There are, however, two large alluvial areas where the water is not suitable for irrigation. These are the two major pastoral centres known as Xihu ("west lake") and Beihu ("north lake"). These areas have water that is just adequate for stock.

There are 172 households specialising in sheep production in the county. On average, each specialised household has around 110 sheep.

- *Sheep breeds*

The sheep in the county are almost entirely of the Mongolian coarse wool variety.

- *Animal productivity (1989)*
  Total wool production (tonne)   102.30
  – fine wool   0.00
  – semi-fine   1.48
  – coarse wool, etc.   100.82

14.2.4 Specific Sheep Breeding Programs

The Animal Husbandry Bureau in Dunhuang County is responsible for the State Veterinary Service Centre which in turn is responsible for the Animal Improvement Station. The role of the Animal Improvement Station is to facilitate the improvement of breeds of animals in the county. In past years, the Station has concentrated mainly on improving pigs. Only a small amount of work has been undertaken on sheep improvement.

The Mongolian coarse wool sheep are said to be particularly useful for mutton production, they give very little trouble as far as disease is concerned and they are well suited to the local climate.

In 1958, fine wool rams were introduced from Xinjiang. These rams were crossed with the local Mongolian ewes. This breeding program, however, proved to be unsuccessful as fine wool rams were unsuited to the harsh local environment and the very tough and fibrous vegetation.

Following the failure of this fine wool breeding program, the Station elected to improve the local breeds by introducing six Shandong Short-tail rams from Shandong Province in 1985. The Shandong Short-tail sheep are multiple birth meat sheep, useful for mutton production and carpet wool production. Shandong Short-tail sheep are said to produce between two and four lambs per lambing. Initially, the six rams imported from Shandong Province were cared for by the Dunhuang State Farm. This State Farm acts as a sire breeding station for goats but not for sheep. After a short time at the State Farm, the Shandong rams were sold to specialised households in the county.

14.2.5 Specific Animal Health Programs

The Animal Examination Station is another station under the Animal Husbandry Bureau in Dunhuang County.

The role of the Animal Examination Station is as follows:
(i) to ensure that farmers are not selling diseased animals to the general public;
(ii) to monitor the level of disease and parasitic infestation in the animal production base; [The two main parasites of concern in Dunhuang County are bot flies and liver flukes. Liver flukes are a particular problem in and around the Xihu and Beihu areas.]
(iii) to provide advice on the prevention and treatment of disease and parasite infestations in the animal production base; [The Station does this through the use of formal training courses and visits to the various townships and villages in the county. Training courses are usually held in winter in the county town, and last for around one week. Between 70 and 80 farmers and village technicians are said to attend these courses each year.] and
(iv) to undertake research into the prevention and treatment of various diseases and parasites, taking into account local environmental conditions and local cultural

practices. [In this connection it was said that farmers are usually unwilling to treat their animals when they are suffering from diseases or parasites which are known to the farmers. The farmers would prefer instead to kill the animal and eat the meat, rather than attempt treatment.]

Staff working in the Animal Examination Station do not perform actual preventative or therapeutic treatment of disease and parasites in animals in the county. Actual treatment is left to the township veterinarians and the village-level disease prevention staff.

14.3 Socio-Economic Development

14.3.1 Income Growth

Nominal net per capita rural incomes in Dunhuang increased greatly during the 1980s (Table 14.3). However, as demonstrated by the graphs in Figure 6.2 and discussed in Section 6.6.1 of Chapter 6, real incomes have declined since 1985.

Table 14.3 Average Net Per Capita Rural Income for Dunhuang County, 1970 to 1989

Year	Nominal net rural income	Real net rural income*
	(¥/capita)	(¥/capita)
1970	101.0	104.3
1971	112.0	116.7
1972	124.0	129.4
1973	133.0	138.0
1974	111.0	114.6
1975	105.0	108.1
1976	121.0	124.2
1977	136.0	137.0
1978	146.0	146.0
1979	170.0	166.7
1981	223.5	206.8
1982	239.6	216.4
1983	396.0	351.1
1984	580.0	506.6
1985	669.0	568.4
1986	674.0	526.2
1987	703.0	517.7
1988	769.0	527.8
1989	832.0	482.0

*Deflated by "Overall Retail Price Index" in State Statistical Bureau of the PRC, *China Statistical Yearbook 1991*, p.199. The base year is 1978.
Source: Dunhuang County Animal Husbandry Bureau

14.3.2 Changes to Commodity Prices

In 1989, the Dunhuang County Supply and Marketing Cooperative (SMC) reduced its wool purchase price by 46% relative to 1988 prices (from ¥13 per kg greasy to ¥7 per kg greasy). This resulted in a very sharp reduction in the proportion of wool being sold to the SMC. Farmers were clearly unhappy with the new price set by the SMC and opted to sell their wool through other marketing channels, in particular private dealers and to a lesser extent, direct sale to carpet wool processors.

In Gansu Province, the monopoly control of the SMC organisation over the wool and cashmere market ceased in 1985. As indicated in the Sunan case study, the provincial government now provides indicative prices which the county level SMC may or may not choose to follow. Between 1985 and 1990, the Dunhuang SMC was entirely free to set its purchase prices at whatever level it desired. However, in 1990 the provincial government introduced a floor price of ¥6 per kg greasy for coarse wool grade I. The reason for this policy offered by the Dunhuang SMC officials was to discourage farmers from slaughtering their sheep and hence reducing the future supply of coarse wool.

The relative prices of coarse wool and cashmere moved in favour of cashmere in the late 1980s and this factor (rather than the absolute level of wool prices) is probably the most important determinant of wool supply in Dunhuang County (see Table 14.4).

Table 14.4 Prices Paid for Animal Fibres in Dunhuang County, 1982 to 1990

Year	Average prices paid for coarse (local) wool	Prices paid for first grade cashmere[1]	Prices paid for camel hair
	(¥ per kg greasy)	(¥ per kg)	(¥ per kg)
1982	na	14	na
1983	na	14	na
1984	2.8	14	na
1985	6	40	15
1986	8	80	na
1987	10	100	na
1988	13	120	na
1989	7	80	15
1990	6	40	10

[1]First grade cashmere comprises 80% of the total SMC purchases. The SMC purchases around 90% of the total cashmere production in Dunhuang County.

Other key factors determining the supply of wool are the prices of mutton, goat meat and beef. While the data in Table 14.5 is incomplete, it appears that the prices paid by the government purchasing agency moved in favour of goat meat and beef in the mid-1980s.

14.3.3 Timing of Introduction of the Household Production Responsibility System

In Dunhuang (as in the rest of Gansu), most of the pasture land, desert land and forestry land is owned by the State. The remaining lands (cropland, cutting land and some forestry land) are owned by collectives. In 1984, all of the collective-owned cropland, cutting land and some of the forestry land was contracted out to individual households. However, none of the State-owned lands were contracted out.

Table 14.5 Procurement Prices Paid for Sheep Meat, Goat Meat and Beef in Dunhuang County, 1978 to 1990

Year	Sheep meat	Goat meat	Beef
	(¥/kg)		
1978	1.06	0.84	1.70
1979	1.06	0.84	1.70
1980	1.06	0.84	1.70
1981	1.18	1.12	1.70
1982	1.18	1.12	1.70
1983	1.18	1.12	1.70
1984	1.18	1.20	1.70
1985	1.18	1.20	3.60
1986			4.40
1987			5.00
1988			5.60
1989			5.60
1990			5.60

The county officials indicated that there were plans in 1990 to contract out the areas of grazing land surrounding the agricultural settlements in the county. Although the State-owned grazing lands have not been contracted out, households are required to pay an animal tax for the use of State-owned pastoral land.

14.3.4 Level of State Assistance

Since Dunhuang is not recognised as a poor county, it does not receive any special poverty alleviation assistance from higher levels of government. However, there are funds made available by the Dunhuang County Bureau of Civil Affairs for the assistance of poor households. These funds, which are provided by the County government, are administered by the Credit Cooperatives but not by the Agricultural Bank of China.

14.3.5 Growth in Output of Major Pastoral Commodities

Dunhuang county is not a major producer of pastoral commodities. The production of wool increased more than threefold during the 1980s but the total output remains extremely modest (Tables 14.6 and 14.7). Almost all the wool grown in Dunhuang is coarse/local wool, the best of which can be made into excellent carpets.

In Section 6.1 it was suggested that it might be possible to expand sheep and goat raising activities in oasis-based counties such as Dunhuang by integrating these activities with agriculture. County officials pointed out that high value cash crops such as melons, grapes, fruit and vegetables were the most profitable and hence common types of agriculture in Dunhuang and other similar oasis-based counties in China. Sheep and goat raising does not integrate well with the growing of these crops. Furthermore, raising small ruminants is a relatively unprofitable form of intensive animal husbandry compared with pigs, poultry and rabbits. Even in the purely pastoral areas of the county, raising sheep for mutton and wool was not sufficiently profitable to encourage significant new investment. In particular, according to the County Animal Husbandry Bureau staff, past experience had shown that fine wool sheep were not suited to the environment in Dunhuang.

Table 14.6 Total Wool, Fine Wool and Semi-fine Wool Production in Dunhuang County, 1985 to 1989

Year	Total wool	Fine wool	Semi-fine wool
		(tonne)	
1985	57.85	0.36	0.27
1986	73.55	0.00	4.00
1987	92.00	1.00	1.00
1988	101.53	0.73	2.00
1989	102.30	0.00	1.48

Table 14.7 Wool, Cashmere and Goat Hair Production in Dunhuang County, 1975 to 1989

Year	Wool	Cashmere	Goat Hair
		(tonne)	
1975	18.75	0.05	12.8
1976	18.40	0.05	10.9
1977	18.70		9.2
1978	21.60		10.7
1979	27.20	0.15	10.7
1980	31.80	0.35	5.9
1981	33.95		12.4
1982	37.00	0.30	12.9
1983	46.50	0.35	10.1
1984	48.15	0.15	8.9
1985	57.85		7.9
1986	73.55	0.10	10.2
1987	92.00	0.32	11.0
1988	101.50	2.15	14.5
1989	102.30	1.75	16.7

Source: Dunhuang Animal Husbandry Bureau

Chapter Fifteen

Tianzhu County (Gansu)

Tianzhu is a Tibetan Autonomous County located about 100km north of Lanzhou, the capital of Gansu Province (Map 15.1). Tianzhu is the most easterly county of Wu Wei Prefecture which is the first prefecture encountered in the Hexi Corridor travelling from Lanzhou towards Xinjiang.

Approximately one-third of the population belong to minority nationalities, with the majority being Tibetan. Most of these people live in the alpine pastoral areas of the county and depend upon fine wool or improved fine wool sheep for their livelihood.

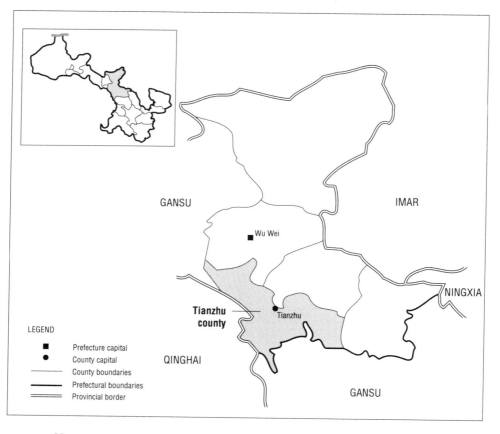

Map 15.1 Tianzhu County in Wu Wei Prefecture of Gansu Province

15.1 Resource Endowment

The Central government classifies Tianzhu as a pastoral county and the Gansu provincial government includes Tianzhu in the list of seven pastoral counties in Gansu. However, nowadays the proportion of rural income which comes from pastoral activities is in the range 25 to 50%. Therefore, at the local level, Tianzhu County is regarded as a semi-pastoral county.

Neither the Central government nor the Gansu provincial government regard Tianzhu as a poverty county.

15.1.1 General Features

There are large areas of agricultural land around the county capital Tianzhu Town but the eastern parts of the county are high mountain pasture areas. A major mountain range, known as the Mou Mou Mountains, extends across the northern and north eastern border of Tianzhu County. This mountain range, which is a spur of the Qilian Mountains, exceeds 5,000m above sea level at several points.

During the "Great Leap Forward" era (1958 to 1961), large areas of the best pasture land in Tianzhu were ploughed up and sown to crops. However, the conditions were not suitable for the growing of crops and the light soils quickly became badly eroded. Increasing grazing pressures on the remaining pastures since 1961 have intensified the erosion problem. Almost all of the once lush pasture land in the low hills east of Tianzhu Town is now in a badly eroded and degraded state.

Research based on taking random samples from clearly defined areas of pasture land over a number of years revealed that the dry matter productivity of the average pasture in Tianzhu was declining at 2.4% per year. Of course, degradation involves changes in the composition of useful grasses as well as a reduction in dry matter. In the Tianzhu study, it was found that as degradation increased over time, the dominant native species, *Steppa breviflora* was replaced by the less useful grass *Asteraceae frigida*. *Steppa breviflora* was said to be a very useful native grass species for animal production.

15.1.2 Land Use

Only a little over half the land area in the county is usable pasture or arable crop land.
- Total land area   10,724,700 mu   (or 7,149.8 km²)
- Total arable land   368,000 mu
- Total rangeland   7,584,000 mu
  - usable pasture land   5,871,100 mu

While no precise figures are available, a high proportion of the usable pasture land would be degraded, most of it badly degraded.

15.1.3 Population

- Total, rural, and urban population (1990)
  - total   209,000   (100%)
  - rural   183,000   (87.56%)
  - urban   26,000   (12.44%)

- Ethnic composition
  In 1990 the minority population (mainly Tibetan) was 32% of the total population.
- Population density   29.23 people/km²
- Administrative structure
  Tianzhu has 1 town and 22 townships.

15.1.4 Livestock Population

- Sheep and goats   429,000
- Large livestock (yaks, cattle, horses,
  mules, donkeys, camels)   134,000

15.2 Improvements to Production Conditions in the Pastoral Systems

15.2.1 Improved Sheep and Wool Production

Song Shan State Farm and Sheep Stud is one of the three studs in Gansu recognised by the provincial government as breeding sheep which meet the Gansu Alpine Fine Wool Sheep breed standard. It is located 17km east of Tianzhu Town. For details on Song Shan Stud, including information about the breeding program and the role imported Australian merino rams have played at this stud, see Longworth and Williamson (1993).

Every pastoral township in Tianzhu County has a ram care station so that it can supply rams to the village AI centres. The rams at these stations come mostly from Song Shan Stud. In the pastoral areas, virtually all sheep are fine wool or improved fine wool sheep and almost all the ewes are mated by AI.

In Tianzhu County, the program of upgrading to fine wool sheep has progressed the furthest of all counties in the Wu Wei Prefecture. Most sheep in the pastoral townships have now been mated to pure Gansu Alpine Fine Wool rams for at least four generations.

Since most F_4 and better crosses to pure-bred Gansu Alpine rams will produce almost completely homogeneous wool (that is, wool with virtually no hair), the F_4 and better generations are usually called "fine wool sheep". However, most F_4 sheep would not meet the general overall breed standards set for Gansu Alpine Fine Wool Sheep (such as evenness of fibre diameter and length over the body, wool colour, degree of crimp, wool length, etc.). It takes about six generations of crossing with pure-bred rams to achieve all these traits.

Agricultural areas of Tianzhu usually do not have ram care stations nor even AI centres in many instances. Most of the sheep are mated by natural means. Some of the rams used will have come from Song Shan Stud. Many of the sheep raised in the agricultural areas are semi-fine wool or improved semi-fine wool sheep. There are also a lot of local coarse wool Tibetan and Mongolian sheep in the agricultural areas.

Wool production data for Tianzhu County are only available for 1988 and 1989. This information is presented in Table 15.1.

Table 15.1 Wool Production in Tianzhu County, 1988 and 1989

Type of wool	1988		1989	
	(tonne)	(%)	(tonne)	(%)
Fine wool and improved fine wool	656.9	67.67	546.0	63.34
Semi-fine wool and improved semi-fine wool	114.4	11.78	175.0	20.30
Coarse/local wool	199.3	20.53	141.0	16.35
Total all wool	970.6	100	862.0	100

15.2.2 Improved Cattle Breeds

Tianzhu County is famous for the Tianzhu White Yak. This breed is very popular in Wu Wei Prefecture where there are at least 40,000 head of this type of yak.

15.3 Socio-Economic Development

No data were collected on socio-economic development in Tianzhu County except for the following details in relation to the construction of a wool scouring plant.

In 1987, Tianzhu County Supply and Marketing Cooperative built a scouring plant and set up a company to operate it. Although the plant, which has a capacity to scour 2,000 tonne of greasy wool per year, was expected to commence operations in 1988, it has processed very little wool to date.

The company established to manage the scouring plant purchased over 1,000 tonne of greasy wool in 1988 for which it was reported to have paid the extraordinarily high average price of ¥20 per kg (greasy). The company was forced to pay extremely high prices to obtain wool in competition with private individuals and other government units who were active as wool-buyers at the peak of the wool boom in China in 1988. After purchasing the wool, the company managers apparently realised that it would be impossible to scour the wool and then sell it at a profit because the price of wool had declined. Consequently, they quickly on-sold about 40% of their stock of greasy wool to other processors, presumably without losing too much on the deal. The remaining 600 tonne of extremely expensive greasy wool has been gradually sold at well below cost, mostly without being scoured. The SMC company is now bankrupt and the scouring plant in Tianzhu City remains idle. For a general critique of county-level wool scouring plants which was written before the authors became aware of the fiasco associated with the Tianzhu wool scouring plant, see Brown and Longworth (1992b).

Chapter Sixteen

Cabucaer County (Xinjiang)

Cabucaer County is a Xibe Autonomous County on the extreme western border of Xinjiang Uygur Autonomous Region (XUAR). It is located in Yili Prefecture on the southern side of the Yili River. Yining, the prefectural capital, is located on the northern side of the Yili River just across from the northern part of Cabucaer County (Map 16.1).

Almost two-thirds of the inhabitants of Cabucaer belong to more than a dozen different minority nationalities, with Uygur (26%), Kazak (20%) and Xibe (14%) people being the most numerous. While a large part of the county is devoted to grazing, less than one in ten households are classified as being pastoral households. Sheep, and in particular fine wool sheep, are the major source of income for these pastoralists.

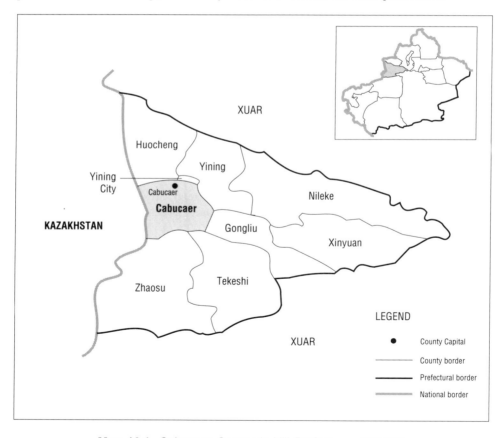

Map 16.1 Cabucaer County in Yili Prefecture of XUAR

16.1 Resource Endowment

Cabucaer County is considered a semi-pastoral county in Xinjiang. About 28% of the total value of agricultural, forestry, fishery and sideline output in Cabucaer is generated by animal husbandry activities. However, at the national level, Cabucaer is not included in the 266 pastoral and semi-pastoral counties described as the pastoral region of China.

Being close to Yining (a city of about 1 million) provides the people living in the agricultural areas in the north of Cabucaer County with many economic opportunities. In general terms, Cabucaer appears to be a relatively wealthy county.

16.1.1 General Features

The northern part of the county has extensive areas of irrigated agriculture. Most of the population in the county live in this northern agricultural belt located along the Yili River.

Between the irrigation areas and the high mountain range to the south and south east are dry stony/sandy loess plains. There are few permanent settlements on these vast plains. Most of the townships and villages not situated near the Yili River are located in the foothills of the high mountain range in the south and south east of the county.

Precipitation varies widely from reasonably high in the foothills of the mountains to less than 200mm on the loess plains.

Of the 6.645 million mu in Cabucaer County, 31,000mu are high mountains and 16,000mu are ice mountains to the south east and south of the county.

A wide range of soil types exists across the county. The soils are richer in the fertile agricultural plains whilst being infertile and stony/sandy in the semi-desert loess plains areas.

16.1.2 Land Use

In contrast with counties such as Dunhuang and Tianzhu, a high proportion of the land area in Cabucaer is usable.

The land use in Cabucaer County in 1990 was as follows:
- Total land area   6,645,000mu (or 4,430.0km^2)
  - usable land area   5,540,000mu
  - arable land   648,000mu
  - forestry land   136,700mu
  - "cutting" land (for making hay)   98,946mu
- Total rangeland   5,027,633mu
  - summer pasture   727,200mu
  - winter pasture   1,465,000mu
  - spring and autumn pasture   2,737,600mu
  - artificial pasture   78,633mu
  - semi-artificial pasture   14,000mu
  - forage land   5,200mu
- Total degraded land   278,000mu

Land degradation in Cabucaer County is not officially recognised as a major problem, as less than 5% of the land is said to be degraded. However, the Director of the Pasture Management Station indicated that in 1950 dry matter production per mu per year ranged from 280 to 400kg. By 1990, dry matter production had fallen to a range of 120 to 200kg per mu. Direct observation suggests the county has a major problem in as far as pasture land degradation is concerned. Vast areas of the extensive loess plains used for spring and autumn grazing appear badly degraded.

16.1.3 Population

In 1990, the total population of Cabucaer was 146,615. Of these people, about 78% are involved in agriculture, about 8% are pastoral households and the remainder are urban dwellers.

The ethnic composition of the population in Cabucaer County in 1990 was as shown in Table 16.1.

Population density   33.1 people/km^2

Table 16.1 Ethnic Groups in Cabucaer County, 1990

Nationality	Population (no.)	Proportion of total population (%)
Uygur	38,590	26.32
Han	52,808	36.02
Kazak	28,751	19.61
Hui	5,127	3.50
Kirgiz	343	0.23
Mongol	308	0.21
Xibe	19,782	13.49
Russian	93	0.06
Manchu	67	0.05
Daur	11	0.01
Uzbek	109	0.07
Tatar	15	0.01
Others	661	0.45
Total	146,615	100.00

16.1.4 Administrative Structure

The county is divided into 13 townships and two State farms. The Cabucaer State Farm and Sheep Stud, which is one of the case study fine wool sheep studs in Longworth and Williamson (1993), is one of these State farms. The other State farm is an agricultural farm with some cattle. Both these State farms belong to the prefectural government and, therefore, they are at the county level in the administrative hierarchy. Of the 13 townships, none are completely pastoral, but some are semi-pastoral. The 13 townships are sub-divided into 56 villages, only two of which are defined as being purely pastoral villages.

16.1.5 Livestock Population

The livestock population for Cabucaer county on 31 December 1991 was:
- cattle   38,329
- horses   13,275
- sheep   241,135
- donkeys   6,216
- goats   21,794
- camels   400

16.2 Improvements to Production Conditions in the Pastoral Systems

16.2.1 Improved Pastures

Very little pasture improvement is undertaken by households in Cabucaer County. A significant proportion of existing pasture improvement is in the form of irrigated lucerne pastures, since dryland pasture improvement is not considered a viable form of investment in most parts of the county. The total artificial pasture amounts to 78,633mu, of which 21,087mu is devoted to lucerne.

Between 1984 and 1990, the total area of pasture land fenced was around 142,000mu. Of this total, 40,000mu (which was aerial planted) was financed by the county government and the remaining 102,000mu was constructed by households.

The National Rangeland Law of 1985 is considered the basis for local measures aimed at controlling grazing animal numbers. "Pasture utilisation certificates" were introduced in July 1989 by the County government. These certificates are similar to those in use in Balinyou and Alukeerqin (see the Appendices to Chapters 8 and 11) in that they spell out maximum allowable stocking rates for each type of pasture contracted to the household in question.

Despite the issuing of these certificates to all the pastoral households in the county, the county AHB officials claimed that there were no provisions for enforcing the stocking rate regulations. Perhaps this is another example of a "policy mirage". Officially, appropriate steps have been taken to prevent overgrazing but, in practice, nothing has changed at the "grass roots".

16.2.2 Supplementary Feeding

Corn and barley are extremely important sources of supplementary feed for sheep, goats and cattle in Cabucaer County. The production of corn and barley per mu is extremely high (500kg and 200kg per mu respectively). Total consumption of grain by livestock in the county is about 6,600 tonnes per year.

16.2.3 Improved Sheep and Wool Production

- *Sheep production systems*

Sheep-grazing in Cabucaer County is a four-season system. Summer pasture is in the high mountains to the south east and south, spring and autumn pasture is on the plains between the mountains and the agricultural areas, and winter pasture is around the agricultural areas near the Yili River or in the low hills between the plains and the high mountains.

The animals are mostly owned by households although village collectives still own a small number of animals. Village collectives organise AI services, shearing and the movement to summer pastures, collect pasture use fees, and organise pasture improvement and road construction, etc. Most farmers bring their sheep to one of the shearing sheds attached to the county SMC purchase stations to be shorn, usually commencing in early June. Each of the six purchase stations operates two or three shearing sheds. At these sheds, the sheep are shorn by professional shearers using machines. The wool is then graded by the SMC technician and checked by the AHB technician. That is, the farmers' wool is carefully graded at shearing time. [Some farmers prefer to shear their sheep by hand shears because they think that the machines will damage their sheep or the wool.]

The shearers charge ¥0.60 per sheep. There is also a further ¥0.40 per sheep for electricity. The shearer's charge may rise or fall with the price of wool. The SMC at the township level pays some of this shearing cost (perhaps 70%) to encourage farmers to use this system for shearing.

Shearing professionally on clean boards improves the quality of the fleece and reduces contamination of the wool. The centralisation of shearing also makes grading more effective and efficient.

- *Proportion of sheep artificially inseminated*

Of the total 146,481 ewes in the county, only about 142,000 were mated in 1991. Of those mated, 53,574 (or 36.6%) were mated using AI. Virtually all of the ewes mated through AI were fine wool sheep.

The level of artificial insemination was low in 1991 owing to drought. In a typical year, the county AHB officials would expect about 45% of all ewes to be mated by AI. There is considerable variation in the use of artificial insemination between villages. In pastoral areas, almost 100% of households use AI. In agricultural areas, perhaps 55% of households use AI.

- *Sheep breeds*

Of the approximately 240,000 sheep in the county, about 80,000 (33.33%) are fine wool sheep and about 110,000 (45.83%) are improved fine wool sheep. The remaining 50,000 sheep (20.83%) are Kazak native sheep.

There were said to be no semi-fine or improved semi-fine wool sheep in the county.

- *Animal productivity*

No specific information on wool production per head is available. Overall production of greasy wool was 770 tonne in 1991.

- *Specific sheep breeding programs*

The SMC personnel (and other officials as well) claimed that farmers are buying and using poor quality rams much more than they have in the past. Perhaps the rams are actually better quality meat-producing rams. The farmers (especially in agricultural areas) are not keen to breed for fine wool quality at the expense of both the quality and quantity of mutton on their sheep.

There are no ram breeding stations in Cabucaer.

Most fine wool rams used in the county come either from Gongnaisi Stud which is the oldest and probably the best fine wool stud in China and which is located in Xinyuan County of Yili Prefecture (Map 16.1), or from Cabucaer State Farm and Fine Wool Sheep Stud.

The history of fine wool sheep-breeding in Cabucaer County began in 1955 when all the sheep were local Kazak sheep. In 1955, Xinjiang Fine Wool rams from Gongnaisi were used both by AI and in natural mating to begin upgrading the seven flocks which existed on the Cabucaer State Farm at that time. Since 1958, all sheep flocks on the State farm have been mated by AI.

In the 1950s and 1960s, the pasture was much better and the area of pasture available to the State Farm was larger than at present and sheep improvement progressed rapidly. The State Farm was recognised at the county level as a commercial sheep farm in 1957 (hence this is taken as the year in which the sheep farm was formally established). In 1960, it became a sheep stud under the Prefectural AHB.

Until 1963, the management of the State Farm were mainly concerned with improving the local Kazak sheep breed by infusing Xinjiang Fine Wool blood. From 1963 to 1966, they concentrated on upgrading all sheep to the Xinjiang Fine Wool Sheep standard. From 1967 to 1972, they raised the quality of the Xinjiang Fine Wool Sheep flock by culling older sheep, sheep with short wool, and low-yielding sheep (<3kg of greasy wool). From 1972 to 1976, they established the nuclear breeding group by selecting ewes with longer staple and denser fleeces.

During the 1966 to 1974 period, the average fleece weight (greasy) was about 5.00kg with a clean yield of 40%. The wool was too yellow, the crimp was poorly defined, the staple was too short, and the fleeces were not dense enough.

Since 1974, the breeding program at Cabucaer Stud has been aimed at improving these characteristics. Now the average staple length is 8.37cm with a clean yield of 53% and the grease colour is much better.

After 1974, an experiment was conducted at the Stud using 25% and 37.5% Australian merino rams from Gongnaisi. As a result it was concluded that the 37.5% rams were superior. In 1986, six Australian merino rams were introduced. Later, another two Australian rams, supplied by the Xinjiang AHB, were acquired. Over the 1974 to 1986 period, 52 Chinese Merino rams were introduced from Gongnaisi, and from 1986 to 1991, a further 32. All these Gongnaisi rams had Australian merino blood in them.

The whole Cabucaer Stud nuclear flock of 4,200 ewes now meets the Chinese Merino breed standard. The remainder of the fine wool sheep on the State farm satisfy the Xinjiang Fine Wool breed standard. For more detail on Cabucaer Stud see Longworth and Williamson (1993).

In recent years, the fine wool stud breeding program has been handicapped by the poor condition of the pastures and the wool price recession. As a result the incomes of the herdsmen and the technicians involved in the fine wool growing activities on the State farm have suffered a significant drop in income relative to the wool boom period of 1985 to 1988.

The management of Cabucaer State Farm is exploring fat lamb production, presumably as an alternative to commercial fine wool production. In 1991, the farm received 30 Suffolk ewes and some rams imported from Australia. These sheep did well and the managers of the farm are planning to expand their fat lamb producing enterprise greatly in the future.

- *Ram purchases for AI*

The Cabucaer County Animal Husbandry Bureau purchases rams for AI mating as follows:

(a) Rams are bought from Gongnaisi Stud at prices set by the Provincial AHB.
  – Specially prepared AI rams (i.e., rams given special care after birth including extra feed) cost ¥450/head
  – Special Grade rams cost ¥350/head
  – Grade I rams (for heat testing and natural mating) cost ¥290/head
(b) Rams are also bought from Cabucaer Stud but the County AHB officials interviewed did not know much about these purchases or the prices paid. While this seems to suggest that the County AHB buys relatively few rams from the Cabucaer Stud, other county officials claimed that the AHB buys 50% of its rams from Gongnaisi and 50% from Cabucaer.

- *Specific animal health programs*
  Sheep in the county are vaccinated for Leptospirosis and "Lamb Diarrhoea".

16.2.4 Improved Goat Breeds

There are no improved or fine cashmere goats in the county.

16.2.5 Improved Cattle Breeds

Improved cattle in the county number 5,383. The main improved breed in the county is the Xinjiang Brown breed of cattle. These cattle have been bred using a cross between the local Xinjiang breed and the Arlaton breed imported from Russia.

16.3 Socio-Economic Development

16.3.1 Income Growth

The available net per capita income data available are presented in Table 16.2. On the basis of this information, it would seem that pastoral incomes rose much faster than rural incomes in general between 1984 and 1988 but since then pastoral incomes (especially in real terms) have tended to fall.

16.3.2 Changes to Commodity Prices

The markets for wool, meat and cashmere were declared open in 1992 in Cabucaer County. In principle, therefore, there will no longer be any government control over the prices of these commodities. However, it is uncertain how these open markets will operate during periods of depressed prices. There are "instructional" prices set by the Price Bureau, and the persons interviewed at the Cabucaer County Policy Research Division indicated that during periods of depressed prices, the government is planning to support certain wool prices. For example, the planned support price for fine wool grade I in 1992 was ¥8 per kg. This applies to wool which is of 64 count or better and 6 to 7.9cm in length. Wool which is of a length less than 6cm will receive 50% of the support price. There are no plans to support the price for local wool produced in the county. The main reason given for this lack of support was that a large proportion of local wool produced is used for household consumption. The actual average prices paid in 1991 for fine wool Grades I and II were ¥8.40 and ¥8 per kg respectively.

Table 16.2 Average Net Income Per Capita in Cabucaer County, 1979 to 1991

Year	Income per capita for rural population		Income per capita for pastoral population	
	Nominal	Real*	Nominal	Real*
	------ (¥/year) ------		------ (¥/year) ------	
1979	119.77	117.4	104.10	102.1
1980	207.77	192.2	167.50	154.9
1981	254.23	229.7	n/a	n/a
1982	413.64	366.7	n/a	n/a
1983	477.78	417.3	n/a	n/a
1984	497.67	422.8	193.40	164.3
1985	513.36	400.7	397.30	310.1
1986	555.84	409.3	423.10	311.6
1987	567.07	389.2	515.47	353.8
1988	596.34	345.5	585.62	339.3
1989	n/a	n/a	583.48	286.9
1990	n/a	n/a	583.47	280.9
1991	n/a	n/a	528.19	n/a

*Deflated by "Overall Retail Price Index" in State Statistical Bureau of the PRC, *China Statistical Yearbook 1991*, p.199. The base year is 1978.

16.3.3 Timing of Introduction of Household Production Responsibility System

The household responsibility concept was first established in Cabucaer between 1958 and 1965 and was known as the "three contracts and one reward system". (See also Section 3.5.4.) The three contracts involved were firstly for animal numbers; secondly for lambing survival rates; and thirdly for specified commodity production, in particular wool and lamb production. The reward specified that if these targets were met, a specified number of work points would be allocated. Households could earn extra work points by exceeding the targets.

There was no household responsibility system operating during the Cultural Revolution (1965 to 1975). The system used during that period was termed "Xue Da Zai" or "learn from Da Zai". This in effect meant that no responsibility system should be applied because all workers were equal and the State and collectives should pursue an egalitarian path.

The concept of household responsibility was reintroduced for a second time between 1975 and 1983 and functioned in a manner very similar to the earlier system which operated between 1958 and 1965. The major difference between the two systems was in the reward offered for exceeding the targets specified in the contracts. Under the second household responsibility system, pastoral households which exceeded their target for lamb production were allowed to retain 60% of over-target lambs for private use. Similarly for wool production, households which exceeded target production were allowed to retain 60% of over-target wool production for sale through the State marketing system.

Eventually, the current household production responsibility system was introduced in 1984 with the sale of animals to individual households. Land contracts were introduced in 1987 and were fully implemented throughout the County by 1989. The major reason for the three-year period of implementation was said to be that it took this length of time to survey the pasture land and divide it into various categories according to quality considerations. Winter and summer pasture lands are said to be fairly precisely defined.

However, spring and autumn pastures are less well defined and, in a large number of cases, are still grazed in common.

The length of the land contract for the current responsibility system is said to be 30 years. For cropland, the contract length is 20 years. The people being interviewed had no idea what would happen when the contracts expired. The only answer offered was that the County government would follow Central government policy (i.e. if the contract system was abolished following expiration of the contracts, households would lose their land).

16.3.4 Level of State Assistance

Cabucaer County is regarded as being a "deficit" county within the national agricultural banking system (i.e. the operations of its Agricultural Bank are subsidised by "surplus" counties in the prefectural group of counties). The losses incurred by the Agricultural Bank of China (ABC) in Cabucaer County are a direct result of Central government policy which requires that the ABC loan money to the county Grain Bureau at rates which are not considered profitable. Grain is a Category I commodity and, as such, loans to purchase grain attract a subsidised interest rate.

It seems that while grain is a subsidised commodity, the subsidy is not being totally borne by the State. Instead it would appear that part of the burden of the subsidy has been shifted onto the wealthier counties which in Yili Prefecture are the pastoral counties. In this regard, pastoral counties in this prefecture are subsidising the operations of predominantly grain-growing counties such as Cabucaer.

16.3.5 Development of Pastoral Product Processing

A major development for the pastoral economy in Cabucaer County has been the establishment of a factory for the processing and export of meat to Kazakhstan.

16.3.6 Growth in Output of Major Pastoral Commodities

The major pastoral commodities produced in Cabucaer County are wool, cashmere and meat (beef, sheep and goat meat). The output of these commodities during the period 1970 to 1991 is shown in Table 16.3.

Table 16.3 Wool, Cashmere and Meat Production in Cabucaer County, 1970 to 1991

Year	Wool	Cashmere	Meat (sheep, goats, cattle)
	(tonne)	(tonne)	(tonne)
1970	309.80		
1975	544.77		
1980	556.26		
1983	662.70	1.51	
1984	544.75	1.42	2,006
1985	653.65	1.90	2,610
1986	705.50	1.78	2,044
1987	857.00	2.36	2,205
1988	781.00	1.93	2,090
1989	806.00	3.13	2,126
1990	822.00	2.28	1,957
1991	770.00	2.21	2,147

Chapter Seventeen

Hebukesaier County (Xinjiang)

Hebukesaier County is one of the most remote places in the world. It lies in the north east corner of Tacheng Prefecture which is on the north western border of China between the Xinjiang Uygur Autonomous Region (XUAR) and Kazakhstan (Map 17.1). The position on Earth remotest from the sea is at Lat. 46°16.8'N, Long. 86°40.2'E in XUAR just south of Hebukesaier County. This point is 2,648km from the nearest sea.

Hebukesaier, a Mongolian Autonomous County, is the only autonomous county in Tacheng Prefecture and one of only six such counties in the XUAR. Mongolians constitute 35% of the population but there are nearly as many Kazaks and people of other ethnic groups as there are Mongolians, so that almost two-thirds of the inhabitants of Hebukesaier belong to minority nationalities.

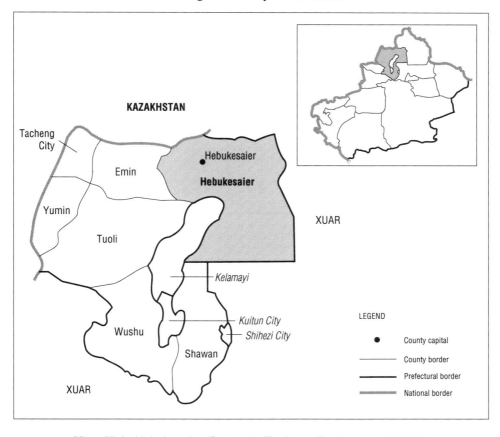

Map 17.1 Hebukesaier County in Tacheng Prefecture of XUAR

17.1 Resource Endowment

Hebukesaier is recognised by all levels of government as a pastoral county. However, it is not designated by either the Xinjiang or the Central governments as being a poverty county. The only natural resources available to the people living in this sparsely populated remote area are the rangelands and the short rivers fed by melting snow which make irrigation possible in some areas.

17.1.1 General Features

There are substantial mountain ranges both to the north and to the south of Hebukesaier Town, the county capital. The summer pastures are in the northern mountains, the spring/autumn pastures are on the plains in between the mountains, and the winter pastures are in the hills and mountains to the south.

Pastoral conditions are harsh with a cold dry climate of long cold winters (average 151 days) and short summer seasons (average 82 days). Rainfall averages 98mm per year but as the data in Table 17.1 demonstrate, the annual average rainfall varies considerably from year to year. Since 1970, the lowest rainfall recorded was 64.5mm in 1978 while the highest was 168.4mm in 1983.

Table 17.1 Annual Rainfall in the Capital of Hebukesaier County, 1970 to 1991

Year	Annual rainfall	Year	Annual rainfall
	(mm)		(mm)
1970	94.8	1981	116.9
1971	77.9	1982	67.3
1972	165.7	1983	168.4
1973	133.3	1984	112.5
1974	70.0	1985	70.4
1975	131.3	1986	94.4
1976	116.7	1987	86.5
1977	76.1	1988	110.6
1978	64.5	1989	150.6
1979	78.9	1990	113.2
1980	82.7	1991	89.7

17.1.2 Land Use

Considerably less than half the land area of Hebukesaier is regarded as usable pasture while the area suitable for agriculture represents only about 0.2% of this vast county.

- Total land area   45,600,000mu (or 30,400km^2)
- Total rangeland (pasture)   23,559,000mu
- Total usable pasture area   19,000,000mu
  - summer pasture   1,640,000mu
  - spring and autumn pasture   6,090,000mu
  - winter pasture   5,850,000mu
  - usable pasture for horses   5,470,000mu
  - "cutting" land   37,100mu

- Arable land   92,000mu
- Forestry land (1991)   1,067mu
- Total degraded land (1984)   1,767,000mu
  - heavily degraded   67,000mu
  - medium to lightly degraded   1,090,000mu

The 1984 figures for the area of degraded pastures presented above (which were the latest official data available) come from a pasture resource survey conducted in that year by the Tacheng Prefectural AHB. However, the Hebukesaier County AHB Pasture Management Station officials said that in 1992 almost all winter and spring/autumn pastures were degraded. Heavy degradation was attributed to locust and rodent damage whilst light degradation was attributed to overstocking. According to the Pasture Management Station officials, pastures in Hebukesaier County were overstocked by at least 100,000 sheep equivalents in 1992.

The overall impact of overstocking and the consequent rangeland degradation in Hebukesaier has been to reduce dramatically the productivity of the native pastures in the county. For example, according to Pasture Management Station officials, pastures in Hebukesaier in 1950 yielded on average between 280 and 400kg of dry matter per mu per annum. By 1990, however, production had fallen to between 120 and 200kg.

17.1.3 Population

- Total urban and rural (1990)
  - total   43,506   (100%)
  - urban   15,132   (34.78%)
  - rural   – agricultural   15,842
            – semi-pastoral   6,832
            – pastoral   5,700
            Sub-total rural   28,374   (65.22%)
- Ethnic composition (1990)
  - Han   14,850
  - Uygur   1,081
  - Kazak   12,015
  - Mongolian   15,117
  - Others   443
- Population density (1990)   1.43 people/km^2

17.1.4 Administrative structure

There is one town and five townships in the county. One of the townships is agricultural, two are semi-pastoral and two are pastoral. In addition, there are four State farms in the county.

17.1.5 Livestock Population

In 1990, sheep were easily the most numerous herbivorous animals but in terms of sheep equivalents, they constituted less than half the total stocking rate.

- Total number of livestock (1990)   379,100
- Sheep   240,000
- Goats   78,700
- Cattle   32,700
  - dairy cattle   2,100
- Horses   19,700
- Donkeys   800
- Mules   300
- Camels   5,700

The long-term trends in total sheep equivalents and in sheep numbers are presented graphically in Fig. 17.1. Clearly there were large increases in both until about 1965. For the last 25 years sheep numbers have remained relatively constant but there has been a steady increase in the total stocking rate (owing to increases in goats and cattle) since 1978.

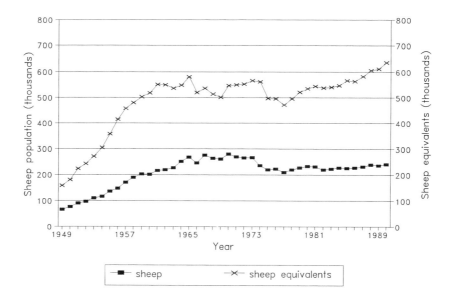

Fig. 17.1 Total Stocking Rate in Sheep Equivalents and Sheep Numbers in Hebukesaier County, 1949 to 1990

17.2 Improvements to Production Conditions in the Pastoral Systems

17.2.1 Improved Pastures

Total area of improved pastures planted in the 1970 to 1991 period:
- cutting land (for hay making)   37,124mu
- artificial pastures   34,308mu
- semi-artificial pastures   nil

Fencing around pastures began in 1970 as earth walls but in the 1980s stone walls were used. Between 1970 and 1991, a total of 182,362mu of pasture had been fenced.

- *Specific pasture improvement programs*

Farmers are constructing artificial pastures in order to compensate for any decline in overall productivity due to overstocking of the natural rangelands. The Pasture Management Station in Hebukesaier County declined to indicate whether or not an increase in artificial pastures would only add to the overstocking problem rather than reduce it. When artificial pastures are being established, barley or wheat are usually planted in the first year followed by lucerne and pasture grasses in the second year.

- *Specific pasture management programs*

The pasture monitoring division of the Hebukesaier County AHB Pasture Management Station is responsible for:
- implementing the National Rangeland Law;
- utilisation of pasture lands;
- preventing fires;
- pasture protection (preventing the digging for medicinal herbs, stone removal, etc.); and
- collecting pasture management fees and penalties.

The penalties which can be imposed on offenders are as follows:
- driving on pastures: <¥100
- digging medicinal herbs: confiscate digging tools and herbs and penalise 2 to 3 times the value of the herbs
- lighting fires in a dry season: ¥40 - ¥50
- use of pasture without permission: penalty of 4 times the value of production from the pasture
- exceed stocking rates: herdsmen must destock to specified levels within 15 days.

Xinjiang, like IMAR, has a sub-law of the National Rangeland Law. It came into effect on 1 September 1989. Pasture Utilisation Certificates apply to each household in Hebukesaier County and specify land area, boundaries, stocking rates, etc. Pasture management fees have been collected since 1991, two years after the introduction of the Pasture Utilisation Certificates. According to the Policy Research Division with the Hebukesaier government, the annual management fee charged is dependent upon the quality of the land being used (i.e. for Class I land the fee is ¥0.3 per mu, class II land ¥0.2 per mu, class III land ¥0.15 per mu, class IV land ¥0.1 per mu, and for class V ¥0.05 per mu). Interviews at the village and household level revealed that rather than the fee being set according to land quality, it was being collected on a per animal basis. The rate per year was ¥0.8 per animal. Interestingly, for the households interviewed, the amount they had to pay on this basis amounted to approximately half the total charge which would have been due if fees had been collected according to rates per mu specified by the officials of the county Policy Research Division.

The Pasture Utilisation Certificates specify stocking limits for summer, winter, spring and autumn pastures. However, the Pasture Management Station has not checked to ensure that herdsmen have complied with the limits stated in the certificates. Once again, as in Cabucaer County, policies are in place but nothing is done at the grass roots to ensure that they are properly implemented.

Limits on stocking rates have been based on pasture surveys conducted between 1981 and 1984. The possibility (certainty?) that the pasture has degraded further since 1984 was not taken into consideration when limits on stocking rates were specified in 1990. If there has been further degradation of the pastures, then the stocking rates specified in the Pasture Certificates may have been set too high. Furthermore, the incidence of high rainfall or drought is not taken into account in determining the stocking rates.

As mentioned earlier, local officials consider that in 1992 the county was overstocked by at least 100,000 sheep equivalents. The Xinjiang government has directed fine wool growing counties such as Hebukesaier to increase their fine wool sheep numbers by 2% per year. Such a blanket directive from the XUAR government takes no account of the particular circumstances in counties such as Hebukesaier. In all fine wool growing areas of Xinjiang (and in China generally), it will be self-defeating to increase further the total number of grazing animals. Fine wool sheep can only be increased by reducing the number of sheep equivalents represented by other types of grazing animals. The data in Fig. 17.1 suggest that sheep have become relatively less popular over the last decade. How this trend is to be reversed so that more of the rangeland resources can be devoted to fine wool sheep has not been spelt out by the Xinjiang government. Indeed, in most areas the pastures are already overstocked. Therefore, there is an urgent need to reduce the total number of grazing livestock. Under these circumstances, the directive to increase fine wool sheep numbers would seem to be in contravention of the 1985 National Rangeland Law and the 1989 Xinjiang Autonomous Region Sub-Law relating to the National Rangeland Law. Both these Laws require the introduction of policies which reduce the grazing pressure on pastures.

The Pasture Management Station officials originally claimed that all the usable pastures had been marked and contracted out to herdsmen. However, this claim was challenged by village leaders and by farmers. In many cases it was argued that, while the land had been contracted out, boundaries for the contracted land had not been identified and the land continued to be grazed in common. When questioned further about the contracting of pastures, the Pasture Management Station officials agreed that perhaps groups of households used designated pasture land, but it was unusual to find large numbers of households using a particular area of pasture. These pasture-using groups were said to vary in size from four to eight households.

The inconsistency between the original statements of the county officials and the real situation as described by the village leaders and individual farmers is an example of a common problem in China. Officials at the county level will often state that official policy has been implemented successfully in their county. Yet, as described above, subsequent interviews at the township, village, and household levels demonstrate that the policy in question has not, in fact, been implemented or if put in place, there have been significant "adjustments to local conditions". The only way to appreciate the true situation at the "grass roots" is to conduct surveys below the county level (i.e. at the township, village and household level). Otherwise policy failures, which are all too common, will go undetected.

17.2.2 Supplementary Feeding

Sheep are fed both hay and grain during the long winter. The hay is typically produced from cutting land contracted to the sheep-raising household. Grain, on the other hand, must be purchased from the State Grain Bureau under the control of the Ministry of Domestic Trade (formerly the Ministry of Commerce and Ministry of Raw Materials). In recent years, the price of grain has often risen sharply. For example, in mid-winter 1992, the Grain Bureau raised the price of maize from ¥0.38 per kg to ¥0.48 per kg. This increase and similar sharp price rises for other grains reflect moves by the Central government to bring State grain prices more in line with free market prices. However, as grain prices rise, pastoralists will become increasingly reluctant to incorporate feed grains into livestock diets. Since the prices paid for pastoral commodities such as wool do not offer significant premiums for quality, there is little financial incentive for farmers to feed their over-wintering animals more than survival rations.

17.2.3 Improved Sheep and Wool Production

- *Sheep production systems*

Sheep are raised in the agricultural, semi-pastoral, and pastoral areas of Hebukesaier. As elsewhere in China, the production systems in these areas are varied, with the specialist pastoral households having much larger flocks and most of the fine and improved fine wool sheep. The distribution of households raising sheep in Hebukesaier by flock size is shown in Table 17.2. More than 10% of the households have in excess of 150 head.

The majority of the sheep are raised in the pastoral areas which cover most of this extremely large county. As mentioned earlier, the summer pastures are in the mountains to the north while winter pastures are in the hills and mountains to the south. Spring/autumn grazing occurs on the plains between the two mountain ranges. Flocks are moved the 20 to 80km distance from winter to summer pastures in the spring and back again in the autumn. On average, the sheep spend about 82 days on the summer pastures and 151 days on the winter pastures. The remaining 132 days are spent travelling backwards and forwards across the spring/autumn pastures on the plains.

Table 17.2 Distribution of Sheep-Raising Households in Hebukesaier County by Flock Size, 1991

Flock size	Households	
(no.)	(no.)	(%)
< 60	1,095	32.3
60 – 80	724	21.4
80 – 100	725	21.4
100 – 150	499	14.7
> 150	342	10.2
Total	3,385	100.0

Source: Hebukesaier Animal Husbandry Bureau

- *Proportion of sheep artificially inseminated*

It is not known how many sheep are actually artificially inseminated in the county. Despite having issued a target whereby at least 85% of fertile ewes should be artificially inseminated, the Animal Husbandry and Veterinary Station had not checked village AI centres to ensure that this target has been achieved.

- *Specific sheep breeding programs*

The program to improve fine wool sheep in Hebukesaier County commenced in 1952 with the establishment of one breeding station and an AI centre capable of inseminating 4,500 ewes. Between 1952 and 1967 rams were introduced from Gongnaisi Stud and Tacheng Stud. These rams were crossed with local Mongolian and Altay ewes. Table 17.3 provides actual data for 1967, 1977 and 1987 together with estimates for 1992 on the changing composition of the Hebukesaier sheep flock. While there appears to have been a steady increase in the number of genuine fine wool sheep up to 1987, the estimates for 1992 seem extremely optimistic. Another feature of the data is the sharp decline in the number of local sheep between 1967 and 1977 which was reflected in the size of the total flock. During the 1980s (discounting the 1992 estimates), there was a swing back to native breeds but only a modest increase in fine and improved fine wools.

Despite providing the optimistic estimates for 1992, officials of the Hebukesaier AHB claimed that at present (1992) farmers did not regard it as worthwhile to improve their sheep for fine wool production. Consequently, farmers were using low grade rams rather than pure-bred fine wool rams. This viewpoint was supported by farmers who were interviewed. They claimed that fine wool sheep were not well adapted to the poor pastures and cold climate which constitute the grazing environment in Hebukesaier County. Moreover, both individual private farmers and production team leaders on the State farm surveyed reported that meat sheep were much more profitable than wool sheep. For example, an Altay lamb is able to attain a liveweight of around 40kg within 12 months with a sale price of between ¥80 and ¥100.

Table 17.3 Long-Term Changes in the Composition of the Hebukesaier County Sheep Flock, 1967 to 1992

Type of sheep	1967		1977		1987		1992*	
	Number of sheep	Proportion of flock	Number of sheep	Proportion of flock	Number of sheep	Proportion of flock	Number of sheep	Proportion of flock
	(no.)	(%)	(no.)	(%)	(no.)	(%)	(no.)	(%)
Fine wools	2,230	0.8	14,411	6.9	23,757	9.8	72,000	30.0
Improved fine wools	120,448	43.5	154,022	73.9	156,868	64.8	120,000	50.0
Local sheep	154,207	55.7	40,031	19.2	61,616	25.4	48,000	20.0
Total	276,885	100	208,464	100	242,241	100	240,000	100

*Estimates only.
Source: Hebukesaier County Animal Husbandry Bureau

In Hebukesaier County, there exists a policy which prohibits herdsmen from crossing improved or fine wool ewes with local breed rams. The policy is aimed at retaining the integrity of the improved and fine wool sheep flock, and is said to be advocated by the XUAR government. However, the county Animal Husbandry and Veterinary Station did not appear to vigorously enforce it. The strongest action that officials from the Station are prepared to take with regard to herdsmen who choose to ignore the breeding policy is to "have a talk with them and to encourage them to conform".

The County Animal Husbandry and Veterinary Station has introduced a number of semi-fine wool rams from other counties in Tacheng Prefecture and from Kazakhstan to cross with local ewes in acknowledgment of the difficulty of raising fine wool sheep in this part of China.

There is said to be no money for the Station to conduct research on various aspects of fine wool production. Very little money seems to be allocated to the station in order to facilitate the breeding program.

17.3 Socio-Economic Development

17.3.1 Income Growth

Nominal net rural incomes per capita increased substantially during the 1980s but levelled off after 1989 (see Table 17.4). Real incomes reached a peak in 1987 and have fallen significantly since that year.

Table 17.4 Average Net Rural Incomes Per Capita in Hebukesaier County, 1970 to 1991

Year	Nominal income	Real income*
	(¥/year)	(¥/year)
1970	151.67	156.7
1971	146.90	153.0
1972	143.40	149.7
1973	114.01	118.3
1974	95.49	98.5
1975	117.55	121.1
1976	127.18	130.6
1977	97.12	97.8
1978	123.51	123.5
1979	142.04	139.3
1980	139.50	129.0
1981	138.87	125.4
1982	157.58	139.7
1983	293.62	256.4
1984	383.20	325.6
1985	471.87	368.4
1986	552.39	406.8
1987	618.64	424.6
1988	683.33	395.9
1989	755.70	371.5
1990	746.97	359.6
1991	752.19	n/a

*Deflated by "Overall Retail Price Index" in State Statistical Bureau of the PRC, *China Statistical Yearbook 1991*, p.199.
The base year is 1978.

It seemed that households living on the State farm surveyed had markedly lower per capita incomes than the households in a pastoral village in the same general area. If this is the case, it is unusual since State farm households usually enjoy relatively good incomes.

17.3.2 Commodity Prices

In remote isolated counties such as Hebukesaier, the traditional administered pricing system still applies to many goods and services which are traded in free markets in the more advanced parts of China. For example, in Hebukesaier the County Price Bureau is responsible for setting the price of daily-use goods such as soap and detergents, essential clothing, most tools, electricity, etc. while major bulk commodities such as grain, coal and cement are sold at prices set by the provincial Price Bureau.

The markets for beef and mutton have been more or less "open" throughout the XUAR since 1985 but the wool market was only officially opened in 1992. Therefore, although there has been an increasing degree of "slackness" in policing the restrictions on free trade in wool in recent years, until 1992 the SMC was the only official buyer of wool from private households. As discussed in Section 7.6.1, since about 1986 State farms have not been compelled to sell to the SMC.

In Hebukesaier, the County SMC has adhered to the "instructive" purchase prices established by the provincial Price Bureau. For example, the full list of greasy wool prices in 1991 was as follows:

Fine wool	– Special Grade	9.73 ¥/kg
	– Grade I	8.76 ¥/kg
	– Grade II	8.00 ¥/kg
Improved wool	– Grade I	7.62 ¥/kg
	– Grade II	5.72 ¥/kg
Skin (fine) wool		4.00 ¥/kg
Skin (improved) wool		2.50 ¥/kg

To give the local SMCs some flexibility, in 1990 and 1991 the provincial Price Bureau permitted the purchase of wool at prices which were within a ±10% band around the official prices listed above. Despite this increased freedom in terms of pricing, the SMC in Hebukesaier (and in other parts of the XUAR as well) lost market share. They purchased 360 tonne of greasy wool in 1989, 197 tonne in 1990 and 180 tonne in 1991. It seems that in 1990 and 1991 the textile mills (illegally) purchased the best wool from private households as well as from the State farms.

At the time of the survey just prior to the commencement of the 1992 wool-buying season, the Hebukesaier County SMC had taken advantage of the newly announced free market for wool in the XUAR to enter into contracts with the four State farms guaranteeing them the 1991 official prices plus 21%. While this pricing strategy was likely to lift the quantity of wool handled by the County SMC in 1992 to about 460 tonne, it may have proved to be too generous because most people in the XUAR expected the free market wool prices in 1992 (including grade price differentials) to remain more or less at the level of the official prices set for 1991.

17.3.3 Development of Pastoral Product Processing

The county has no plans to build a scouring plant or small woollen mill. Production is simply too small to justify such a large outlay of investment capital.

17.3.4 Growth in Output of Major Pastoral Commodities

The major pastoral commodities produced in Hebukesaier County are wool, cashmere, mutton, goat meat and beef. The available production data for these commodities are presented in Table 17.5. [There was no data available for the different grades of wool or for the different kinds of meat.] In the case of wool, about 250 tonne or almost half the total output was grown on the four State farms in 1991.

Wool production increased considerably in the 1970s, reaching a peak in 1983. Since 1983, wool output seems to have plateaued but cashmere production has more than doubled and meat output has almost doubled. This reflects the change in the wool/cashmere and the wool/meat price relativities. As elsewhere in the pastoral region of China, the pastoral households in Hebukesaier have demonstrated their capacity to change their output mix in response to price incentives.

Table 17.5 Wool, Cashmere and Meat Production in Hebukesaier County, 1970 to 1991

Year	Wool	Cashmere	Meat
		(tonne)	
1970	281.0	14.0	842.0
1971	280.0	14.0	841.0
1972	282.3	14.0	840.0
1973	340.3	13.9	858.2
1974	339.3	12.9	1,023.7
1975	317.7	9.5	1,759.1
1976	331.0	7.5	1,896.0
1977	318.0	7.7	1,715.3
1978	394.5	9.4	1,461.0
1979	447.5	8.9	1,820.0
1980	449.9	12.9	1,563.0
1981	443.8	13.6	1,768.6
1982	443.5	8.0	1,803.7
1983	487.0	7.6	1,497.0
1984	425.8	8.6	1,801.5
1985	467.0	10.0	2,216.9
1986	486.9	7.9	2,082.0
1987	432.2	11.2	2,207.0
1988	467.0	10.0	2,044.0
1989	470.0	11.5	1,933.2
1990	490.0	15.0	2,164.0
1991	506.0	18.0	2,666.0

PART IV

THE BIG PICTURE

Chapter Eighteen

Constraints to Development in Pastoral Areas

A subtle interplay between public policy, technology and the environment exists in the pastoral areas of China. The case studies presented in this book provide evidence that the improvements in the incomes of traditional pastoralists achieved in the 1980s are not sustainable unless major constraints to progress are addressed. If these issues are not effectively dealt with in the near future, the remarkable achievements since 1978 will be lost; the ecological damage to the rangelands, already substantial, will become disastrous; and the minority nationality pastoralists will become even more alienated from mainstream Chinese society. The ecological, economic and political stability of a vast sector of China is at risk.

After 1978, as with the rest of China, people living in the pastoral region experienced changes (reforms) in many fields of public policy. At the same time, a whole new spectrum of biological, mechanical and informational technology became available to Chinese pastoralists. The stimuli and incentives created by the policy changes interacted with the available new technology to produce significant short-term gains. Paradoxically almost, the policy-technology interplay which generated these benefits also represents the major threat to the long-term sustainable development of the pastoral areas. In the simplest possible terms, much of the progress achieved since 1978 has been at the expense of the rangelands. That is, the chief natural resource available has been "mined" to achieve short-term improvements in living standards. In the long term, such a development strategy is not sustainable.

The first step to remedying the situation is to understand why the post-1978 mix of policies, technology and the environment has not produced sustainable development. In particular, it is important to appreciate the factors which have led to overstocking and hence to increasing degradation of the rangelands. Of course, as pointed out in many places in this book, degradation and destruction of natural pastures are not new phenomena in China's pastoral region. While there are some particularly serious examples which date from the middle of last century, rangeland degradation on a broad scale has become an increasingly serious problem since 1949. The policy reforms after 1978, however, have created a new milieu.

The situation which has existed in the pastoral areas of China since 1978 is far too complex to be portrayed in a simple diagram. Nevertheless, Fig. 18.1 captures the major factors involved. Three broad sets of policy-related issues, namely population pressures, market distortions and institutional uncertainties, have been interacting with the adoption of technology in a unique environmental setting to generate development outcomes. It is the interplay between these various elements of "the big picture" which is of critical concern. To a large extent, the policy mix has encouraged rangeland degradation and technology has provided the means by which this has been achieved.

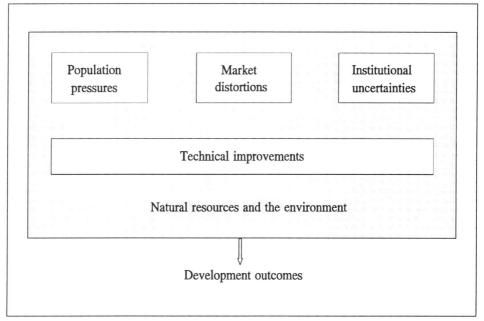

Fig. 18.1 Major Factors Influencing Development Outcomes in the Pastoral Areas of China

18.1 Population Pressures

The human population density in the pastoral areas of northern and north western China is much lower than in the eastern and southern parts of the country. Nevertheless, the general lack of resources and the rapid increase in the population in these areas since 1949 has placed increasing pressure on the rangeland which is the principal natural resource available, and ultimately on the potential for sustainable development in these ecologically fragile environments.

18.1.1 Link Between Population Pressure, Poverty and Rangeland Degradation

The process by which population pressure contributes to rangeland degradation is, in a general sense, well understood. That is, as population pressure rises, per capita incomes decline, and day-to-day financial pressures grow, leaving farmers with little choice but to increase their intensity of range use to levels which eventually begin to destroy the natural pastures. This process leads to second-round effects whereby as the natural resource base declines, day-to-day financial pressures increase further, in turn forcing farmers to further intensify their use of the rangeland.

Thus farmers in pastoral areas find themselves in an increasingly untenable position. The pressures of rising population and falling living standards force them to adopt progressively more intensive management strategies in an attempt to satisfy their short-run needs. However, in the longer term, these strategies contribute to the destruction of their environment and ultimately, in so doing, the basis for much of their livelihood.

In the case of China, there is a growing pool of empirical evidence demonstrating the strong link between population pressure, poverty and environmental degradation.

Rozelle and Huang (1993), for instance, note the strong association between high rates of population growth, a worsening of poverty, and environmental degradation in agricultural areas. Similar associations are also evident in pastoral areas. For example, prefectures in the IMAR with the highest proportion of impoverished pastoral and semi-pastoral counties and generally high population densities, such as Chifeng City, Zhelimu and Yikezhao League, also have the highest level of rangeland degradation (Table 18.1). Conversely, prefectures with lower population densities and without any impoverished counties such as Hulunbeier, Bayanzhuoer and Alashan, and Xingan with only one poor county, have lower levels of rangeland degradation.

Table 18.1 Incomes, Poverty, Population Density and Rangeland Degradation in Pastoral and Semi-Pastoral Counties of the IMAR

City or league (prefecture)	National income per capita in 1988[1]	Number of pastoral and semi-pastoral counties with NSP status	Number of pastoral and semi-pastoral counties with PSP status	Total number of pastoral and semi-pastoral counties	Average population density for pastoral and semi-pastoral counties in each prefecture	Percentage of total pasture area degraded in each prefecture
	(¥)	(no.)	(no.)	(no.)	(person/km²)	(%)
Chifeng City P.	630	6	1	7	29.73	84
Hulunbeier L.	1,223	0	0	7	9.89	14
Xingan L.	654	1	0	4	22.12	25
Zhelimu L.	861	1	3	7	46.04	67
Xilinguole L.	1,098	0	4	10	3.87	49
Wulunchabu L.	586	2	2	4	14.71	26
Yikezhao L.	914	4	3	8	13.57	74
Bayanzhouer L.	1,167	0	0	4	9.92	32
Alashan L.	1,910	0	0	3	0.60	9

[1]"National income" is defined as the sum of Net Output Value of Agriculture, Industry, Transportation, Construction, and Commerce, the five material production sectors of the economy. The coverage of National Income in Chinese statistics excludes the value added in "non-material production sectors". China's National Income is approximately equivalent to the Net Material Product (NMP) used by the United Nations.
Sources: *Inner Mongolia is Advancing Courageously*, 1989 and Tables 5.2 and 5.4 in Chapter 5 above

Data for Alukeerqin and Hebukesaier, two of the pastoral counties surveyed, also provide strong support for the suggestion that increasing human populations contribute to increasing grazing pressures and degradation, and eventually, to falling productivity and real incomes. In both the Alukeerqin case study (Chapter 11) and the Hebukesaier case study (Chapter 17), it was pointed out that the stocking rates for these counties now exceed sustainable levels and that pasture degradation is becoming an increasingly serious problem. In these two counties, as can be seen from Fig. 18.2, there appears to be a close link between the trend in rural population and herbivorous livestock numbers. For four decades in the case of Alukeerqin, there has been a remarkable correspondence between the two populations. While data are only available for the last twenty years for Hebukesaier, again the graphs exhibit similar trends. The incomes of a high proportion of the rural population in these counties are dependent to varying degrees on grazing animals. Although the proportion of the rural population classified as pastoral or semi-pastoral in both counties has declined since the 1950s, it was still 77% for Alukeerqin and 44% for Hebukesaier at the end of the 1980s. In addition, many agricultural households in these counties also raise herbivorous animals, especially sheep, to supplement their income from cropping. Pastoral activities, therefore, have a major impact on average rural incomes in these two counties.

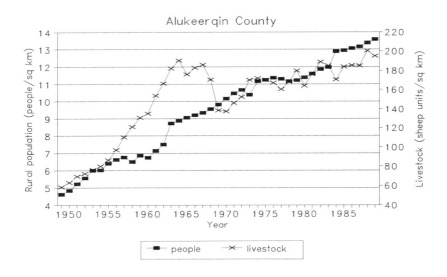

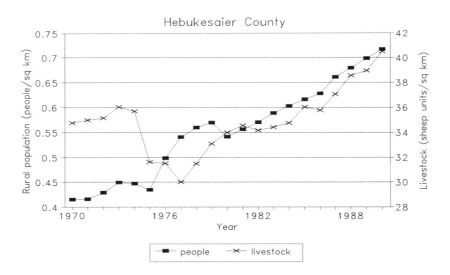

Fig. 18.2 Long-Term Changes in Rural Population Densities and Livestock Densities for Two Pastoral Counties

Alukeerqin and Hebukesaier counties are located in very different parts of the pastoral region of China. Alukeerqin, which lies in the Eastern Grasslands of IMAR, currently supports a stocking rate five times greater, and a rural population density almost 20 times greater than Hebukesaier, which is a semi-desert county almost 2,000km further inland in north western XUAR. Nevertheless, the experiences in these two counties strongly support the intuitive argument that, in pastoral areas, increasing rural populations lead to increasing stocking rates which may eventually exceed sustainable levels, causing degradation of the rangelands. Furthermore, serious degradation can occur at markedly different rural population densities depending upon local environmental conditions.

Another important point illustrated by the graphs in Fig. 18.2 is that the relationship between the two populations seems to have become stronger over the last decade or so. This suggests that as rural population densities have grown over time, there is increasingly less scope for expansion of livestock numbers beyond the immediate needs of local populations. By contrast, however, in earlier times there appears to have been considerable scope for major changes in livestock numbers which, as illustrated in Fig. 18.2 for Alukeerqin, have not always been as closely linked to the growth in rural population. In particular, the introduction of the commune system in 1958, combined with the fact that the rangeland resource was comparatively undegraded at that time, enabled livestock numbers in Alukeerqin to be boosted well beyond the immediate needs of the local population. However, as the rangelands have become progressively more degraded, and hence by implication more expensive to exploit, the rate of increase in livestock numbers has become more closely aligned to changes in the local pastoral population.

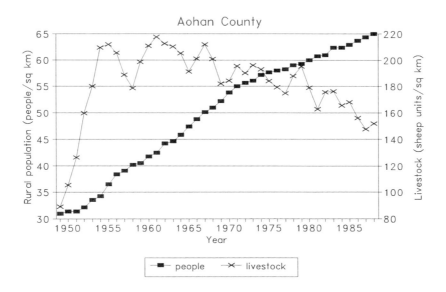

Fig. 18.3 Long-Term Changes in Rural Population Densities and Livestock Densities in an "Agricultural" County

The observation that livestock and rural populations in pastoral counties have become increasingly linked over time demonstrates that in the absence of alternative sources of income, rising population pressure in pastoral areas is likely to contribute strongly to further upward pressure on livestock numbers in these areas. Of course, the key point which arises out of this observation is that the link between the growth in the density of rural and livestock populations is only valid where there are little or no alternative sources of income available to local residents. In other counties where rural residents have major alternative sources of income, there may be little or no relationship between changes in the rural population density and stocking rates. Almost certainly, this arises because the immediate needs of the population are addressed by the agricultural sector rather than the pastoral sector in the county. Aohan, another case study county which is located in the same prefecture of the IMAR as Alukeerqin, illustrates this situation.

Although the dramatic increase in livestock populations in the 1950s and 1960s in both Aohan and Alukeerqin resulted in large areas of pasture land being degraded in both counties, the response in Aohan County has been distinctly different to that in Alukeerqin County. As demonstrated by Fig. 18.3, livestock numbers have been allowed to decline in Aohan as the rangeland resource has become progressively more degraded and the rural population has become more dependent on agriculture. By contrast in Alukeerqin, the local population did not (and still does not) have the luxury of comparatively plentiful agricultural opportunities to offset falls in the productivity of their rangeland resource. Hence, as rangeland degradation continues to worsen in pastoral areas, population pressure will become increasingly important as a factor influencing intensity of range use in these areas.

18.1.2 Policies Contributing to Population Pressures

The increasing population pressure on the rangelands of China since 1949 can be traced to three basic policy initiatives of the Central goverment: the expansion of cultivation in the pastoral region; the introduction of the household registration system; and the granting of family planning concessions to minorities. Taken together, these three national policies have imposed major constraints on sustainable economic development in pastoral areas and, as a result, they have perpetuated regional poverty and attendant environmental decline.

- *Expansion of cultivation*

In many of the more remote parts of the pastoral region (such as the XUAR), the area under cultivation was expanded in conjunction with programs to resettle large numbers of people of the Han Nationality from elsewhere in China. In the less remote parts of the pastoral region, such as in Chifeng City Prefecture in the IMAR where there had been large inward migrations of Han people last century, increased agricultural production was promoted primarily on the basis of the existing agriculturally-oriented population, although there were some small-scale resettlement programs in these areas as well.

The decision to expand cultivation in pastoral areas has had an enormous impact on the resources available for use by traditional pastoralists. For instance, in Chifeng City Prefecture, the official statistics indicate that around 15.52 million mu or 21% of total rangeland in the prefecture was lost to agricultural production between 1953 and 1979. However, the real reduction in pasture resources is likely to be considerably larger for a number of reasons. First, the rangeland "reclaimed" for agricultural use would almost certainly have been the better quality (most productive) in the prefecture.

Hence, actual losses in terms of total pastoral resources are likely to have been much larger than the official reduction in rangeland area would suggest. Secondly, the official statistics relating to the area of land reclaimed only include "successful" (i.e. currently farmed) areas. Large amounts of poorer quality rangeland, which were destroyed by cultivation as a result of the "grain first" policies introduced during the "Great Leap Forward" (1958 to 1961) and the Cultural Revolution (1966 to 1976) but which are no longer being cultivated, are not recorded in the official statistics. In Aohan County in Chifeng City Prefecture, for example, the recently completed National Rangeland Survey revealed that around 22% of the rangeland in the county (1.77 million mu) was abandoned agricultural ground (Yong et al., 1990). Thirdly, it has been suggested that the official figures may significantly understate the actual area of agricultural land successfully reclaimed. Officials with the Pasture Management Station of Chifeng City Prefecture, for instance, indicated that as much as 5.22 million mu or 28.8% of the total *actual* reclaimed agricultural land in that prefecture is not recorded in the official statistics. This discrepancy is said to arise because, in the past, compulsory State acquisition quotas for grain were determined on the basis of arable land area. Thus, it was in the interest of farmers (and local officials) to understate the area of agricultural land available, thereby enabling a greater surplus of grain to be retained within the local economy.

The three reasons just given for why the real loss of pasture resources in Chifeng City Prefecture is likely to exceed greatly the official estimate of rangeland lost to cultivation, would obviously apply throughout the pastoral region. Furthermore, the area of rangeland lost in the more remote parts of the region is likely to be even greater than in Chifeng City Prefecture because Chifeng was not an area primarily targeted for resettlement programs in the 1950s and 1960s.

Finally, it must be noted that the destruction of rangeland is not specifically confined to land reclaimed for agricultural use. In most cases, pastures surrounding the newly developed agricultural areas become severely degraded. Many agricultural households in pastoral counties raise a few sheep and/or goats as a sideline activity. Typically, these farmers randomly and intensively graze their animals on the surrounding rangeland as a secondary source of income. Hence the *actual* areas of rangeland destroyed around agricultural settlements are likely to be several times larger than the areas of pasture lost to cultivation.

In summary, therefore, the impact of expanding cultivation in pastoral areas has been to force the people of minority nationalities who make up the traditional pastoral communities to seek their livelihoods from smaller and considerably less productive (marginal) areas of rangeland. As population has increased in these more marginal areas, the pressure to overgraze the remaining rangeland has worsened. As a result, the productivity of the pastures has declined, leading to a further second-round squeezing of incomes, and so on.

- *Introduction of household registration system*

The second policy which has substantially added to population pressure in the pastoral region was the introduction of the Household Registration System (HRS). The HRS is the major institutional mechanism used to regulate interregional and rural to urban population flows in China. This system was introduced in 1955 and became established in its current form in the early 1960s (Wu, 1991). Individuals are registered in the system as belonging to one of four possible groups: urban/agricultural; urban/non-agricultural; rural/agricultural; and rural/non-agricultural.

Those people registered within a specific residential grouping are permitted to move within that grouping from one region to another, providing they are able to obtain the necessary permission from the relevant local authorities and are able to find friends and relatives to support their activities in their new location. In circumstances where an individual wishes to move from one residential grouping to another, special permission must be obtained from government authorities. Obtaining this permission is especially difficult for people wishing to shift into the urban/non-agricultural residential group and is usually only forthcoming in circumstances where an individual has passed national university tests, or through promotion in government and party administrations.

The principal means of enforcing the HRS include enforcement by law as well as exclusion from any welfare benefits and subsidies enjoyed by each of the residential groups. In particular, rural or urban/agricultural residents who have settled illegally in urban/non-agricultural areas are not entitled to significant housing and staple food subsidies enjoyed by residents in these areas (Yang and Tisdell, 1992). In this regard, illegal migration or transfer to an urban/non-agricultural area would present significant financial difficulties as housing in particular is tightly controlled and managed by State authorities. Indeed, Anderson (1990) notes that it is not so much the lawful barriers presented by the HRS *per se* but rather the linkage between the system and the high cost of accommodation which presents the real barrier to population movement.

A significant concession embodied in the HRS is the right to migrate on a temporary basis. People wishing to migrate from one resident grouping to another may apply to the relevant local authorities for a temporary resident's permit. This permit entitles these people to work as a member of that residential group for a specified time. However, in cases where the temporary resident moves to an urban/non-agricultural area, the temporary resident is still not able to enjoy the welfare benefits and subsidies enjoyed by registered urban/non-agricultural residents. Therefore, although lawful, this mode of population transfer is extremely costly for migrants who often must leave their families as well as their accommodation. In addition, they must feed themselves at prices significantly higher than those enjoyed by permanent residents of the urban/non-agricultural grouping.

In terms of its impact on development in the pastoral areas, the HRS prevents the otherwise natural out-migration from these areas. That is, given the rising population pressure and low and declining relative incomes in pastoral areas, one would expect people to migrate to regions or sectors in the economy where economic growth was higher. However, under the HRS, the opportunities for market-based adjustments to the problem of population pressure are extremely limited. Therefore, in the absence of compensating policy mechanisms, retaining the HRS for pastoral areas will intensify population pressure, thereby raising the possibility of increasing poverty and further environmental damage. Williamson and McIver (1993) use a two-sector, three-factor version of the specific factors model to demonstrate the theoretical basis for this intuitive conclusion.

- *Family planning concessions for minority nationalities*

The third area of policy related to population pressure in the pastoral region is the long-standing decision of the Central government to allow ethnic minorities substantial family planning concessions. These rights have been granted as part of a more general policy of allowing minorities a limited measure of ethnic autonomy. By comparison to the family planning rights assigned to the ethnic Han majority (who are limited to one child in urban areas and two in rural areas), the concessions traditionally granted to people belonging to the ethnic minorities have been extremely generous. For example, minority couples resident in the more central pastoral provinces such as Qinghai, Gansu and IMAR were permitted to have between three and four children, while in the more remote parts of the country such as XUAR and Tibet, minority couples were permitted to have as many children as they desired (Zhang, 1991).

Although as Mackerras (1994) points out, many autonomous governments formally moved during the 1980s to adopt much less lenient family planning arrangements, the traditional concessions have had a significant impact on the rate of growth in the populations of minority nationalities relative to the total population. In this regard, Zhang (1991) notes that the proportion of ethnic people to the total population increased from 5.82% in 1964 to 8% in 1990. Between 1982 and 1990, the total minority population of China increased by 35.5% compared with 12.1% for the whole population (Table 18.2). If this rate of increase continues for the next decade, the minority population will rise from 91 million in 1990 to 132 million by the year 2000.

Table 18.2 Changes in Selected Minority Populations in All China and the 12 Pastoral Provinces, 1982 and 1990

Ethnic group	1982 census		1990 census		Change 1982 to 1990	
	Total population in All China	Population in 12 pastoral provinces	Total population in All China	Population in 12 pastoral provinces	Total population in All China	Population in 12 pastoral provinces
	('000)				(%)	
Mongolian	3,417	3,341	4,807	4,653	40.7	39.3
Hui	7,227	4,499	8,603	5,318	19.0	18.2
Tibetan	3,874	3,770	4,593	4,468	18.6	18.5
Korean	1,766	1,754	1,921	1,909	8.8	8.8
Manchu	4,304	4,080	9,821	9,444	128.2	131.5
Miao	5,036	363	7,398	548	46.9	51.0
Yi	5,457	1,527	6,572	1,788	20.4	17.1
Tujia	2,835	597	5,704	1,089	101.2	82.4
Uygur	5,963	5,951	7,214	7,197	21.0	20.9
Kazak	908	907	1,112	1,111	22.5	22.5
Xibe	84	83	173	170	106.0	104.8
Yugur	11	11	12	12	9.1	9.1
Ewenki	19	19	26	26	36.8	36.8
Sub-total of above	40,901	26,902	57,956	39,733	41.7	47.7
All 55 Minorities	67,295	n/a	91,200	n/a	35.5	n/a
Total Chinese population	1,008,200	-	1,130,500	-	12.1	-

Source: Statistics of Ethnic Minorities of China, 1949-1990. Statistical Press of China, Beijing, 1991

The first 10 minorities listed in Table 18.2 are the groups with a total population in 1990 in excess of 1 million people and who reside in the 12 pastoral provinces. (See Table 3.8 in Chapter 3.) The other three ethnic groups included in Table 18.2 are much less numerous but almost all of these people live in the pastoral region. Some of the population changes recorded between 1982 and 1990 and shown in Table 18.2 are remarkable. With the exception of the Koreans and the Yugurs, all ethnic groups listed in the table grew much faster than the whole Chinese population. Furthermore, for most of these groups, the growth rates recorded in Table 18.2 were also above those experienced for earlier periods.

The normal demographic explanations for the high rates of population growth in the 1980s are an increase in life expectancy (owing to improved health care in remote areas and the absence of widespread famines) and a higher than normal birth rate. However, some authorities such as Mackerras (1994) claim that a significant proportion of the increased growth rate during the 1980s in some, if not all, minority populations was the result of two non-conventional demographic factors. First, it is believed that a large number of individuals who previously described themselves as Han re-registered as members of a minority group between 1982 and 1990. The Manchus in particular are believed to have had their ranks swollen as a result of this phenomenon. Secondly, an amnesty was declared in the late 1980s for people who, for whatever reason, were not officially registered. The existence of such unregistered individuals had not previously been recognised in population statistics. Remote ethnic communities, such as those in parts of the pastoral region, may have included many unregistered people who took advantage of the amnesty to have their existence formally recorded. Even after allowing for these two unusual factors and the improvements in life expectancy, the evidence points to a "baby boom" in pastoral areas since 1978.

There are undoubtedly many reasons for the jump in birth rates during the 1980s. The State was probably able "to encourage" birth control more effectively in remote, sparsely settled areas when the population was organised in communes. Since the early 1980s with the introduction of the household production responsibility system (HPRS), families are less subject to State control. Furthermore, the demise of the communes removed a "social safety net", making children more important to the future security of the family. In addition, under the HPRS, each additional child represents an investment in the future labour force of the household. Such an investment may be an economically rational allocation of the resources currently available to the family. The longer-run implications of these rapidly growing pastorally-based populations for employment, incomes and ultimately for rangeland degradation, are extremely serious, especially if the household registration system continues to be strictly enforced in pastoral areas.

18.2 Market Distortions

The household registration system discussed above is a major impediment to a properly functioning labour market in China. But this policy is only one of many market distortions which constrain development in pastoral areas.

Gradually, or in Chinese terms "step-by-step", the pastoral sector of the economy is being subjected to market forces. Unfortunately, many policy changes have been uneven, erratic and unco-ordinated. As a result, the incentive structures put in place have combined with product and factor market distortions to generate outcomes which are undesirable in the short term but potentially disastrous for the ecology of pastoral areas in the longer run.

Some may wish to argue on so-called "Second Best" grounds that reducing or removing distortions in product and factor markets in pastoral areas, while major distortions continue to exist elsewhere in the economy, will not necessarily lead to an improvement in economic welfare. However, at least in relation to the market distortions specifically referred to in this section, there are clearly major potential gains to be had in terms of both economic efficiency and environmental sustainability if the distortions can be removed.

18.2.1 Product Market Distortions

Under a completely centrally planned economic system, prices can be set to achieve income distribution and other goals without considering their impact on the allocation of resources. Price structures which generated socially acceptable income distributions under central planning are unlikely to exhibit price relativities which lead to an efficient allocation of resources under a free market regime. Therefore, as the Chinese economy shifts from being a centrally planned system to a mixed economy with an increasing proportion of economic activity guided by market forces, the well-known conflict between equitable and efficient prices has emerged in many sectors.

Finding a solution to this conflict without disrupting social harmony is a major challenge for Chinese policy-makers. In the short run, there is likely to be a strong tendency to maintain (or even introduce) policies which reinforce "traditional" price relativities and hence income distributions. To the extent that the historical price structure is not consistent with the price relativities which would exist in a free market, such policies will create market distortions, producers will receive the wrong price signals, resources will not be used efficiently, and total social welfare outcomes will [be sub-optimal. The provision of] ool and other pastoral products illustrates the [conseq]uences which arise if distortions in the product

[market are not addressed. In the case] of China, the introduction of the household [responsibility system has] provided pastoral households with the freedom [to make] significant changes in the output mix during the [past two decades. Man]y case studies in this book provide convincing [evidence that herders c]an and do respond to changing price relativities. [It is essential,] therefore, that the relative prices of outputs such [as wool and mutton, for examp]le, reflect actual relative values and [input prices such] as winter feedstuffs, drenches, and even labour

[Tradition]al marketing and pricing arrangements which [are e]ntirely inappropriate on efficiency grounds once [herders have freed]om to adjust production according to market [signals. Tradi]tional arrangements protected the incomes both [of herders who prod]uce coarser and lower grade wool and of the [employees of the s]tate monopoly wool marketing authority. Both [groups, especially pr]oducers, are extremely numerous and politically [powerful. As d]iscussed in more detail in Section 4.5.2, income distributional and political considerations have made it extremely difficult to introduce reforms designed to make the traditional wool marketing and pricing system more economically efficient.

Two specific aspects of wool marketing which illustrate the problems involved are the lack of price premiums for quality and the practice of purchasing wool from farmers on a greasy weight basis according to a very crude grading system. A third major distortion occurred when the market for cashmere became free but wool prices remained administratively determined in most pastoral areas. These three product market distortions have had serious implications for the use of the rangelands, sheep improvement programs and the development of higher-value fine wool production which is one of the few feasible means by which the incomes of pastoralists may be improved without necessarily increasing the grazing pressure on the rangelands.

- *Lack of premiums for quality*

Prior to 1985, wool was classified as a Category II commodity. The State set production quotas and established prices paid for quota wool. Production in excess of quota could be sold to the State procurement agency (in the case of wool, this was the Supply and Marketing Cooperative) usually at State procurement prices. Trading in Category II commodities by non-authorised individuals or units was prohibited. As discussed in Section 4.5.2 and in several of the case studies, wool prices remained at roughly the same levels for about 20 years until the mid-1970s when they rose a little. After a modest downward adjustment in 1981, wool prices were once again constant until 1985.

Not only did the general level of wool prices remain virtually constant for many years but the relativities between the prices for the various grades of wool also remained fixed. These premiums and discounts were effectively written into the National Wool Grading Standard which assigns each grade a "quality" index relative to a reference grade (Longworth, 1993b). The National Price Bureau traditionally set the price for the reference grade and prices for the various grades were determined by applying the fixed and administratively determined "quality" indices to the price of the reference grade.

The quality relativities incorporated in the grading system may have originally represented a crude estimate of the relative value to the textile mills of the various grades of wool. However, over time, changing textile technology and shifts in the relative demand for and supply of the various grades would undoubtedly have altered the premiums and discounts which should apply.

The Central government declared that wool was no longer a Category II commodity in 1985. The provincial and county case studies include discussions on what happened following this decision. In summary, despite a period of relative chaos in 1986 to 1988, commonly referred to as the "wool war" period, the traditional administratively-determined premiums and discounts continue to dominate the prices to which wool-growing households respond.

While the details are discussed in Section 4.5.2, the magnitude of the distortion this situation involves can be appreciated by the fact that, traditionally, producers of both semi-fine wool and fine wool received virtually the same reward. Semi-fine wool is wool of 25.1 to 40µm while fine wool is ≤ 25µm in fibre diameter. These two types of wool would usually be processed differently and have very different end uses. They are virtually different commodities. Perhaps even more importantly, the premium for genuine fine wool relative to cross-bred fine wool (or improved fine wool as it is called in China) has traditionally been extremely modest.

The overall inappropriateness of the traditional wool pricing structure in China can be gauged from the fact that raw wool of all types is classified into only six grades for pricing purposes (see Table 4.5 in Chapter 4, for example). There is obviously going to be enormous variation within each grade because wool is an extremely heterogeneous commodity. In Australia, by comparison, there are potentially over 3,000 different recognised wool types.

The crude grading system and the lack of premiums for better quality wool has severely retarded the development of fine wool production in China since the introduction of the household production responsibility system (HPRS). Farmers have almost no price incentive to upgrade their sheep to fine wools. Furthermore, as pointed out in the case studies, fine wool sheep may be more costly to raise, more susceptible to drought and other environmental risks, and produce less mutton of an inferior quality (according to traditional tastes). Under these circumstances, giving the households the freedom to allocate their resources according to market forces under the HPRS without allowing the market to reflect true relative values is imposing a serious constraint on the production of better quality wool. Fine wool has a significantly higher social value relative to semi-fine and coarser wools. In many pastoral areas, the real premium for fine wool may be more than sufficient to compensate households for the higher production costs (etc.) associated with raising fine wool sheep. Properly priced, therefore, better quality wool has the potential to improve the incomes of pastoralists and contribute significantly to the development of the pastoral region.

- *Purchasing on a greasy wool basis*

Closely tied to the lack of premiums for quality and the crude grading system are the issues surrounding the practice of buying wool from farmers on a greasy wool basis. In Section 11.3.2 in the Alukeerqin case study, details are provided about how the buyers at the purchasing stations operated by the SMC determine the price for a particular lot of wool. In general, any foreign matter not shaken out of the wool is bought at the same price as the wool itself. Farmers, therefore, have an incentive to incorporate as much dust, animal wastes, grass seeds, etc. as possible into their fleeces. It is not surprising that the average yield of clean wool obtained from scouring raw wool is so low in China. Clean yields below 40% are normal in China while the corresponding figure is over 65% in Australia.

- *Wool versus cashmere*

One of the remarkable policy developments related to the marketing of pastoral products post-1985 was the decision by virtually all governments in pastoral areas to allow free trade in cashmere. Cashmere prices rose eightfold between 1984 and 1989 while State purchase prices for wool little more than doubled (Fig. 11.2 of Chapter 11). In areas where a free market existed for wool as well as cashmere, peak wool prices were about four to five times higher than 1984 price levels. Consequently, the incentives for households to raise more goats at the expense of sheep, especially fine wool sheep, were much less in areas which experienced free-market prices for both cashmere and wool.

In pastoral areas where wool prices were tightly controlled such as Balinyou and Alukeerqin Counties, the swing to goats after 1985 was remarkable. (See, for example, Fig. 11.1 in Chapter 11.)

• *Links with rangeland degradation*

All three examples of serious distortions in the markets for pastoral commodities discussed above have led to greater rangeland degradation than might otherwise have been the case.

The lack of premiums for better quality wool encourages farmers to aim for quantity and not quality. There are many facets of this issue. For example, farmers who want to grow better quality wool are likely to stock their pastures less heavily because they will be aiming to increase the output per head (as will be the case if they raise more fine wools). They will also want to overwinter fewer sheep for any given amount of winter feeding resources available to them, since they will aim to avoid tender wool by feeding each sheep more adequately during the winter. At present, there are no price penalties for producing tender wool, so there is no incentive to feed wool-growing sheep sufficiently well in winter to avoid a break developing in their wool fibres. In general, therefore, incentives for quality *ceteris paribus* are also incentives for more conservative rangeland management practices.

Similarly, if farmers were penalised for incorporating dust in the wool, they would have an incentive to avoid dusty environments. At present, the marketing arrangements encourage wool-growing households to consider dust as one of the "inputs" in the production of wool. Indeed, one could envisage a "dust production function" from which an optimal input of dust could be determined and, by implication, an optimal amount of pasture destruction to provide the dust needed to "produce" the wool.

Finally, goats are acknowledged as being more damaging to natural pastures than sheep. For any given number of small ruminants, pasture degradation is more severe the higher the goat/sheep ratio. Allowing a free market for cashmere while distorting wool prices (downward) has encouraged farmers to replace sheep with goats. The result has been much more pressure on the rangelands.

These examples demonstrate the subtle but all important linkages between product market distortions and rangeland degradation.

18.2.2 Factor Market Distortions

The basic factor or resource available in pastoral areas is natural pasture. Herbivorous animals, as Liu (1990) points out, are biological factories which convert this resource into products useful to people such as meat, milk, wool, leather, etc. Natural pastures are a renewable resource if managed appropriately. If ownership of this resource is entirely in the public domain, the social value assigned to the resource will have a major influence on how it is managed. Even when the rangelands are privately owned, social attitudes will condition management practices.

Traditional pastoralists revered their rangelands. The Mongolians and the other Peoples who for millennia herded their animals in the areas now referred to as the pastoral region of China, placed a high social value on the natural pastures. On the other hand, ethnic Hans are agricultural people. Historically, the Han and the traditional nomadic inhabitants of the pastoral areas had very different social attitudes to rangelands. Indeed, it is said that Genghis Khan once ordered the death of every living Han because he considered the whole race guilty of the crime of violating the grasslands for agricultural purposes (Grousset, 1967).

As already discussed, the migration of large numbers of Han into pastoral areas after 1949 led to much of the best rangeland being "reclaimed" for cultivation.

Furthermore, the political dominance of the Han greatly influenced the value local governments placed on natural pastures. To a large extent, the Han people considered (and perhaps still consider) that the reclamation of rangeland for cultivation was converting idle, valueless land into productive agricultural uses. These traditional culturally-based attitudes were reinforced by the application of communist/Marxist orthodoxy after 1949. Marxist Thought, as interpreted by the Russian Communist Party and subsequently adopted by the Chinese Communist Party, held that natural rangeland had little or no value as a resource since it embodied no labour (Zhang, W., 1990). Hence, from a Marxist ideological stance, exploitation of rangeland, even to the point of serious degradation, was an acceptable practice. In addition, the "grain first" philosophy which dominated agricultural policy in China, especially during the 1950s and 1960s, generally devalued pasture land.

Present day ethnic minority communities in pastoral areas have lost much of their traditional culture. They have been under the influence of Han-dominated governments for at least 45 years and for much longer in many parts of China. The almost godlike status attached to grasslands by their ancestors has gone for ever and in its place is a kind of commercial pragmatism. The danger arising from this move to a more commercially-motivated society is that if, within a given economic and policy framework, private households are provided with commercial incentives to degrade the rangeland, then social barriers to such actions are likely to be minimal.

As indicated in the preceding section, the introduction of the household production responsibility system has given private households the freedom to respond to commercial incentives. It was also pointed out in relation to product markets that policy-induced distortions have seriously undermined the social desirability of many of these incentives. As a result, there are likely to have been significant losses in terms of economic efficiency but worse still, the distortions have encouraged practices which are harmful to the ecological system. Similar distortions also exist in the land, labour, capital and farm input markets.

- *Land market*

Private ownership of land is prohibited in pastoral areas of China. As explained in Section 4.4.1 and in the county case studies, most rangeland previously under the control of communes has now been contracted out to households or groups of households. These "leases" are not transferable by sale but in some counties they can be inherited and/or sub-let. When sub-contracting is permitted, it is only permitted on a very short-term basis such as for a season. Furthermore, the sub-contractor must belong to the same administrative village as the original contractor.

The complete lack of any form of market either for ownership rights or for use rights creates serious distortions not only in relation to how land resources are allocated between competing uses, but also in how competing capital investment possibilities are prioritised.

Setting aside for the moment the issue of tenure uncertainty (see Section 18.3 below), the rangeland under contract to the private household represents a totally illiquid "asset". Other assets such as livestock and transport machinery (tractors, trailers, trucks, etc.) are relatively freely traded and, therefore, represent a much more flexible investment. Under these conditions, private households will under-invest, relative to the socially optimal level of investment, in improvements irrevocably tied to pasture land such as fences, wells and improved pastures. As discussed in most of the county case studies, all levels of government implicitly recognise this "market failure" because they adopt countervailing policies, including subsidised credit, to stimulate investment in rangeland preserving (e.g. fencing) and rangeland developing (e.g. pasture improving) activities.

Perhaps the absence of a land market may be even more important in regard to how it influences disinvestment strategies. To the extent that a private household, for whatever reason, decides to run down its stock of capital, the inability to sell land means that there is no market penalty for "mining" the land. That is, in the case of rangeland, the household will tend to over-exploit the land because, from the viewpoint of the household, the residual value of the land does not accrue to the household. Of course, imposing penalties for overstocking, lengthening the term of the land contracts, removing tenure uncertainty, and allowing inheritance are all measures designed to reduce the likelihood that households will choose to "mine" their rangeland. Permitting rangeland contracts or use-rights to be freely traded would be a major improvement. Assigning private titles to land and allowing a completely free market to operate for rangeland ownership rights, while considered far too radical at present, may eventually be adopted as the most effective means of encouraging private households to take better care of their pastures.

- *Labour market*

As mentioned earlier, the household registration system is a major constraint on the efficient allocation of labour in China. It is an especially serious market distortion in pastoral areas. The vast distances and general remoteness of many of these localities prevent even the most mobile members of the workforce from making temporary workplace-related moves. At the same time, permanent relocation is strictly policed and for the most part impossible. Of course, Han people in the remote areas often have relatives living in other parts of China where job prospects are better. These relatives are permitted to "sponsor" temporary working migrants from the pastoral areas. On the other hand, few people who belong to an ethnic minority would have similarly well placed relatives.

It could be argued that the policy-induced distortions in the labour market which prevent people leaving the pastoral areas are the single most important constraint on the future prospects of these areas. Furthermore, as pointed out in Section 18.1.2, the recent population explosion in pastoral areas will exacerbate the problem in the near future.

Another distortion in the labour market which may have contributed to the baby boom of the 1980s is the historical ban on hired labour. Nowadays, pastoral households are more or less free to employ non-family members but the extremely unpleasant experiences of those who employed others in the 1947 to 1957 period (see Section 3.5.2) and especially during the Cultural Revolution are still fresh in the minds of the older generation. Understandably, few households explicitly employ outsiders for long periods. Of course, within any one village, certain individuals (usually old men and/or young boys) may be regularly "employed" to shepherd sheep and goats but such arrangements are not likely to be negotiated in the context of a free labour market. Households prefer to supply their own labour and hence the incentive to have larger families to cope with a growing demand for labour.

- *Capital market*

Until the introduction of the household production responsibility system in the early 1980s, virtually all investment in pastoral areas was financed by the State. Nowadays, although the State remains the major source of capital, a growing proportion of investment funds is controlled by private households. The Agricultural Bank of China (ABC) is the dominant financial institution but the Agricultural Credit Cooperatives (usually closely integrated with the ABC) and to a lesser extent other major banks such as The Commercial and Industrial Bank of China, provide some competition for the ABC in most pastoral areas. A reasonably healthy capital market has emerged even in the most remote areas to mobilise household savings and to provide commercial credit to private farmers.

Unfortunately, the use of funds available through the capital market has been badly distorted by tenure and other institutional and policy uncertainties. Investment in private housing rather than productive assets is one common manifestation of these distortions and the lack of enthusiasm for investments tied to pasture land such as fences, wells and re-seeding is another important example which has already been mentioned. Perhaps with the passage of time and as the people gradually become less concerned about tenure uncertainties, these misallocations of private investment funds will diminish.

Misallocation of the investment funds available to the State also presents a major problem in many pastoral areas. Three examples of this issue of particular concern are public investments in pastoral-product processing facilities, the use of poverty alleviation funds provided by domestic governments, and the application of foreign aid funds.

- The introduction of the fiscal responsibility system (FRS) in 1983 provided the primary stimulus for local governments to become more involved in value-adding activities. The FRS refers to a package of economic reforms commonly referred to as "fiscal reforms" (World Bank, 1988; Watson et al., 1989). In this book, these reforms are referred to as the FRS to highlight the similarity, in terms of economic impact, between the fiscal reforms and the introduction of the household production responsibility system. Like the household production responsibility system, the FRS was aimed at decentralising decision-making by making economic agents directly responsible for their actions in terms of any profits and losses they incurred. Provincial, prefectural and county governments were assigned revenue targets by the Central government, with revenue over and above these targets being retained by the respective levels of government.

Since the inception of the FRS, many county governments in the pastoral region have shifted resources away from primary production-related industries such as animal husbandry, and into first-stage and latter-stage processing of animal products. Compared to the primary pastoral production of live animals and animal fibres, the value-adding industries represent potentially significant tax revenue advantages for county governments. For instance, the national tax on greasy (raw) wool is 10% of the farm-gate purchase price, whereas the tax on profits from processing the wool (scouring, etc.) vary on a progressive scale from 10% for the first ¥1,000 earned to 55% for amounts greater than ¥200,000. In addition, value-added taxation levied at various stages in wool processing range from 14% for tops, blankets and carpets to 20% for knitted goods and woollen clothing. (For more details on taxation in China, see World Bank, 1988.)

The shift in focus from primary production to value-adding activities is not without risk. The remoteness and industrial backwardness of the pastoral region suggest, at least at the county level, that the region's comparative advantage lies in primary pastoral production rather than in the processing of animal husbandry products. Furthermore, the fundamental natural resource available in the region, the rangelands, underpins not only the primary pastoral production but also the local first-stage processing activities. Given the shortage of capital in these regions, shifting the focus towards value-adding activities may divert investment funds away from conservation of the rangelands and so undermine the input base of the value-adding activities.

The total quantity of capital available to the pastoral region is regulated in accordance with the national credit plan (World Bank, 1991). At current interest rates, the supply of credit made available under the credit plan is considered to be insufficient to meet the demand for credit in the pastoral region. Furthermore, given the high opportunity costs incurred in allocating existing levels of capital to these remote pastoral areas (compared with the returns to capital in the faster-growing, industrial regions of the eastern seaboard), it is considered unlikely that the pastoral region's credit limit will be increased significantly in the foreseeable future. The incentives created by the FRS combined with the tighter credit limits may have motivated county governments to make inappropriate investment decisions.

The growth in local first-stage processing of wool in pastoral counties illustrates these potential problems. For instance, prior to the introduction of the FRS in Chifeng City Prefecture of IMAR, all wool was scoured at the large prefectural mill. By 1991, the fiscal reforms had encouraged all six pastoral counties in Chifeng City Prefecture either to build their own scouring plant or to be in the process of establishing one. That is, value adding to wool represented one of the only industrial opportunities available in these remote pastoral areas and hence one of the few investments by which the county governments could generate fiscal revenue.

Brown and Longworth (1992b) point out that the construction of county wool-scouring facilities has had two major unintended negative effects. First, whereas wool-scouring capacity was almost fully utilised throughout the pastoral region prior to the FRS, by the early 1990s less than one-third of the available scouring capacity was being utilised. Given the significant economies of size and throughput in wool scouring, the low utilisation of scouring capacity imposes substantial organisational inefficiency and added costs on the wool-processing industry. Indeed, in the late 1980s, rather than providing a source of fiscal revenue, many of the county scouring plants were drawing on county revenues by requiring subsidies to cover the losses incurred as they operated at low capacity utilisations.

The second major negative effect of the growth of county wool-scouring facilities, and one which may have more insidious long-term effects on regional development, is the low quality of the scoured wool produced by these county plants. The counties have had no previous experience in wool scouring, an operation which is technically demanding and one which, if done incorrectly, can seriously damage the wool fibres. Moreover, the extent of the damage typically emerges only at later processing stages. The intense competition between fibres in the Chinese textile market ensures that a lowering of the quality of scoured wool will impact adversely on the wool industry and hence development in the pastoral region.

In addition to the apparently adverse impact of the FRS in many pastoral counties, the "success" of the FRS in other parts of China also appears to have had a considerable impact on the development of pastoral areas. In particular, the FRS seems to be making it increasingly difficult for the Central government to transfer capital from the richer southern and eastern parts of the nation to poor areas in the pastoral region and elsewhere. Most notably, the FRS has provided both the incentive and the means for local authorities in the richer southern and eastern parts of the country to shift investments into activities which do not require profit from these activities to be shared with the Central government. Therefore, as the revenue crisis deepens for the Central government, poorer parts of China such as many of the pastoral areas will find it increasingly difficult to rely on transfers of capital from the Central government for development purposes.

- Special poverty alleviation schemes are one of the major ways in which capital transfers are made to poor counties in the pastoral region. Details of the poverty alleviation program operated by the Central government are explained in Section 3.4.3. The application of this program in relation to the poor pastoral counties is discussed in several of the case studies (see, for example, Section 11.3.4). Provincial governments also have poverty alleviation schemes such as the seven-point program operating in Gansu (Section 6.6.2). Even at the county level, there are frequently special funds made available by the county government to assist with poverty alleviation (e.g. Section 14.3.4).

Unfortunately, the emergence of a capital market at the grass roots may help to frustrate the intended impact of these capital transfers on the rate of investment by pastoral households. Private households which have access to concessional investment credit (or grants) under poverty alleviation programs may simply invest these funds instead of their own. As a result, the households may boost their savings or consumption rather than undertake any additional investment in rangeland conservation or development. Of course, to the extent that the poverty alleviation programs are designed to raise the welfare of the households, displacement of private investment funds by concessional credit may be a most appropriate outcome, at least in the short run. However, if the objective of making the public capital available at concession rates of interest is to increase investment in long-term sustainable rangeland development, then the growth of a more or less commercial capital market is likely to reduce the effectiveness of the scheme under present conditions.

- Another approach to reducing poverty in poor pastoral areas has been for the Central government to arrange for major contributions of foreign capital to stimulate development (Remenyi, 1992). The IFAD North China Pasture and Animal Development Project was one such scheme which assisted eight counties in north east China between 1981 and 1988. A total of ¥168 million was invested under this project with IFAD supplying a little over half these funds. A major objective of the project was to encourage the establishment and fencing of new pastures in certain townships within the eight designated counties. As mentioned in the respective case studies in this book, some of these townships were in Balinyou, Wongniute and Alukeerqin counties. Brown and Longworth (1992a) compared the rate of development of fencing and pasture establishment in IFAD-assisted townships with the rates achieved in similar townships not in the IFAD project. They found little difference. Similarly, they reported that much the same improvement in average net income per capita was achieved in non-IFAD as in IFAD townships over the 1980 to 1981 period. It would seem that the major injection of capital by the IFAD Project had little impact on the rate of investment in fencing and pastures or on incomes.

As with the poverty alleviation programs, perhaps the low-interest IFAD loans were used to finance the intended amount of investment in pastures (etc.) while the private funds otherwise "earmarked" for these investments were re-directed to other uses.

Unfortunately, it was not possible to test this hypothesis because insufficient reliable data were available on household consumption, savings or off-farm investment.

Another possible reason for the apparent failure of the IFAD loans to stimulate extra investment in fencing and pastures was that the availability of other loan funds was restricted by the ABC and other State-controlled credit agencies in townships where households had access to IFAD Project loans. These "saved" loan funds were then made available to households in non-IFAD-assisted townships, thus "spreading the benefit" of the additional capital injected by the IFAD Project. Once again, it was not possible to obtain data with which to test this hypothesis.

The major point which arises from the above discussion in relation to the capital market and the use of capital is that, in terms of effectiveness, fiscal policies which influence the investment strategies of county governments; poverty alleviation programs which provide concessional interest-rate loans to households; and foreign aid projects - all depend for their success upon the existence of appropriate policy-induced incentive structures. In the absence of the right incentives, the emergence of a more or less free capital market may actually contribute to the misallocation of scarce public capital.

- *Farm input market*

In pastoral areas, Seed Companies, Fuel Companies and Supply and Marketing Cooperatives (SMCs) control the distribution of almost all farm inputs not supplied by households themselves. These three State monopolies are effectively under the control of the Ministry of Domestic Trade (formerly the Ministry of Commerce). However, the SMCs in some areas have more or less broken away from the Ministry.

These monopolies are large vertically-organised hierarchies. Each has a head office in Beijing from which tentacles reach down through provincial, prefectural, county and, in most places, to the township and even village level. Under the old central planning arrangements, seeds, fuels and other inputs such as tools, fertilisers, etc. were distributed down through these channels to the communes. Each commune, the basic production unit in the national plan, was allocated its share (or quota) of each input according to the State Plan. Nowadays, the distribution system appears to have retained the major elements of the old planning arrangements at the higher levels. That is, the State allocates inputs down to about the county level and then local administrative and market forces become operative. In principle, the private households have increasing freedom to purchase seeds, fuels, tools, farm machinery, etc. In practice, because the distribution system for most of these inputs remains under the control of State monopolies, major distortions occur relative to what would happen under a competitive free-market system. Shortages, both genuine and contrived, are common, leading to the development of "black markets" and corruption.

The State monopolies frequently resort to "tied deals" to coerce the private households into maintaining traditional trading arrangements. For example, in pastoral areas the SMCs have the official monopoly rights to the distribution of fertiliser which is often in short supply. The SMCs, therefore, are in a position to make fertilisers available only to those farmers who sell their wool to the SMC. Under certain circumstances such as during the "wool war" era (1986 to 1988), the SMCs have adopted a kind of mixed barter/monetary trading system which has allowed them to exchange fertilisers for wool at relative values which reflected the real relative value of these commodities much better than did the official State prices at the time. But normally, the monopoly power of the farm input suppliers creates major distortions in the market for these commodities. These distortions are a significant constraint on progress.

Another specific and, in the context of sheep raising, important facet of the SMC monopoly over wool marketing to which reference has already been made is the lack of premiums for quality. This product market distortion leads to a lack of demand for, or even interest in, a range of wool-quality-improving inputs such as drenches, dips, winter feedstuffs, mineral supplements, and even such things as sheep coats. As already stressed, farmers who sell their wool to the SMC on a greasy weight basis and according to an extremely crude grading system are not penalised for growing tender wool and may even be rewarded for cleverly incorporating as much dust and other foreign matter as possible in their fleeces. These wool growers have no economic incentive to use drenches to control internal parasites, to dip carefully to minimise external parasites, to feed their sheep better (especially in winter) to avoid breaks or tender spots in the fibre, or to adopt management strategies (including putting plastic coats on their sheep) to avoid contamination of the wool. There is clearly a direct link between inappropriate product pricing and the resulting distortion on the demand side of the market for a wide range of modern inputs.

As pointed out many times in this book, the Central government relaxed its policy on wool marketing in 1985 and since that year the wool market has been declared "free" in more and more wool-growing areas. For example, even the XUAR government has permitted a free market to develop in its area of jurisdiction since 1992. Despite these formal steps to free-up the marketing system for wool, most pastoral households not living on State farms continue to sell their wool through the traditional SMC channel. The tied-sales arrangements referred to earlier are one reason, the provision of credit on other purchases is another, and tradition and social mores requiring people to retain traditional patterns of behaviour are also extremely important in isolated communities. For these and other reasons, the SMCs retain a large slice of the private household wool trade in pastoral areas.

From the viewpoint of the SMC organisation, it must do everything possible either to retain its formal monopoly or to dominate the free market. Like all State bureaucracies in China, the SMC is required to pay salaries both to current employees and to retirees. In the past, under the planned economy system, the SMC was not concerned about efficiency to any great extent. The labour productivity of its workforce was low and the workforce was excessively large. Nowadays, with increasing emphasis on efficiency and the consequent pressure on the SMC to reduce its labour force, it is faced with a massive problem of how to continue paying salaries to a growing number of retirees. All large State bureaucracies and firms in China are encountering similar difficulties.

In the case of the SMC, the problem is exacerbated because it is localised. County-level SMCs must look after their own retirees, prefectural SMCs need to generate sufficient funds from their own activities to meet their obligations to their pensioners, etc. Thus at each level in the SMC hierarchy, there is a scramble to retain traditional control over wool marketing and the distribution of inputs to wool-growers.

To the extent that market reforms weaken the SMCs, the welfare of the former SMC employees becomes a responsibility of the local county or prefectural administrations. Therefore, since the local governments in pastoral areas have already been placed under increasing financial pressure by the fiscal reforms, they will not wish to destabilise the SMCs by encouraging a free market for wool and farm inputs such as fertiliser traditionally handled only by the SMCs.

Recent changes in relation to grain-pricing policy is another example of how reforms in one sector of the economy can have unintended adverse effects elsewhere. The Central government has been moving away from completely controlling the grain market in China. As a consequence, as pointed out in several of the case studies, the prices of grain for livestock feed have increased substantially in recent years relative to the price of wool. Sheep-raising farmers, therefore, have tended to feed less grain to their sheep in winter and wool quality has deteriorated. The lack of a market penalty for allowing wool quality to decline, to which reference has already been made several times, encourages such a response from farmers to higher feedstuff prices. In this case, a major reform aimed at removing distortions in the grain market and encouraging more economically rational use of grain has led to less socially desirable outcomes in terms of wool production owing to inappropriate pricing arrangements for wool.

18.3 Institutional Uncertainties

Institutions, broadly defined to embrace public policies, organisations, and property rights, have been subject to sudden and erratic changes in China for more than a century. In particular, policy revisions and reversals, the reorganisation and restructuring of both governmental and Party instrumentalities, and the redefinition of property rights (and obligations) have been commonplace occurrences since 1949. Under these circumstances, economic and political decision-makers can be expected to behave in ways which reflect the high degree of institutional uncertainty they face.

The major reforms introduced in relation to economic policy and property rights after 1978 were initially regarded with scepticism and even suspicion at the "grass roots". Local government officials, Party cadres and private individuals alike, all reacted cautiously to the new arrangements. By the mid-1980s, a steadily growing proportion of the population was taking advantage of the new freedoms to improve their private and, to a lesser degree, their social circumstances. The emphasis, however, was on short-term gains because no-one knew how long the reforms would remain in place.

In the pastoral areas, the introduction of the household production responsibility system (HPRS), not only in place of communes operated on a collective basis but also on State farms, was the most profound institutional change directly affecting private households. The implementation of the fiscal responsibility system (FRS) at about the same time greatly altered the economic incentive structures faced by local governments, especially county governments. The impact of the HPRS and the FRS on development in pastoral areas has been discussed throughout this book. Clearly, both these reforms have been of profound importance both in a positive sense and, unfortunately, in a negative sense. The next section highlights some of the unintended consequences of implementing the HPRS and the associated property rights reforms in pastoral areas, given the strong historical sense of uncertainty which must have existed (and may still exist) regarding the permanence of these reforms. The following section explores the hypothesis that the FRS and other reforms have greatly strengthened the "power of local governments" and encouraged the "policy mirage" syndrome.

18.3.1 Property Right Uncertainties

At the heart of the HPRS as introduced in pastoral areas was the assignment of ownership rights to livestock and use rights to pasture land under what is now called the "double contract" system. The erratic "history" behind the property right changes after 1978 is outlined in Section 3.5 along with a general explanation of the present "double contract" system. The present arrangements, as discussed in the county case studies, were introduced at slightly different times and in different ways in different parts of the pastoral region. Furthermore, the present system, especially the pasture land contract aspect, has continued to evolve over the last decade or so. Modifications have been made much more readily in some areas than others (Williamson and Longworth, 1993).

Nowadays, the term of the pasture use contract and other details concerning the contract varies remarkably from place to place. At present, the Central government delegates to the provinces the power to set the term and other details for land contracts. Provincial governments have established guidelines, but county governments are permitted to vary the conditions which apply to pasture contracts to suit "local conditions". Furthermore, within counties, individual townships and even villages may agree to different conditions to meet "local peculiarities". For example, the IMAR government has set the term as five to seven years. But the IMAR pastoral county of Balinyou (Chapter 8) has established the term as being greater than 15 years. Within Balinyou, two townships were surveyed. One township administration, Bayantala, insists that village collectives under its control contract pastures to households for only five years, whereas the nearby township of Shabutai permits contracts with a term of greater than 15 years.

Another serious source of uncertainty surrounding pasture use contracts in some areas is that while the contract specifies the area assigned to the household, it does not designate the precise location of this pasture land. These "partial" contracts obviously encourage grazing-in-common practices and discourage investments in pasture conservation and improvement by individual households. Even when the contract does refer to a unique piece of pasture land, the village collective (or higher authority) may have the power to re-assign specific areas of pasture within the term of the contract.

Obviously, property rights in general, and use rights in connection with pasture land in particular, must be perceived by pastoral households as most uncertain. Clearly defined and stable property rights are a prerequisite to the adoption of long-term sustainable pasture management strategies by private households. The uncertainty surrounding the future of the HPRS in general, and pasture use contracts in particular, is a major constraint to sustainable development in pastoral areas. However, even where households have become reasonably certain that the HPRS is a "permanent" reform, the relatively short term for which the use rights are contracted out and the relative ease with which county governments and even township administrations can alter the conditions which apply to the contract, mean that households still have a strong incentive to "mine" their pastures. As mentioned in Section 18.2.2, the lack of any real market for the pasture use rights exacerbates the problem.

18.3.2 The "Policy Mirage" Syndrome

Institutional reforms in China since 1978 have greatly decentralised power and responsibility. The discussion of the uncertainties associated with property rights, and in particular the pasture use contract, in the preceding section highlighted the variation from place to place in how property rights reforms are being implemented at the "grass roots". Local government officials and Party cadres, who are paid by their local constituents, have become more aligned with the interests of the local inhabitants as the rigidities in the vertical bureaucracies of which they are part have been relaxed. That is, as local governments become more responsible for their own revenues and expenditures, vertical control over their activities has been greatly reduced.

While these changes have stimulated development in remote pastoral areas, they have also permitted the "policy mirage" syndrome to flourish. That is, at Central government level certain policies are in place and provincial, prefectural, county and even township officials will describe, often in considerable detail, how the policy is working. However, at the village and household level, the policy does not exist. In fact, the real situation at the "grass roots" may be almost exactly the opposite to that "seen" from higher levels in the administration. The higher-level officials are seeing a policy mirage.

Situations illustrating this policy failure problem were observed in relation to the policing of pasture stocking rate limits (see, for example, the discussion in the Cabucaer case study), "compulsory" sheep-breeding regulations (e.g. Hebukesaier case study) and in relation to a number of other policies. Apart from the obvious implications for the assessment of the effectiveness of particular government policies, this deterioration in the degree of official control which has followed the breakup of the communes emphasises the need to get the private incentives right. Formal government regulations can no longer be relied upon, if they ever could, to control socially undesirable private activities such as over-stocking.

Another difficulty with the policy mirage syndrome is that it adds to the policy uncertainty faced by pastoral households. Government officials and Party cadres are moved around on a regular basis. While one township governor may permit certain policies to be adapted "to suit local conditions", the next appointee may not. Under these circumstances, farmers who are currently being permitted to adopt exploitive practices such as overgrazing native pastures have an incentive "to make hay while the sun shines".

18.4 Technical Improvements

In the first three sections of this chapter, it has been argued that population pressures, market distortions and institutional uncertainties have interacted with the fragile ecosystem in pastoral areas to cause rangeland degradation. That is, public policy settings have interacted with the socio-cultural and economic environment to induce private households, and in some areas State farm managers as well, to adopt non-sustainable rangeland management practices of which overstocking is the most important. Technology, on the other hand, has enhanced the capacity of these rangeland managers to develop animal husbandry industries based on native pastures. Whether pastoralists choose to utilise this technology in an exploitive or sustainable fashion has depended (and will continue to depend) upon the policy and institutional framework within which they have operated and the pattern of economic incentives created by this framework. The remainder of this chapter, with the aid of Fig. 18.4, explores some of the potentially serious negative aspects of introducing new technology into a socio-cultural situation in which the economic incentives are distorted by inappropriate public policies and the lack of essential institutional structures.

18.4.1 Pasture Improvement

It seems to be an almost universally accepted orthodoxy in China that one way to reduce the destruction of native pastures is to increase the area of rangeland sown to improved pastures (i.e. introduced plant species). The principle behind this school of thought is that improved pastures will not only raise the overall production capacity of the system (through increased production of primary dry matter) but also alleviate feed shortages in winter and to a lesser extent, feed shortages during times of drought. Thus, with fatter animals in summer and autumn and less starving animals during winter and at times of drought, animal productivity rises, thereby allowing for productivity-based permanent improvements in pastoral incomes. The emphasis placed on the sowing of improved pastures by county officials, which is discussed in most of the county case studies in this book, confirms the widespread acceptance of the value of pasture improvement technology.

The critical difficulty with this scenario is that it assumes an economic and policy environment in which pastoralists are favourably disposed to investment activities with long-term payback periods (such as conservative/sustainable rangeland management practices). However, in a situation where this assumption is no longer valid, a more plausible scenario is depicted by tracing out the impact of technical improvements by means of sown pastures in Fig. 18.4.

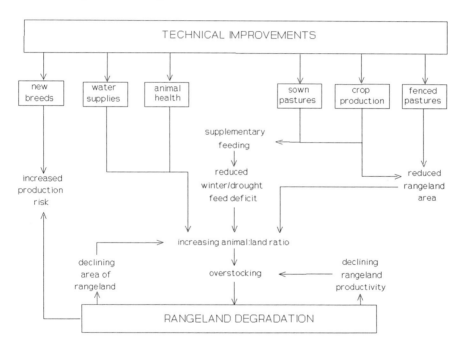

Fig. 18.4 Impact of Technical Improvements on Rangeland Degradation when Policy, Institutional and Incentive Structures are Inappropriate

Sown pastures, which are almost always used as "cutting land" for haymaking, provide low-cost, relatively high-quality hay for use as supplementary feed. However, rather than using the extra supplies of supplementary feed to increase the per animal feeding regime and thus the quality of the animal products produced, the pastoralists choose to utilise the feed to increase total animal numbers, keeping per animal feeding regimes at or near their previous low levels. The increase in livestock numbers results in a higher stocking rate being applied to surrounding rangeland (and, in particular, the winter/spring pastures) raising the animal:land ratio which eventually leads to overstocking and rangeland degradation.

At the same time, secondary effects further worsen the environmental situation. That is, as rangeland degradation increases, total rangeland productivity declines which leads to further overstocking and so on. Furthermore, some of the native pastures may be completely destroyed, reducing the area of available rangeland and directly increasing the animal:land ratio, thereby further adding to rangeland degradation.

Data from those areas where substantial pasture improvement has occurred strongly support the thrust of the scenario just described. For example, the graphs in Fig. 18.5 suggest there has been a strong association between the increasing amounts of hay conserved in Chifeng City Prefecture (almost entirely as a result of an expansion in sown pastures) and the upward trend in herbivorous livestock numbers.

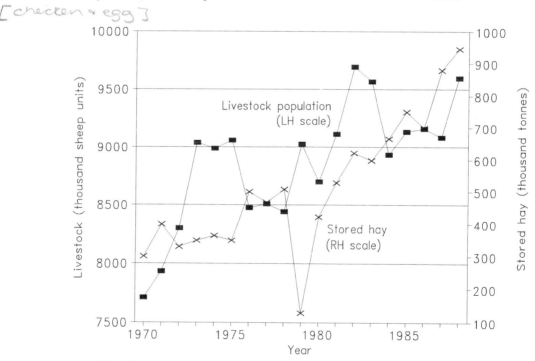

Fig. 18.5 Herbivorous Livestock Population and Amount of Hay Stored in Chifeng City Prefecture, 1970 to 1988

Farmers interviewed in all four counties surveyed in Chifeng City Prefecture invariably reported winter feeding regimes well below recommended supplementary feeding standards such as those set out in Section 9.2.2 for Wongniute. Since most of the interviewees were likely to be above average managers, the average level of winter feeding adopted by farmers in Chifeng City Prefecture would be even worse. Furthermore, the farmers do not ration feed strictly according to biological demand. That is, since ewes are usually mated in autumn, their biological demand for nutrients will increase as winter and their pregnancy progresses. More generally, all livestock will have an increased demand for nutrients in the middle of winter when the ambient temperature can drop to -30°C.

In summary, the evidence is that farmers have not used the increasing amounts of hay made available by sowing large areas of improved pastures, to enhance the quality of their winter feeding regime and hence further improve the level of productivity per animal. Instead, they have elected to overwinter increasing numbers of herbivorous animals, which has put increasing pressure on the native pastures, especially those areas used for winter/spring grazing.

18.4.2 Crop Production

Agricultural products such as maize, millet and crop residues (maize stalks, etc.) are also major sources of supplementary feed and hence are likely to impact on the pastoral system in a similar manner to that suggested for supplementary feed (hay) produced from improved pastures.

Another aspect of crop production illustrated in Fig. 18.4 is that by expanding crop production at the expense of rangelands, there is a significant spillover effect on the animal:land ratio which is applied to the remaining rangeland. Moreover, as was previously explained in Section 18.1.2, because it is often the best rangeland which is "reclaimed" for agricultural use, the real impact in terms of the resulting animal:land ratio is likely to be proportionately higher.

Supplementary feeds sourced from agricultural activities form a much larger proportion of the feedstuffs fed to livestock in winter in agricultural areas than in pastoral areas. This is not surprising, given the limited amounts of agricultural production in pastoral areas and the costs of importing such feedstuffs. However, it may also reflect a greater interest in quality and per head productivity by agricultural households. One reason for this is that the opportunity cost of labour is likely to be higher, increasing the marginal cost of raising an extra animal. Thus, from an economic perspective, there is an incentive to concentrate on increasing the output and revenue per head from the existing livestock rather than expanding the flock.

Another way of looking at the situation in agricultural areas is to define grazing as providing the supplementary feed in spring, summer and early autumn, when agricultural by-products and grain are in short supply.

That is, the herbivorous animals are raised in agricultural areas primarily to utilise crop wastes and by-products, and natural pastures which grow in spring, summer and autumn are a convenient, but not the only, source of supplementary feed during these seasons. This perspective helps explain why in the pastoral region rangelands are generally much more heavily degraded in agricultural areas than in pastoral areas. Indeed, in these agricultural areas it is the availability of the "supplementary" rangelands, especially in winter and spring, which are now limiting livestock numbers rather than the supplies of feedstuffs required for winter feeding. In many respects, therefore, in counties such as Aohan in the IMAR and Cabucaer in the XUAR, herbivorous livestock production systems have "mined" the natural pastures to the point where it is the rangeland resource rather than winter feedstuffs which is now the principal limiting factor in the production system.

18.4.3 Fencing

As with pasture improvement, the county case studies demonstrate that fencing is widely accepted as an excellent way to attack pasture degradation. In some counties such as Wushen (see Chapter 12), fencing has greatly improved the situation. But in most other areas, fencing has been limited to crop land, improved pastures and the best of the native pastures (i.e. most "cutting land" is fenced). Fencing, therefore, reinforces the impact of pasture improvement and the expansion of crop land on the remaining native pastures (Fig. 18.4). Unfenced pastures are almost always more overgrazed and degraded than the fenced areas in the same vicinity.

At present, it is impractical to consider fencing the extensive rangelands. The cost of shepherding *vis-à-vis* the cost of fencing; the lack of market rewards for controlling internal and external parasites by rotational grazing and flock separation; the lack of suitable fencing materials; and, perhaps most importantly of all, the lack of agreement on the location of boundaries, are all major impediments to the more general and widespread use of fencing in the pastoral region of China.

18.4.4 Animal Health and Water Wells

The environmental consequences of improved animal health programs in the absence of an appropriate economic and policy environment are essentially the same as those arising from improved access to the supplies of low-cost supplementary feeds. That is, both ultimately lead to an increase in the animal:land ratio. However, the manner in which this process takes place differs significantly in the sense that, whereas improved supplies of supplementary feeds facilitate an increase in the numbers of animals, improvements to animal health programs ensure that the population of extra livestock is not subsequently reduced through naturally-occurring outbreaks of parasites and disease. Thus the technologies which assist in improving animal health are complementary to and strengthen the effectiveness of technologies which promote improved supplementary feed supplies. This complementary relationship between animal health and supplementary feed is particularly important given the propensity for farmers to utilise low per animal feeding regimes and the subsequent higher incidence of parasites and disease normally associated with poor diets in animals.

Improved access to drinking water of better quality is likely to have an impact similar to measures which control parasites and disease (Fig. 18.4). To the extent that new wells are located in rangeland areas previously protected from livestock for some or all of the year owing to the lack of water, they will facilitate the exploitation of these areas. Experience in other arid areas such as sub-Saharan Africa suggests that the longer-term impact of providing better livestock watering facilities can be disastrous.

18.4.5 Livestock Improvement

Along with pasture improvement, upgrading local breeds of livestock by infusing new genetic material is a cornerstone of the "technology will fix it" syndrome in the minds of Chinese policy-makers. Local officials are especially enamoured with this approach to developing animal husbandry. The keenness on the part of Chinese officials concerned with sheep to acquire Australian merinos to upgrade their fine wool breeds is one of the better known instances of this phenomenon (see Longworth and Williamson, 1993).

Livestock improvement has the potential to raise incomes. For example, the point has been made several times in this book that upgrading nondescript local sheep to good quality fine wools, something which could be achieved in less than 20 years, is perhaps one of the best potential ways of permanently raising the incomes of traditional pastoralists.

Unfortunately, however, breed improvement programs may also have detrimental impacts on development. For instance, if the new breeds have characteristics which make the animal unsuited to the local environment, there could be disastrous consequences. There is strong evidence that fine wool sheep, for example, are not well suited to the harsh environment in much of the pastoral region. During the commune era, collectives and State farms were required to use fine wool sires. Consequently, the number of sheep carrying a heavy dose of fine wool genes increased sharply. However, as discussed in many of the case studies, once farmers were given more freedom to choose their own breeding lines in the early 1980s, the problem of "genetic regression" appeared. That is, farmers rejected the fine wool rams and mated their ewes to more environmentally suited sires (see, for example, the discussion in Section 7.5.1).

Rather than contributing to higher incomes and hence regional development, breed improvement has the potential not only to raise production costs more than returns, but also to significantly increase production risks.

Animals which are less hardy require better feeding, housing and animal health care if they are to be productive. This need for additional care and attention means higher costs which, as pointed out many times, may not be compensated for by sufficiently higher prices for the better quality meat or fibre which the improved animals are capable of producing. Animal improvement programs can only lead to better incomes if the markets generate appropriate premiums for improved products.

Increased production risks are the second major potential problem retarding the acceptance of new breeds. Less hardy sheep, for example, are more likely to die as a result of a natural disaster such as drought, severe snowstorm, "black disaster", etc. Pastoral households in northern China are often heavily dependent on their flocks both for income and for sustenance. The loss of their animals could imperil the very existence of the family. The breakup of the communes with the attendant disappearance of much of the community-wide sharing of environmental risks has made the newly independent households more cautious about production risks.

Under those circumstances, the swing back to goats in Alukeerqin (Chapter 11) and to coarse wool sheep in Hebukesaier (Chapter 17), for example, may reflect much more than a response to changing relative prices. Livestock improvement, therefore, remains very much a two-edged sword.

Chapter 19

The Way Ahead

Effective survey-based research in a large and diverse part of the world such as the pastoral region of China requires a selective focus. The study from which this book developed concentrated on sheep and wool; on people who raise the sheep and especially on minority nationalities; on degradation of the rangelands on which the animals and ultimately the herdsmen and their families depend; and on identifying the constraints to further sustainable development in pastoral areas. A detailed overview of these four interrelated facets of China's pastoral region has emerged which is based on interviews with sheep-raising households, village heads, township officials, and key people in government and other organisations at the county, prefectural, provincial and national level.

There are enormous advantages in being able to interact with relevant individuals at all decision-making levels. Grass roots observations of facts and opinions frequently identified major gaps and inconsistencies in information collected further up the Chinese bureaucratic hierarchy. As a result, this book presents a comprehensive picture of a remote but fascinating and increasingly important part of the world. Unique first-hand information is presented and analysed in a holistic system-wide framework. Many new insights emerge about the problems faced by the residents of this often almost forgotten part of China. Concrete suggestions are made about ways to improve the long-term economic circumstances of the traditional pastoral communities and others living in the pastoral areas. The information, insights and suggestions contained in this book are not summarised in this chapter. Instead, the four facets of China's pastoral region on which attention has been concentrated are placed in a global context and some brief comments are made about possible future developments in relation to each of these topics.

19.1 Sheep and Wool

During the 1980s, the Chinese woollen textile manufacturing capacity expanded rapidly and China emerged as a major force, both as a buyer in the international market for raw and semi-processed wool, and as a seller in the world woollen textile and garment trade (Anderson and Park, 1988; Anderson, 1990a; Zhang, C., 1990). The need to import large amounts of wool not only for re-export after processing but also to supply domestic consumers with more wool-based goods such as knitting wool, textiles, ready-made apparel and blankets, lowered wool self-sufficiency and added to the pressure on foreign currency reserves. Ministry of Agriculture officials and others long concerned about the need to develop the domestic Chinese wool-growing industry suddenly found new allies in the Central government bureaucracy as policy-makers involved in trade issues became interested in domestic raw wool production.

With a national flock of 113 million sheep in 1990, China ranks with Australia, New Zealand and the former USSR as one of the four most important sheep-raising parts of the world. Prior to the study reported in this book, relatively little was known about the domestic-supply side of the Chinese wool market. There had been no previous comprehensive industry-wide investigation of the grass roots situation in the pastoral areas of China. A great deal of the information about wool production and marketing presented in this book has not previously been available, even in Chinese. Apart from being of major relevance to understanding the development issues facing the pastoral areas, the material in this volume also sheds a great deal of light on the future wool production potential of the Chinese flock. Both Chinese policy-makers concerned about the domestic wool textile industry and related trade issues and foreigners interested in the international markets for wool and wool textiles, have a keen interest in the future prospects for wool production in China, especially in regard to apparel wool or fine and improved fine wool as it is called in China.

China produced almost no wool of this type in 1950. After four decades of breeding and upgrading, the output of fine and improved fine wool (i.e. wool of $\leq 25\mu m$ in mean fibre diameter) reached 109,000 tonne in 1991. For the reasons set out in Section 18.2.1, the yield of clean wool obtained after scouring is much lower in China than in other major wool-growing countries. Therefore, the size of the Chinese wool clip needs to be assessed on a clean scoured basis when making international comparisons. In clean scoured terms, the amount of fine and improved fine wool grown in China in 1991 was about 46,000 tonne or roughly one-thirteenth of the amount of wool of $25\mu m$ or less produced in Australia in that year. Furthermore, while a high proportion of the Australian wool in this category would have been genuine "fine wool" in Chinese terms, probably only about one-quarter of the fine and improved fine wool grown in China in 1991 was wool of this type. The other 75% was wool which the Chinese would describe as "improved fine wool" but which the international wool trade would term "crossbred" or "comeback" wool.

Fine and improved fine wool production in China in 1991 was more than 56% above the output achieved in 1981 (Section 4.3.2). Chinese Ministry of Agriculture officials and others concerned about the pastoral region for strategic and other reasons have long seen the upgrading of sheep for fine wool production as one major means of improving productivity and hence incomes in pastoral areas. While this strategy has been successful in some places, and upgrading will continue to make a major contribution in selected parts of the pastoral region, there are many localities in which farmers are rejecting official pressure to continue upgrading their flock towards finer woolled sheep. The reasons for this phenomenon are considered in detail in the case studies in this book but there are essentially two major factors at work. First, fine woolled sheep are not well adapted to the harsh environments under which many traditional Chinese pastoralists must operate and secondly, other small ruminant products have become relatively more profitable, in particular cashmere, mutton and goat meat.

China's meat industry is being encouraged to obtain more of its supplies from ruminant animals (cattle, sheep and goats) in a bid to reduce the pressure on grain supplies as consumer demand for meat rises even faster than incomes. At present, 78% of the meat eaten in China is pork. Half the world's pig population lives in China and these pigs consume a huge amount of grain. Beef, mutton and goat meat currently contribute less than 9% of the national meat supply while poultry, rabbits, and other small animals make up the rest. In addition to the burgeoning domestic market demand for meat, the XUAR is enjoying a rapid growth in livestock and meat exports to the new Central Asian Republics, especially Kazakhstan. Throughout the pastoral region, therefore, there is increasing emphasis being given to breeding sheep especially for mutton production. In most areas, specialist mutton production matches the feed requirements of the sheep flocks to the growth pattern of natural pastures much better than wool-growing production systems. (See the discussion in Section 7.5.3, for example.)

In the past, fine wool production has been increased in pastoral areas both by running more sheep and by increasing the wool cut per head. Since the total stocking rate of grazing animals now exceeds sustainable levels in almost all pastoral localities, more fine wool sheep can only be raised at the expense of other herbivorous animals. Relative product prices in the early 1990s favoured mutton and cashmere rather than fine wool and this situation is expected to persist for some time. Hence, even in those areas where fine wool sheep are well adapted, it is unlikely that the total number of fine and improved fine wool sheep will increase significantly in the foreseeable future, although the upgrading process will continue and the proportion of this type of sheep which are genuine fine wools will increase.

As more sheep are upgraded to genuine fine wools and as management practices improve, the average amount of wool grown by each sheep will increase. Therefore, even if the total number of sheep classified as fine and improved fine wools does not increase, more wool of this type will be produced in the traditional pastoral areas. However, output in the 1990s will expand much less rapidly in these areas than during the 1980s.

While the potential to increase sheep numbers in traditional pastoral localities is severely constrained unless the number of other grazing livestock is reduced, there is said to be considerable capacity to raise more sheep without cutting back on other animals in three other parts of China. The first of these areas is the agricultural localities of the north eastern pastoral provinces (Liaoning, Jilin and Heilongjiang). These parts of the country have a coarse grain surplus, and a considerable amount of soybean meal and cake is produced. In addition, there is a tradition of raising small ruminants (sheep and goats) and a considerable wool-industry infrastructure already exists (studs, technicians, etc.). However, given present and likely market prospects and the growing interest of policy-makers in expanding ruminant meat production, any expansion in sheep raising in the north eastern pastoral provinces is likely to concentrate on mutton rather than wool production.

The same conclusion can be drawn with even more conviction in regard to the eastern agricultural provinces (Jiangsu, Zhejiang, Shandong, Henan and Anhui) which are the second area in China considered capable of supporting a much larger sheep population. These provinces have a ready market for better quality wool because there are many large wool textile mills in this part of China. Nevertheless, they are unlikely to continue to expand into fine wool production. The demand for meat is growing strongly in this part of China and these provinces traditionally raised sheep for meat rather than wool. There is little wool-industry infrastructure in these provinces (except for parts of Shandong) to support a push into wool growing and there is very little pasture land.

The third part of the country often claimed to have great potential in regard to wool production is the mountain pasture areas in five southern provinces (Sichuan,

Hubei, Hunan, Guizhou and Yunnan). As discussed in detail in Section 4.2.2, these areas are not generally suited to fine wool production and, to the extent that a sheep industry develops in this part of China, it will be based on the production of mutton and semi-fine wool.

19.2 Minority Nationalities

Of the 56 officially recognised ethnic groups in China, the Han people with 92% of the population play an overwhelmingly dominant role as the "majority nationality". The Han first emerged from the gradual integration of a number of related tribes and ethnic groups during the Han Dynasty (206 BC to 220 AD). Originally from the Yellow River basin, the Han people now occupy all of the country, although in many parts of China's pastoral region they are not the most numerous ethnic group.

Of the 55 minority nationalities, 30 are represented in the 12 pastoral provinces/autonomous regions and 20 of these ethnic groups are not found elsewhere in China. Of the 30 minorities living in the pastoral region, the Mongolian, Hui, Tibetan, Korean, Manchu, Miao, Yi, Tujia, Uygur and Kazak ethnic groups are the most numerous.

Not all minority groups living in the pastoral region are traditionally associated with the raising of sheep and other grazing livestock. However, most groups historically occupied remote parts of the country and even today these "homeland" areas are relatively isolated from mainstream Chinese society. Of the 10 case study counties, Wushen (90% of the pastoral households are Mongolian), Sunan (Tibetan and Yugur people represent half the population), and Hebukesaier (almost two-thirds of the population are Mongolian or Kazak) are examples of remote localities with a long history of occupation by the forebears of the current residents. Cabucaer, a case study county in Yili Prefecture of the XUAR has 60% of its population drawn from three ethnic minorities (Uygur, Kazak and Xibe). In this case, however, the Xibe are relatively recent settlers. Their ancestors marched across northern China from their traditional homeland in north east China to garrison the north western border for the Qing government in the mid-18th century (Ma, 1989).

China, of course, is not especially unusual in having a large number of ethnic minorities living within its borders. Indeed, the term "fourth world" has been coined to describe enclaves of indigenous populations within nation states of the first, second or third world. The growing international concern for the rights of the residents of the fourth world precipitated the decision by the United Nations to declare 1993 the Year of the Indigenous People.

In a formal sense, the rights of the minorities appear to be better protected in China than in many countries. Since 1949, as briefly discussed in Section 3.4, the Central government has promoted autonomous rule, encouraged national representation and stimulated development in minority autonomous areas. However, as with many policies in China, there may be a significant gap between the formal provisions in the Constitution and national legislation etc. as described by Chinese authors such as Ma (1989) and grass roots reality.

Although there are elaborate formal provisions for autonomy and self-determination, in practice the evidence suggests that the long-term objective of the Central government is to integrate the minorities into mainstream Chinese society especially in a political and economic sense. As Mackerras (1994) spells out in considerable detail, some of the largest minority groups are already almost completely integrated into the Han-dominated society. On the other hand, smaller ethnic minorities living in remote parts of the pastoral region, such as the Tibetans in particular, continue to resist integration even to the extent of public violence.

Economic reforms in China since 1978 have led to remarkable economic development in most of China. As the case studies in this book demonstrate, even in some of the remotest counties predominantly inhabited by ethnic minorities, real incomes and hence living standards improved dramatically. However, the way ahead for these people looks much less promising in an economic sense unless the constraints to further development discussed in Chapter 18 are seriously addressed. Should these communities experience a deterioration in their economic circumstances not only in relative terms but also absolutely, the pressure for political reforms in these localities will intensify.

19.3 Rangeland Degradation

There is an urgent need to raise the profile of rangeland degradation as a global environmental issue. Papers presented at the XVII International Grassland Congress held in New Zealand and Australia in 1993 suggested that at least 25% of the earth's land surface is affected and the future livelihood of up to a fifth of the world's population is under threat. The continued loss of the natural savannas on a global scale has implications for mega-environmental concerns such as global climatic change and loss of genetic diversity which are at least as serious as those associated with the destruction of rainforests.

"There is nothing more humble than grass" (Cribb, 1993). Although natural savannas played a crucial role in the evolution of humankind, urban civilisations have always neglected these important terrestrial systems. Much of the middle east was turned to desert by ancient civilisations which destroyed their grasslands. The same is occurring now in the Sahelian savannas, the North American prairies, the Pampas of South America, the Australian outback, and in the pastoral region of China.

The World Bank and others have published desktop studies summarising the macro-picture in China. For example, Rozelle and Huang (1993) quote a 1992 World Bank study which estimates that 33 million hectares of natural grassland were turned to desert prior to 1980 and that a further 150,000 hectares per annum became desert during the 1980s. In addition to the area already desertified, at least another 1.3 million hectares of natural pasture is becoming degraded each year so that the total area of degraded pasture now exceeds 30 million hectares or one-tenth of China's usable pasture as defined by the World Bank.

The aim of this book has been to dig deeper. The first-hand observations and otherwise unavailable local data presented herein greatly refine the definition of the extent of the problem, at least in the areas covered by the case studies. For instance, while the World Bank study states that 10% of the usable pasture area is degraded, the research reported in this book suggests that virtually all natural pastures in China are now affected to some extent, and it is critically important to define carefully the degree to which any particular area of pasture is degraded before passing judgement on the seriousness of the problem in the locality under discussion. Several of the case studies in this book spell out in detail precisely what the various levels of pasture degradation being discussed mean in terms of pasture ecology, ground cover, productivity, etc. (See, for example, Section 11.1.2.) The local data on the extent of pasture degradation are presented and discussed in the light of these detailed definitions. The overall conclusion which emerges is that pasture degradation is not a new problem but it has now reached crisis levels in many pastoral areas. The number of grazing livestock exceed the sustainable stocking rate almost everywhere.

While overstocking is the obvious direct "cause" of the problem, the real issue is what induces the owners of the livestock to adopt management strategies which are so destructive. That is, the real causes of degradation are behavioural and not readily discernible.

Throughout this book, a holistic approach has been adopted to the analyses of issues, hence the case study approach. In regard to the underlying causes of pasture degradation, the major blame is laid at the feet of policy. In Chapter 18, the case is made that the policy environment in China since 1949 has created the incentives and uncertainties which have induced pastoralists to behave in an exploitative manner. Furthermore, the availability of modern technology has facilitated and intensified the "mining" of natural pastures in China's pastoral region.

To solve or even ameliorate the degradation problem will require a concerted system-wide rethink about the relevant public policies at both the national and the local level. The behavioural responses of the private pastoralists, State farm managers and even local governmental and party cadres need to be modified to ensure that more sustainable micro-level management practices are adopted on a wide front. This is a massive and extraordinarily difficult task and not one which is unique to China's pastoral region.

While in general terms the situation in north and north western China is similar to that in other parts of the globe, which Barbier (1991) collectively refers to as "drylands", in detail there are many unique features of the pasture degradation problem in China which need careful analysis before meaningful solutions are likely to emerge. Furthermore, local conditions, as demonstrated in the county case studies presented in this book, vary a great deal across China's pastoral region. Some of the local socio-cultural features in China have the potential to facilitate a shift to more sustainable grazing practices. On the other hand, some local factors exert a strong negative influence.

One example of positive local conditions is the re-emergence of traditional clan-like management units in some of the more remote and especially environmentally fragile grazing areas. Prior to collectivisation in the late 1950s, the sparse populations roaming these areas managed their herds and flocks on a communal basis. Each clan or sub-tribe "owned" certain grassland areas and collectively regulated the size of their livestock inventories to preserve these lands for posterity. In many respects, at least initially, the establishment of communes in these areas in the late 1950s did not represent a major change to the traditional decision-making framework. However, over time, pressure both to abandon the nomadic way of life and to increase production (initially to satisfy State Plans but more recently to generate income for an expanding population) diluted traditional approaches to the management of livestock. During the 1980s, the breakup of the communes recreated the need for voluntary communal management. In some minority nationality communities where the traditional culture has been maintained, the old clan system of management has been re-established and represents a major step towards solving the well known grazing-in-common problem. Furthermore, under these circumstances, regulations imposed by the State and aimed at preventing overgrazing are likely to be relatively effective because they are consistent with the local socio-cultural environment.

In a country like China with a long tradition of bureaucratic control, the most common approach to preventing private activities which are judged to be detrimental to the public good is to introduce regulations. Of course, persuading local officials to enforce regulations which are inconsistent with the short-run aspirations of their constituents is another matter. Furthermore, regulations which inhibit the short-run expansion in physical output from the area for which the official is responsible will not be in the career interests of a bureaucrat who expects to be rewarded for meeting production targets, etc. As pointed out several times in this book, while the "policy mirage" syndrome was common enough when the Chinese economy was completely centrally planned, the decentralisation of power to local governments during the 1980s has encouraged grass roots cadres to give even greater weight than previously to local attitudes to regulations imposed from above. Under these circumstances, regulations aimed at solving the pasture degradation problem will not be successful unless the appropriate policy-induced economic incentive structures are also in place.

Creating a policy environment which generates the required incentives (social as well as economic) to encourage private decision-makers to make socially desirable resource allocation decisions is a world-wide natural resource management problem. In the context of China's pastoral region, changes to present policy settings which would greatly improve the situation are discussed in Chapter 18 and elsewhere in this book.

19.4 Sustainable Development

The intuitively obvious concept that development, broadly defined, needs to be sustainable in perpetuity has attracted a great deal of attention in recent years. Although the idea of "sustainable development" had its origins in United Nations debates of the 1960s, it gained increasing credibility during the 1980s and finally came of age as a concept when the World Commission on Environment and Development released the so-called Brundtland Report in 1987. Since then there have been many intensive debates at conferences and in the professional literature. (See, for example, Peters and Stanton, 1992 and the references cited therein.) It would seem that in many real-world circumstances it is not always clear precisely what sustainable development really means. On the other hand, in some special situations such as in relation to pastoral activities in the pastoral region of China, sustainable development would appear to be unambiguously defined as preserving a balance between the number of grazing livestock and the productivity of the natural pastures. Of course, this balance or equilibrium need not be at the maximum biological sustainable yield nor is the socially optimum level of pasture utilisation likely to be at the biological maximum. Indeed, to the extent that the pastures can recover from periods of overgrazing once the grazing pressure is reduced, long-term sustainable use may involve short periods during which the pastures are overstocked and hence are being degraded. Therefore, even in the context of pastoralism in north and north western China, defining precisely what sustainable development means is not straightforward. The practical approach more or less implicitly adopted in this book has been to regard development which preserves or expands the renewable natural resource base on which pastoral communities depend as being sustainable development. Other authors such as, for example, Southgate et al. (1990) have stressed the enormous challenges which must be overcome if sustainable development even in this restricted sense is to be achieved in the pastoral regions of many developing countries.

In the context of China, the most urgent issue which needs to be addressed is how to match the size of the human population to the available resources in pastoral areas. As argued in Chapter 18, population and labour market policies currently inhibit the outmigration of surplus people. At the same time, there is little likelihood of any substantial increase in the flow of capital into the remote pastoral areas for the purposes of enlarging the resource base of the pastoral industries. The returns are low relative to alternative investment opportunities, and the risks are high. Of course, as discussed in many places in the book, for strategic and other reasons, the Central, provincial and even local governments provide public funds for poverty alleviation, some of which are made available to households at concessional interest rates for investment in pastoral activities.

A new approach and one which is not canvassed to any extent in this book is that entirely new sources of livelihood could develop in many pastoral areas, hence relieving the human population pressure on the rangelands. Tourism, for example, could rapidly become a major source of employment and income in certain pastoral areas. For example, of the 10 case study counties surveyed, Wushen County in the IMAR, Dunhuang and Sunan Counties in Gansu, and Cabucaer County in the XUAR are localities with scenery, features of historical and cultural interest and other attractions which create an enormous potential for both domestic and international tourism. Many of the more remote pastoral counties in the XUAR have reserves of coal and some have oil and mineral deposits. Indeed, parts of China's pastoral region appear to be on the verge of a major minerals boom. Should tourism and mining expand in China's pastoral region, the additional economic activity will lead to new employment opportunities, new markets, and better infrastructure (airports, roads, rail, telecommunication, health services, education, etc.). To the extent that the traditional pastoralists can "plug into" these new industries, the pressure on the traditional resource base will be eased. However, in other parts of the world, traditional communities have tended to be marginalised as high technology, capital and skill intensive industries such as tourism and mining develop in pastoral areas.

The problems associated with the incorporation of indigenous ethnic groups living in remote parts of the country into the national development process are not unique to China. Other large countries with rapidly growing economies, such as Indonesia, face similar difficulties. Like China, Indonesia has a great diversity of ethnic groups dispersed over a vast geographic area. Furthermore, as in China, the population densities and the availability of natural resources and income levels vary greatly from one part of Indonesia to another. While in terms of climate, culture, history and many other characteristics, the remote islands of Indonesia would appear to be vastly different to China's pastoral region, in principle, the problems of how to achieve sustainable community development are essentially the same. The traditional populations resident in these areas, together with settlers who have moved in to these previously relatively underpopulated localities from elsewhere in the country, are now exerting excessive pressure on the natural ecosystem. Further development along the lines which have occurred in the recent past is not sustainable in the longer term.

The single most important factor in the development of the pastoral areas of China since 1949 has been the steady expansion in the number of grazing livestock, especially sheep. Not only has the national sheep flock (of which almost 90% is in the pastoral region) expanded from around 25 million in 1949 to over 112 million in 1990, but also there has been a major improvement in terms of the wool production per head. The increased average productivity has occurred because the number of fine wool and improved fine wool sheep has increased from a few thousand in 1949 to over 32 million in 1990. Traditional pastoral communities which are heavily dependent for their livelihood on sheep have improved their economic circumstances. At the same time, the long-standing threat of the loss of natural pastures has become progressively more serious, and it is now the most obvious constraint to further progress.

Further sustainable development of China's sheep and wool industry, and hence of the minority nationality communities which depend heavily on sheep raising for their economic livelihood, requires that ways be found to overcome the increasingly severe physical constraints being imposed by continued rangeland degradation. This book explores in depth the crucial interlocking elements involved in trying to find a sustainable way forward. The case studies presented in Parts II and III demonstrate that there are no panaceas. Local climatic, cultural, socio-economic and historical factors vary greatly across northern and north western China. Technical improvements (i.e. the application of new technology) alone will not lead to further sustainable development. Indeed, as argued in Chapter 18, it may exacerbate the problem. Major policy changes are needed at all levels of government to create the right incentives and institutional structures, not only to encourage private livestock-raising households, State farm managers, and local officials to adopt more sustainable management practices, but also to create a milieu in which local communities are prepared to accept and to enforce restrictions on short-term individual behaviour in the interests of long-term socially desirable outcomes.

References

Anderson, K. (1990a) China and the multi-fibre arrangement. In: Hamilton, C.B. (ed.), *Textile Trade and the Developing Countires: Eliminating the Multi-fibre Arrangement in the 1990s*. The World Bank, Washington, D.C. pp.139-58.

Anderson, K. (1990b) Urban household subsidies and rural out-migration: the case of China. *Chinese Economy Research Unit Working Paper*. Economics Department, The University of Adelaide, 90/3, 1-11.

Anderson, K. and Park, Y. (1988) China and the international relocation of world textile and clothing activity. *Pacific Economic Paper* 158, Australia-Japan Research Centre, ANU, April 1988.

Anon. (1985) *China ABC*. New World Press, Beijing.

Anon. (1987) *The Population Atlas of China*. Population Census Office, State Council of the People's Republic of China and the Institute of Geography of the Chinese Academy of Sciences, Oxford University Press.

Anon. (1989) *Outlines of Economic Development in China's Poor Areas*. Office of the Leading Group of Economic Development in Poor Areas Under the State Council, Agricultural Publishing House, Beijing.

Anon. (1991) *History of the Chinese Communist Party - A Chronology of Events (1919-1990)*. Compiled by the Party History Research Centre of the Central Committee of the Chinese Communist Party, Foreign Languages Press, Beijing.

Barbier, E.G. (1991) Natural resource degradation: policy, economics and management. In: Winpenny, J.T. (ed.), *Development Research: The Environmental Challenge*. Westview Press, Boulder, Colorado.

Brown, C.G. and Longworth, J.W. (1992a) Multilateral assistance and sustainable development - the case of an IFAD Project in the pastoral region of China. *World Development*, 20(11), 1663-74.

Brown, C.G. and Longworth, J.W. (1992b) Reconciling national economic reforms and local investment decisions: Fiscal decentralization and first-stage wool processing in Northern China. *Development Policy Review*, 10(4), 389-402.

Carrad, B., McIntyre, K., Obst, J., and Shelton, M. (1989) Improving pastures in China's south west - The Yunnan Livestock and Pasture Development Project. *AIDAB Evaluation Series*, No. 7, p.18.

Chandra, S. (1993) Assessing the performance of aid projects in agriculture: the case of Gansu grassland agricultural systems research and development project, China. *Paper presented to the 37th Annual Conference of Australian Agricultural Economics Society*, February 9-11, Sydney, Australia.

Chen, L.Y. and Buckwell, A. (1991) *Chinese Grain Economy and Policy*. CAB International, Wallingford.

China Daily (1991) Ethnic minorities enjoy freedom. 5 November, p.4.

China Daily (1992a) Ethnic autonomy helps uphold national unity. 31 January, p.4.

China Daily (1992b) More attention on minority area urged. 14 January, p.1.

Cribb, J. (1993) The battle to save rainforests get regular publicity but meanwhile a crucial resource is vanishing. *The Australian*, 8 March, p.13.

Grousset, R. (1967) *Conqueror of the World* (Translated by Denis Sinor and Marian MacKellar). Oliver and Boyd, London.

Guangming Daily, P.R.C. (ed.) (1990) *The People's Republic of China*. Lotus Publishing House, Hong Kong.

Henzell, E.F., Blair, G., and Syme, J. (1987) *Review of the Institute of Crop Breeding and Cultivation (Beijing) and Grasslands Research Institute (Huhehot)*. ACIAR (Mimeo), Canberra.

Khan, A.R. (1984) The responsibility system and institutional change. In: Griffin, K. (ed.), *Institutional Reform and Economic Development in the Chinese Countryside*. M.E. Sharp, New York.

Lehane, R. (1993) China and wool - a giant constrained. *Partners in Research for Development*. ACIAR, Canberra, 6(May), 2-9.

Lin, X. (1990) Development of the pastoral areas of Chifeng City Prefecture. In: Longworth, J.W. (ed.), *The Wool Industry in China: Some Chinese Perspectives*. Inkata Press, Melbourne, pp. 74-92.

Liu, Y. (1990) Economic reform and the livestock/pasture imbalance in pastoral areas of China: a case study of the problems and counter measures in Balinyou Banner. In: Longworth, J.W. (ed.), *The Wool Industry in China: Some Chinese Perspectives*. Inkata Press, Melbourne, pp. 93-105.

Longworth, J.W. (ed.) (1989) *China's Rural Development Miracle: With International Cmparisons*. University of Queensland Press, St. Lucia.

Longworth, J.W. (ed.) (1990) *The Wool Industry in China: Some Chinese Perspectives*. Inkata Press, Melbourne.

Longworth, J.W. (ed.) (1993a) *Economic Aspects of Raw Wool Production and Marketing in China*. ACIAR Technical Reports No. 25, Australian Centre for International Agricultural Research, Canberra, pp.36-49.

Longworth, J.W. (1993b) Wool marketing in China: A system in transition. In: Longworth, J.W. (ed.), *Economic Aspects of Raw Wool Production and Marketing in China*. ACIAR Technical Reports No. 25, Australian Centre for International Agricultural Research, Canberra, pp. 36-49.

Longworth, J.W. and Brown, C.G. (1994) *Wool Marketing in China: A System in Transition*. Inkata Press, Melbourne (forthcoming).

Longworth, J.W. and Williamson, G.J. (1993) *Fine Wool Sheep Breeds and Breeding in China: The Impact of the Australian Merino*. Inkata Press, Melbourne (forthcoming).

Ma, Y. (ed.) (1989) *China's Minority Nationalities*. Foreign Languages Press, Beijing.

Mackerras, C. (1994) *China's Minorities: Integration and Modernization in the Twentieth Century*. Oxford University Press, Hong Kong (forthcoming).

Minson, D.J. and Whiteman, P.C. (1989) A standard livestock unit (SLU) for defining stocking rate in grazing studies. *Proceedings of XVI International Grasslands Congress*, Nice, France, pp.1117-1118.

Niu, R. and Chen, J. (1992) Small farmers in China and their development. In: Peters, G.H. and Stanton, B.F. (eds.), *Sustainable Agricultural Development: The Role of International Cooperation. Proceedings of the Twenty-First International Conference of Agricultural Economists*. Dartmouth, Aldershot, pp.619-643.

Peters, G.H. and Stanton, B.F. (eds.) (1992) *Sustainable Agricultural Development: The Role of International Cooperation. Proceedings of the Twenty-First International Conference of Agricultural Economists*. Dartmouth, Aldershot.

Prosterman, R.L. and Hanstad, T.M. (1990) China: A fieldwork-based appraisal of the 'household responsibility system'. In: R.L. Prosterman, M.N. Temple, and T.M. Hanstad (eds.), *Agrarian Reform and Grassroots Development: 10 Case Studies*. Lynne Ryenner, London, pp. 103-36.

Remenyi, J. (1992) *A Poverty Alleviation Framework for AIDAB in China*. Centre for Applied Social Research, Deakin University, Geelong.

Rozelle, S. and Huang, J. (1993) Poverty, population and environmental degradation in China. *Paper presented at an International Symposium on Sustainable Agriculture and Rural Development*, May 25-28, Beijing, China.

Sicular, T. (1988) Agricultural planning and pricing in the post-Mao period. *The China Quarterly*, 116, 670-705.

Southgate, D., Sanders, J. and Enhui, S. (1990) Resource degradation in Africa and Latin America: Population pressure, policies, and property arrangements. *American Journal of Agricultural Economics*, 75(5), 1259-1277.

Spence, J.D. (1990) *The Search for Modern China*. Century Hutchinson, London.

Standing Committee of the Sixth National People's Congress (1985) *Rangeland Law of the People's Republic of China*. Adopted at the Eleventh Session of the Standing Committee of the Sixth National People's Congress, June 18, 1985.

Tu, You-Ren (1992) *Sheep, Goat and Wool Production in China*. Inner Mongolian Academy of Animal Husbandry Sciences, Huhehot, Inner Mongolia (unpublished).

Walker, K.R. (1989) 40 years on: provincial contrasts in China's rural economic development. *The China Quarterly*, 119, 448-480.

Wang, X. (1989) The current situation and future development of the Balin Grassland. Balinyou County Government, Balinyou, (unpublished).

Watson, A. (1989) Investment issues in the Chinese countryside. *The Australian Journal of Chinese Affairs*, 22(1), 85-126.

Watson, A. and Findlay, C. (1992) The "wool war" in China. In: Findlay, C. (ed.), *Challenges of Economic Reform and Industrial Growth: China's Wool War*. Allen and Unwin (in association with The Australia-Japan Research Centre, ANU), Sydney, pp. 163-80.

Wen, S. (1989) Development of the Cooperative Economy in China. In: Longworth, J.W. (ed.), *China's Rural Development Miracle: With International Comparisons*. University of Queensland Press, St. Lucia, Australia.

Williamson, G.J. and Longworth, J.W. (1993) Tenure uncertainty and sustainable development in the pastoral region. *Paper presented at an International Symposium on Sustainable Agriculture and Rural Development*, May 25-28, Beijing, China.

Williamson, G.J. and McIver, R. (1993) Household registration, income inequality and environmental degradation in China's border regions. *Working Paper Series*, University of South Australia, No. 8, 1-28.

World Bank (1988) *China: Finance and Investment*. World Bank, Washington D.C.

World Bank (1991) *China: Financial Sector Policies and Institutional Development*. A World Bank Country Study, The World Bank, Washington D.C.

Wu, H.X. (1991) China's urbanization and rural-to-urban migration: estimates and analysis in a perspective of economic development in pre- and post-reform periods. *Chinese Economy Research Unit Working Paper*, Economics Department, The University of Adelaide, 91/8, 1-44.

Yang, C. and Tisdell, C.A. (1992) China's surplus agricultural labour force: it's size, transfer, prospects for absorption and effects of the double-track economic system. *Asian Economic Journal*, VI(2), 149-81.

Yong, S., Cui, H., and Li, Y. (1990) Pasture vegetation map of Chifeng City, Chifeng City Prefecture Animal Husbandry Bureau, Chifeng City. (unpublished).

Zhang, C. (1990) An overview of the Chinese wool textile industry. In: Longworth, J.W. (ed.), *The Wool Industry in China: Some Chinese Perspectives*. Inkata Press, Melbourne, pp. 31-43.

Zhang, C. (1993) Development of Chinese wool auctions. In: Longworth, J.W. (ed.), *Economic Aspects of Raw Wool Production and Marketing in China*, ACIAR Technical Reports No. 25, Australian Centre for International Agricultural Research, Canberra, pp. 58-62.

Zhang, C., Zhou, L., and Xu, Y. (1991) The easily forgotten half of China: towards sustainable economic development in the pastoral region. *Paper presented at the XXI International Conference of Agricultural Economists*, 22-29 August, Tokyo, Japan.

Zhang, K. (1991) Ethnic baby boom overloads ecology. *China Daily*, 21 October, p.4.

Zhang, W. (1990) Development of market for rural land use in China. *Department of Agricultural Economics and Business Management Miscellaneous Paper*, University of New England, No. 9, 49 pp.

Index

Administrative village 24, 45, 313
Afghanistan 1, 159
Agriculture 9, 18, 32, 33, 54, 65-67,
    86, 96, 97, 118, 126, 127, 129,
    130, 137, 140, 144, 148, 150,
    157, 159, 175, 218, 270, 277,
    278, 286, 301, 304, 328, 329
  abandoned agricultural ground 305
  agricultural population 28, 29, 120
  agricultural settlements 4, 136,
    270, 305
Agricultural Commission 18, 148
Animal husbandry 2, 3, 25, 32, 33,
    35, 48, 57, 61, 65-67, 91, 94-97,
    101, 106, 107, 109, 110, 121,
    128-130, 148, 152, 153,
    156-158, 177, 233, 263, 264,
    271, 277, 291, 292, 293, 315,
    322, 326
  Academy of Animal Husbandry
    Sciences 94, 96, 97
  Animal and Veterinary Station 253
  Animal Examination Station 267,
    268
  Animal health 194, 253, 267, 282,
    326, 327
  Animal Husbandry and Veterinary
    Science Department 18, 66, 67,
    96
  Animal Husbandry Bureau (AHB)
    18, 19, 66, 67, 82, 85, 95, 96,
    123, 129, 130, 148, 157, 168,
    170, 175, 179, 182, 187, 189,
    190, 194, 195, 197, 198, 201,
    202, 204, 205, 210, 212, 214,
    220, 223-226, 243, 248, 253,
    257, 264, 267, 268, 270, 271,
    281, 291, 292
  Animal Improvement Station 19,
    154, 155, 176, 192, 193, 204,
    218, 267
  Animal Medicine and Machinery
    Company 157
  drinking water 326

Army 66, 67, 148
  army reclamation farms 148
Artificial insemination 48, 92, 175,
    176, 178, 192-194, 204, 206,
    217, 218, 241, 252, 253, 280
  AI centres 154, 194, 198, 218, 219,
    253, 274, 292
  fresh semen 176, 206
  frozen semen 92, 176, 178, 193,
    206
Australian Centre for International
    Agricultural Research (ACIAR)
    9-11, 16, 17, 19-22, 27, 37, 52,
    76, 80, 97, 119, 130, 139
Australian International Development
    Assistance Bureau (AIDAB) 55,
    66, 130, 148
Autonomy 1, 21, 39, 40, 99, 306, 331
  autonomous areas 39-41, 331
  autonomous counties 40, 41, 68,
    120, 136
  autonomous governments 40, 307
  autonomous prefectures 2, 40, 120
  autonomous regions 15, 25-31,
    33-35, 37, 38, 40, 43, 53-63, 76,
    97, 106, 116, 148, 149, 331
  autonomous rule 1, 2, 39, 331
  autonomous self-government 39,
    40

Banking system 284
  Agricultural Bank of China 14, 19,
    172, 181, 209, 223, 246, 270,
    284, 315
  Credit Cooperatives 270, 315
  Industrial and Commercial Bank of
    China 257, 315
  national credit plan 316
Black disaster 174, 327

Camel cashmere 91
Capital market 314, 315, 317, 318
Capital transfers 317

Index 341

Cashmere 31, 32, 47, 56, 59, 68, 76, 90, 127, 129, 195-198, 215, 220-222, 224-226, 271, 282, 284, 310-312, 329, 330
 cashmere prices 56, 87, 182, 255, 311
 cashmere production 87, 91, 96, 129, 178, 195, 198, 208, 210, 225, 226, 269, 295
 cashmere yield 178
Cattle 24, 30, 31, 36, 45, 55, 85, 112, 159, 169, 170, 173, 178, 184, 185, 189, 195, 202, 214, 232, 234-236, 240, 250, 256, 258, 265, 266, 274, 275, 278, 279, 284, 329
 dairy cattle 2, 91, 156, 288
 dual-purpose cattle 91, 159
 Shorthorns 91, 178
 Simmentals 91
 Three River cattle 91
 Xinjiang Brown 282
Chifeng City Prefecture 15, 16, 18, 27, 43, 44, 52, 76, 77, 81-86, 88, 89, 91-95, 100, 167, 180, 186, 193, 199, 203, 206, 208, 211, 212, 215, 218, 219, 222, 304, 305, 316, 324
Chinese Academy of Agricultural Science 9, 130
Chinese Academy of Social Science 9
Collective 23, 45, 47, 64, 65, 97-99, 106-108, 131, 154, 172, 183, 190, 203, 209, 231, 253, 256, 259, 269, 320, 321
 collective accumulation fund 98, 131
 collective-ownership 106
 collective property 106
 collectivisation 44, 64, 98, 99, 333
Communes 44-46, 64, 98, 154, 175, 250, 308, 313, 318, 320, 322, 327, 333
 commune structure 24
Communism 21
 Chinese Communist Party 19, 42, 46, 64, 313
 communist/Marxist orthodoxy 313
Contracts 23, 47, 65, 97-100, 107, 108, 171, 181, 183, 197, 216, 230, 231, 233, 240, 244, 245, 251, 256, 259-261, 270, 283, 284, 294, 313, 321, 323
 Certificates of Rangeland Ownership and Rangeland Use 109
 double contract system 181, 321
 Dual Management Land Contract 259
 inheritance rights 99
 joint households 131, 191, 197, 209, 231
 leases 99, 313
 pasture contracts 99, 131, 251, 321
 pasture utilisation certificates 131, 142, 251, 279, 289
 sub-contracting 100, 313
 three contracts and one reward system 283
Cooperatives 14, 19, 44, 45, 49, 68, 69, 96, 158, 270, 315, 318
Credit 172, 190, 209, 216, 223, 270, 314-319
 concessional credit 317
Cultural Revolution 40, 46, 56, 86, 140, 143, 145, 149, 186, 215, 241, 283, 305, 314
 four clean-ups campaign 46

Data problems 22
Degradation 1-3, 5, 10, 33, 47, 58, 81-84, 101, 112, 123, 142, 167-169, 171, 173, 182, 191, 214, 231, 251, 263, 273, 287, 290, 299-301, 303, 304, 308, 312, 313, 322-324, 326, 328, 332-334, 336
 cultivation mistakes 83, 306
 degraded pasture 82, 86, 124, 169, 188, 190, 213, 332
 desertification 108, 121, 240
 grassland degradation 188, 190
 heavily degraded 84, 169, 188, 213, 228, 249, 286, 325
 land degradation 81, 239, 261, 278
 lightly degraded 84, 124, 169, 188, 213, 249, 286
 medium degraded 84, 169, 188, 213
 salinity 30
Department of Animal Science 129
Development 1-5, 9-11, 18, 21, 39, 41, 44, 48, 52, 55, 66, 67, 71, 92, 93, 96, 101, 110, 112,

116, 128, 130, 137, 149, 153, 160, 171, 182, 190, 197, 209, 223, 227, 232, 233, 243, 245,254, 257, 259, 263, 284, 295, 299, 300, 304, 306, 308, 310, 311, 316-318, 320-322, 327, 328, 329, 331, 334-336
  development strategy 299
  economic and policy environment 323, 326
  economic development 2, 4, 11, 32, 42, 133, 134, 162, 179, 195, 206, 219, 243, 254, 268, 275, 282, 293, 304, 332
  economic welfare 17, 101, 309
Disease 23, 63, 194, 218, 227, 253, 267, 268, 326
  disease prevention 23, 194, 227, 253, 268
Dust production function 312
Dynamic programming 18

Eastern Grasslands 4, 81, 82, 211, 242, 257, 303
Economic Commission 257, 258
Economic Management Station 96, 97, 129, 196
Eerduosi Grasslands 4, 16, 81, 82, 237, 238
Ethnic minority 28, 37, 40, 41, 313, 314
  Common Program 40
  ethnic autonomy 39, 40, 306
  ethnic composition 170, 188, 201, 214, 239, 249, 265, 273, 278, 287
  Ethnic Reform Law 42, 43
  minority groups 1, 25, 26, 28, 29, 34, 36-40, 75, 77, 117, 118, 133, 262, 304, 306-308, 331, 332
  minority nationalities 1, 4, 10, 25, 37-41, 43, 53, 75, 77, 120, 140, 247, 265, 272, 276, 285, 305-307, 328, 331
  minority population 28, 37, 222, 274, 307

Factor market distortions 308, 312
Family farm 173
Family pasture fencing program 172
Farm inputs 69, 318, 319
Felt-making 258

Fiscal reforms 134, 315, 316, 319
  fiscal responsibility system 315, 320
Flock size 16, 202, 242, 291

Gacas 26, 27, 228, 229, 231, 233
Gansu 3, 4, 10, 15, 16, 18, 28-31, 33-35, 37-39, 42-44, 46, 51, 53, 55-62, 65, 68, 69, 72, 116-121, 123-134, 142, 148, 164, 238, 247, 248, 252-255, 258, 259, 261, 262, 263, 269, 272-274, 307, 317, 335
Gansu Agricultural University 129, 130
Gansu Grassland Ecological Research Institute 129
Genghis Khan 238, 312
Goats 2, 10, 24, 30, 31, 36, 56, 69, 85-87, 90, 91, 94, 96, 117, 121-124, 126, 129, 143-145, 147, 150, 151, 169, 170, 173, 184, 185, 189, 191, 195, 198, 202, 212, 214, 215, 222, 234-236, 240, 247, 250, 256, 258, 265-267, 274, 279, 282, 284, 288, 305, 311, 312, 314, 327, 329, 330
  dual-purpose goats 178
  Liaoning white cashmere goat 96, 178
  Mongolian white cashmere goat 96
Grain 2, 69, 101, 133, 154, 174, 175, 194, 203, 253, 279, 294, 305, 313, 320, 325, 329, 330
  Grain Bureau 284, 291
  "grain first" policies 305
  grain-pricing policy 320
  subsidised grain 174
Grassland Research Institute 96
Great Leap Forward 5, 20, 140, 273, 305

Han 3, 22, 28, 39, 140, 167, 170, 188, 201, 214, 232, 238, 239, 249, 262, 265, 278, 287, 304, 307, 308, 312-314, 331, 332
Hexi Corridor 118, 121, 123, 124, 127, 128, 248, 262, 272
Household production responsibility system 109, 304, 305, 308, 314
Household Registration System 304,

305, 308, 314

IFAD North China Pasture
   Development Project 171, 182,
   190, 223
Improved pastures 123, 169, 171-173,
   189, 190, 194, 203, 216, 240,
   249, 250, 257, 266, 279, 288,
   313, 323, 325, 326
   aerial sown pasture 23
   aerial spot planting 172
   artificial pasture 23, 169, 171-173,
      189, 198, 216, 217, 239, 240,
      250, 277, 279
   fenced pasture 189, 190, 216, 217,
      231, 250
   semi-artificial pasture 23, 173, 189,
      216, 239, 240, 250, 277
Income 16, 22, 23, 32, 33, 41, 44, 65,
   71, 100-103, 131-133, 150,
   160-162, 179, 181, 192, 195,
   196, 197, 206-208, 209, 212,
   219, 220, 222, 229, 243, 244,
   254, 268, 273, 276, 281, 282,
   283, 293, 301, 302, 304, 305,
   309, 317, 327, 333, 335
   net income per capita 22, 100, 131,
      133, 160, 162, 179, 195, 207,
      220, 254, 283, 317
Information Institute 148
Inner Mongolia Autonomous Region
   (IMAR) 3, 4, 10, 15, 16, 19, 26,
   27-35, 40, 41, 43-46, 52, 58-68,
   75-78, 80-106, 116, 129, 130,
   142-144, 148, 162, 167, 171,
   172, 179-181, 186, 192,
   198-200, 206, 207, 211, 214,
   218, 220, 221, 237, 239, 242,
   244, 251, 257, 289, 301, 302,
   304, 307, 316, 321, 325, 335
Institute of Animal Sciences 130, 148
Investment 85, 149, 169, 171, 172,
   181, 182, 190, 194, 196, 223,
   233, 240, 259, 270, 279, 295,
   308, 313-318, 323, 335
   investment strategies 318
   private investment 259, 315, 317
Kazakhstan 1, 137, 156, 157, 159,
   284, 285, 293, 330
Keerqin Desert 187, 200

Land market 313, 314

Landlords 43, 44
   animal landlords 43
   exploitation index 43
   The Three Nos and Two
      Favourables 44
Lanzhou Research Institute of Animal
   Science 129, 130
Laws 39, 40, 109, 142, 216, 229, 290
   Agrarian Law 260
   Electoral Law 40
   Outline Law 43
Linear programming 18
Liver flukes 267
Livestock Improvement Programs 3,
   91, 153
Livestock Improvement Station 96,
   129

Market distortions 299, 308-310, 312,
   322
Marketing 2, 3, 9, 10, 14, 19, 48-50,
   65, 68-70, 96, 158, 160, 162,
   176-178, 180, 196, 197, 207,
   208, 223-225, 255, 256, 268,
   269, 275, 283, 309-312, 318,
   319, 329
   animal product quota 259, 260
   Animal Products Trade Centre 157,
      164
   black market 104
Marketing organisations
   All China Union of SMCs 69
   Animal By-Products or Native
      Goods Companies 69
   Canton Animal Development
      Company 160
   China General Corporation of
      Animal Husbandry Industry and
      Comm 157
   Food Companies 69, 96, 158
   Supply and Marketing Cooperative
      (SMC) 14, 19, 49, 50, 68, 69,
      70, 96, 104, 105, 109, 154, 158,
      160, 164, 176-178, 180, 196,
      197, 198, 207-210, 220-222,
      224, 225, 244-246, 254, 255,
      257, 268, 269, 275, 280, 294,
      310, 311, 318, 319
Marketing system 69, 162, 283, 319
Markets 17, 47, 68, 70, 71, 104, 159,
   160, 180, 220, 245, 282, 294,
   309, 312, 313, 318, 327, 329,

335
  free markets 47, 71, 160, 294
  market failure 314
  market penalty 314, 320
  marketing information 96
  marketing margin 69, 158
  product market 309, 310, 312, 319
Marxist Thought 313
Meat 1, 2, 23, 24, 31, 32, 60, 85, 90-92, 94, 125, 133, 144, 147, 153, 156, 157, 160, 169, 177, 178, 205, 208, 220, 226, 256, 258, 267-270, 280, 282, 284, 292, 295, 312, 327, 330
  beef 31, 32, 55, 90, 147, 159, 180, 182, 196, 197, 208, 210, 220, 226, 269, 270, 284, 294, 295, 329
  goat meat 24, 31, 32, 90, 147, 180, 182, 208, 220, 226, 256, 269, 270, 284, 295, 329
  mutton 24, 31, 32, 52, 55, 90, 92, 94, 101, 128, 147, 153, 156, 160, 162, 178, 180, 182, 196, 197, 208, 210, 217, 218, 220, 226, 267, 269, 270, 280, 294, 295, 311, 329-331
  pork 31, 32, 329
Meat marketing 256
Meat processing 258
Meat quota 177
Military Commission 148, 149
Ministry of Agriculture 9, 18, 54, 66, 67, 96, 97, 129, 130, 157, 175, 218, 328, 329
Ministry of Commerce 69, 159, 254, 291, 318
Ministry of Domestic Trade 69, 96, 159, 256, 291, 318
Ministry of Material Supplies 69
Ministry of State Farms and Land Reclamation 65, 66
Ministry of Textiles 55
Mu Us Desert 238
Mutual aid teams 44, 45

National Rangeland Law 2, 83, 142, 217, 227, 266, 279, 289, 290
National Rangeland Survey 305
North West Institute of Animal Science 130

Old silk road 118
Output value 32, 33, 35, 43, 149, 150, 257, 301
  gross agricultural output value 32, 35
  gross animal husbandry output value 32, 35
  gross output value 32, 149, 150
  National Standard output value 257

Palatability 122
Pastoral areas
  inhabitants/population 120, 170, 189, 201, 214, 272
  pastoral areas of Gansu 117
  pastoral areas of IMAR 76
  pastoral areas of XUAR 136
  policy issues 33, 42-47, 63-65, 98, 101, 130-131, 142, 160, 162, 197, 210, 223, 299, 325, 333
  sheep production 52-54, 58, 154, 173, 174, 191, 194, 270, 291
  survey problems 21-24
  use of AI 175, 205, 274, 280
Pastoral counties 4, 21, 25, 34-37, 41, 53, 71, 76-78, 86, 100, 101, 117-120, 131, 133, 134, 137-139, 160, 206, 218-221, 238, 245, 246, 273, 277, 284, 301, 302, 304, 305, 316, 317
Pastoral provinces 1, 2, 10, 15, 25, 28-35, 37-39, 43, 52-56, 58-60, 62, 63, 65, 116, 307, 308, 330
Pastoral region 1-5, 9-11, 13, 15, 18, 25, 28-32, 34, 35, 37, 39, 41, 43, 47, 48, 56, 59, 60, 64, 76-78, 80, 93, 94, 100, 101, 119, 127, 134, 138, 167, 175, 181, 196, 197, 214, 246, 262, 277, 295, 299, 303-306, 308, 311, 312, 315-317, 321, 325-327, 328-336
Pastoralism 1, 2, 5, 10, 25, 28, 35, 334
Pastoralists 1, 3, 10, 48, 174, 276, 291, 299, 304, 309-312, 322, 323, 327, 329, 333, 335
Pasture degradation 5, 82, 83, 86, 112, 167, 169, 182, 213, 249, 251, 263, 301, 312, 326, 332, 333, 334
Pasture Improvement Station 19
Pasture land contract 47, 230, 261,

321
Pasture Management Station 96, 129, 142, 171, 197, 233, 249, 250, 266, 278, 287, 289, 290, 305
Pasture protection 233, 289
Policy failure 322
Policy mirage 20, 100, 279, 320-322, 334
Policy uncertainty 322
Population 12, 24-26, 34, 35, 37, 39-41, 56, 75, 77-81, 108, 116-120, 125, 139-141, 143, 146, 149, 150, 162, 167, 179, 187-189, 195, 202, 212, 215, 222, 228, 232, 237-240, 249-251, 261, 262, 264, 265, 272-274, 277-279, 285, 299-308, 314, 320, 322, 324, 326, 328-333, 335
  birth control 308
  demographic factors 308
  family planning concessions 304, 306
  financial pressures 300
  Household Registration 304, 305, 308, 314
  out-migration 306
  pastoral population 21, 196, 283, 303
  population density 28, 29, 77-80, 117, 119, 137-139, 170, 188, 201, 214, 240, 247, 249, 265, 274, 278, 287, 300-302, 304
  population flows 305
  population pressures 299, 300, 304, 322
  population transfer 306
  residential grouping 306
  temporary resident's permit 306
Poverty 4, 78-80, 119, 138, 187, 200, 209, 212, 238, 245, 246, 273, 285, 300, 301, 304, 306
  absolute poverty 42
  anti-hunger projects 133
  concessional interest-rate loans 318
  NSP counties 41, 42
  poor areas 2, 41, 42, 128, 316
  poor counties 41, 42, 77, 100, 119, 131, 133, 160, 317
  poor households 209, 228, 270
  poverty alleviation 4, 41, 80, 119, 131, 133, 139, 160, 168, 181, 182, 197, 222, 223, 245, 246, 257, 270, 315, 317, 318, 335
  poverty counties 41, 42, 197
  PSP counties 41, 42
  relative poverty 42
  Sanxi regions 42
Prices 24, 33, 44, 99, 156, 158-160, 162-164, 182, 195-198, 207, 208, 210, 220-222, 224, 243-246, 253-256, 268-270, 275, 291, 306, 309-312, 327, 330
  instructive prices 158, 255
  liveweight prices 256
  maximum and minimum (reserve) prices 105
  official purchasing prices 104, 105
  Price Bureau 68, 101, 104, 105, 152, 158, 162-164, 180, 208, 220, 221, 224, 244, 255, 256, 282, 294, 310
  price differentials 254, 294
  price incentives 295
  price penalties 312
  price premiums 258, 310
  price relativities 162, 226, 295, 309
  price structures 309
  Prices Commission 255
  (reserve) prices 105
  State procurement prices 310
  wool prices 56, 70, 71, 87, 101, 104, 125, 128, 133, 160, 162, 164, 180, 195, 208, 215, 221, 269, 282, 294, 310-312
  "instructional" prices 282
Product market distortions 309, 310, 312
Production and Construction Corps (PCC) 3, 19, 66-67, 141, 147-153, 155, 156, 158, 160
  Association for PCC Type of Xinjiang Fine Wool Sheep 156
  Xinjiang Production and Construction Corps 19
Property rights 12, 25, 42, 45-47, 63, 320-321
  ownership rights 42, 43, 46, 47, 107, 109, 313, 314, 320
  rangeland ownership rights 314
  tenure arrangements 181
Provisional Constitution 40

Questionnaires 13-16, 22
  survey questionnaires 13
Quotas 45, 68, 159, 180, 183, 191, 258, 305, 310
  animal product quota 259, 260

Rangeland conservation 317
Rangeland contracts 99, 314
Rangeland degradation 1-3, 5, 10, 47, 142, 169, 188, 191, 214, 287, 299-301, 304, 308, 312, 322-324, 332, 336
Rangeland improvement 181
Rangeland management 83, 106, 108-110, 217, 227, 312, 323
  animal:land ratio 324-326
  carrying capacity 81, 121-123, 169, 173, 190, 202, 251
  conservative/sustainable rangeland management 323
  cultivation mistakes 83
  cultural practices 268
  culturally-based attitudes 313
  dry matter productivity 273
  environmental risks 311, 327
  grassland productivity 121
  grazed in common 64, 99, 171, 191, 209, 222, 284, 290
  grazing capacity 109, 198
  grazing pressure 87, 124, 222, 251, 290, 310, 334
  pasture conservation 227, 231, 261, 321
  Proper Stocking Capacity 112, 115
  stocking rate limits 172, 322
  sustainable stocking rate 121, 122, 182, 191, 333
  theoretical carrying capacity 169
  utilized rate of pastures 111
Rangeland management fee
  grazing levy 99
  grazing management 5
  pasture management fee 99, 217, 251
  pasture use fee 99, 100, 171, 173, 251
  user pays 65, 99, 100, 131, 171, 191, 216, 251
Rangeland management office 108, 109
Rangeland management organisation 109, 110

Rangeland Management Regulation 83, 106, 217, 227
Reforms 10, 42, 43, 44, 46-48, 63, 64, 71, 130, 134, 310, 315, 319-321, 323
  Agrarian Land Reform 43
  economic reforms 21, 47, 48, 315, 332
  fragmentation 171
  market reforms 319
  policy reforms 5, 13, 299
  political reforms 332
Registration 109, 304, 305, 308, 314
Research 2, 9-14, 16-22, 48, 66, 67, 83, 93, 96, 97, 110, 129, 130, 148, 152, 153, 203, 267, 273, 282, 289, 293, 328, 332
  collaborative research 9, 10, 20
  desktop research 13, 20
  fieldwork 3, 9, 10, 12-15, 17, 18, 20, 21, 24
  household surveys 21
Resettlement 304, 305
  Han settlement 238
  resettlement programs 304, 305
Responsibility system 13, 42, 45, 46, 48, 56, 63, 64, 85-87, 93, 94, 98, 124, 130, 149, 154, 170, 171, 173, 181, 197, 202, 209, 215, 222, 241, 245, 256, 269, 283, 284, 308, 309, 311, 313-315, 320
  household production responsibility system 13, 42, 45, 46, 48, 56, 63, 64, 85-87, 93, 94, 98, 124, 130, 149, 154, 170, 171, 173, 181, 197, 202, 209, 215, 222, 241, 245, 256, 269, 283, 308, 309, 311, 313-315, 320
  household registration system 304, 305, 308, 314
  household responsibility system 45, 283
  work points 44-46, 130, 209, 283
Risk 15, 299, 315
  production risks 327

Seed Companies 69, 318
Semi-pastoral counties 4, 25, 34-37, 41, 53, 71, 76-78, 100, 117-120, 131, 133, 137-139, 160, 277, 301

Shearing 105, 243, 244, 280
　shearing sheds 280
Sheep breeding 94, 129, 130, 148, 156, 177, 192, 204, 218, 242, 253, 259, 267, 280, 292
　body weight 219, 243
　breed improvement 48, 129, 154-156, 169, 175, 178, 253, 327
　breed improvement programs 129, 156, 327
　breeding policy 293
　breeding programs 48, 51, 129, 177, 192, 204, 218, 242, 253, 267, 280, 292
　breeding rams 128, 156, 176, 259
　breeding specialists 156
　daughter studs 155, 156
　genetic improvement 243
　genetic regression 154, 155, 327
　mass selection techniques 238
　(National) Sheep Breeding Centre 148
　progeny testing 243
　ram breeding station 205
　ram care households 176
　ram holding/breeding stations 175, 176
　selection procedure 243
　sheep improvement program 94, 153, 155, 156, 177, 218, 242
Sheep breeds 48, 49, 51, 55, 94, 126-129, 176, 192, 199, 204, 205, 217, 218, 241, 266, 280
　Aohan Fine Wool sheep 92, 177, 193, 204
　Australian merino 92, 93, 155, 160, 177, 192, 241, 242, 253, 274, 281
　Border Leicester 129
　Charolais 94
　Chinese Merino 51, 55, 92, 153, 160, 192, 193, 281
　coloured wool sheep 252
　Corriedale 55
　crossbred sheep 50, 52, 65, 252
　Dorset Downs 94
　Erdos Fine Wool 51, 92, 238, 241-246
　Fat-tail 153
　Gansu Alpine Fine Wool 51, 127, 128, 252, 253, 274

　Gansu Alpine Merino 51
　Inner Mongolian Fine Wool sheep 92
　Inner Mongolian Semi-fine Wool sheep 92
　Kalequer (Black Lamb Pelt) sheep 159
　Kazak sheep 122, 153, 281
　Keerqin Fine Wool sheep 92
　Lincoln 156
　local sheep 50, 55, 93, 126, 177, 292, 327
　meat sheep 94, 153, 156, 157, 267, 292
　merino 51, 55, 92, 93, 127, 153-155, 160, 177, 192, 193, 218, 241, 242, 253, 274, 281
　Mongolian coarse wool 266, 267
　Mongolian type sheep 94
　North East Fine Wool 51
　Romney 55
　Russian merino 177
　Salisi dual-purpose sheep 175
　Shandong Short-tail 267
　Suffolk 156, 281
　Wulanchabu Fine Wool sheep 92
　Xingan Fine Wool sheep 92
　Xinjiang Fine Wool 51, 55, 127, 153-156, 159, 160, 253, 281
　Xinjiang Lamb Pelt sheep 153
　Xinjiang Merino 51
Sheep coats 319
Sheep equivalents 86-89, 100, 112, 113, 115, 124, 125, 143, 145, 146, 170, 172, 173, 183, 184, 185, 188-191, 202, 214, 215, 217, 228, 230-236, 251, 287, 288, 290
　conversion coefficients 24
　sheep units 24, 173, 191, 261
　standard sheep equivalent 112
Sheep husbandry 54, 130
Sheep Studs 27, 55, 65, 67, 130, 153, 157, 175, 176, 199, 200, 219, 248, 274, 278, 280, 281
　Aohan Stud 92, 175, 194, 201, 204-206, 218
　Cabucaer Stud 281, 282
　Chaganhua 51, 218
　Gadasu Stud 92, 193, 218
　Gongnaisi 51, 136, 153-155, 218, 253, 280-282, 292

Huang Cheng Stud 252, 253, 257, 258
Nan Shan 153, 154
Song Shan 127, 128, 274
Ziniquan 51, 153, 155, 156, 218
Shihezi Agricultural University 151
State farms 49, 63, 65-67, 71, 117, 128, 129, 141, 148, 149, 151, 154-156, 158-160, 170, 173, 189, 253, 257, 265, 278, 287, 294, 295, 319, 320, 327
  Huang Cheng State Farm 130, 248
  PCC State farms 66, 141, 149, 151, 156, 158, 160
  State Farms and Land Reclamation Department 66, 67
  Xinjiang General Bureau of State Farms and Land Reclamation 66, 148
Sumus 26, 76, 191, 192, 232, 240
Supplementary feeding 154, 174, 194, 203, 266, 279, 291, 324
  cut-and-carry 30
  cutting land 81, 99, 120, 171, 174, 183, 185, 191, 200, 203, 209, 231, 233, 266, 269, 288, 291, 323, 326
  daily fodder intake 112, 113
  feeding regime 12, 194, 323, 325
  hay supplements 174
  supplementary feeding standards 324
  winter feeding regimes 324
Survey counties
  Alukeerqin 4, 15, 16, 41, 43, 77, 78, 82, 86, 87, 89, 93-95, 99-101, 103, 186, 197, 198, 206, 211-227, 230, 245, 251, 279, 301-304, 311, 317, 327
  Aohan 4, 15, 27, 35, 41, 51, 77, 79, 87, 89, 92, 94, 100, 101, 103, 175-178, 186, 187, 192, 193, 194, 197, 199-204, 206-212, 214, 215, 217-221, 245, 304, 305, 325
  Balinyou 4, 15, 16, 19, 43, 77, 78, 82, 86-88, 93-95, 99-102, 167-183, 186, 188, 189, 196, 200, 202, 206, 211, 212, 214, 215, 220, 251, 279, 311, 317, 321
  Cabucaer 4, 16, 25, 37, 137, 142, 147, 154, 157-160, 162, 276-284, 289, 322, 325, 331, 335
  Dunhuang 4, 15, 25, 37, 117, 118, 121, 133, 262-271, 277, 335
  Hebukesaier 4, 16, 21, 26, 52, 93, 137, 138, 142, 147, 155, 158, 160, 162, 175, 285-295, 301-303, 322, 327, 331
  Sunan 4, 15, 18, 21, 46, 68, 117-120, 122, 123, 127, 131, 133, 134, 247-259, 261, 269, 331, 335
  Tianzhu 4, 16, 18, 21, 35, 117-120, 122, 124, 127, 128, 134, 272-275, 277
  Wongniute 4, 15, 77, 78, 82, 87, 88, 100-102, 186-198, 200, 202, 211, 212, 215, 317, 324
  Wushen 4, 5, 16, 21, 77, 79, 82, 92, 200, 237-246, 326, 331, 335
Sustainable development 1, 48, 160, 299, 300, 321, 328, 334-336

Tax 22, 23, 69, 71, 98, 133, 158, 164, 173, 209, 224, 238, 270, 315
  pasture construction tax 164
  value-added taxation 315
Technology 5, 10, 48, 96, 110, 128, 133, 176, 183, 193, 194, 253, 299, 310, 326, 328, 333, 335, 336
  technical improvements 322, 323, 336
Textile Corporation 246
Textiles 11, 55, 101, 148, 328, 329
  Bureau of Textiles 101
Tibet 28-34, 38, 39, 57, 58, 61, 62, 65, 307
Training
  Wuxi Training Centre 246

Uncertainty 68, 181, 251, 314, 320-322
  institutional changes 63, 171
  institutional uncertainties 299, 320, 322
  tenure uncertainty 251, 313, 314
Veterinary Institute 148
Veterinary Station 96, 129, 204, 253, 292, 293
Villages 16, 24, 26, 27, 35, 37, 53,

69, 131, 142, 170, 175, 189,
    193, 196, 201, 214, 217, 228,
    229, 231, 233, 239, 240, 249,
    253, 265, 267, 277, 278, 280,
    321
  administrative village 24, 45, 313
  natural or physical villages 24, 27,
    170, 240
  village collectives 64, 98, 99, 129,
    158, 256, 280, 321

Warlords 21
Water supplies 171, 263
Water wells 326
Weed species 83, 124, 169
Wool
  clean wool yield 93, 192, 241
  fibre density 252
  fibre diameter 11, 48, 49, 52, 60,
    62, 70, 96, 127, 218, 219, 274,
    310, 329
  Fibre Testing Bureau 19, 158
  fleece weight 243, 281
  grease 93, 178, 242, 281
  hair fibres 49, 50
  homogeneous wool 50, 51, 274
  stained 258
  staple length 49, 50, 93, 242, 243,
    281
  wool colour 274
  wool fibre 178, 219
  wool grading systems 258
  wool homogeneity 49
  wool length 192, 252, 274
  wool quality 54, 175, 176, 205,
    223, 245, 280, 320
  wool yield 93, 192, 241
Wool grading 49, 241, 244, 258
  grade differentials 162, 163
  grades of wool 24, 163, 164, 205,
    254, 255, 295, 309, 310
  grading skills 246
  grading system 49, 246, 310, 311,
    319
  grading technicians 245
  industrial grading system 246
  National Wool Grading Standard
    49, 70, 105, 162, 163, 180, 196,
    244, 310
  spinning count 127, 156, 219, 252,
    258
  training 55, 96, 97, 129, 152, 153,
    246, 259, 267
Wool marketing 48, 49, 68, 160, 207,
    309, 310, 319
  auctions 71
  institutionalised marketing margin
    69
  Purchasing Standard 49
  purchasing stations 311
  State procurement agency 68, 310
  State purchase quota 164
  wool exports 11
  wool sales 100, 164
  wool supply 269
  wool trade 158, 319, 329
  wool war 48, 68, 69, 104, 158,
    163, 164, 207, 208, 310, 318
Wool markets 68, 104, 180, 220
  category II commodity 68, 70, 310
Wool pricing 49, 70, 71, 100, 101,
    162, 311
Wool processors 14, 269
  wool textile 11, 68, 175, 258, 329,
    330
  wool-processing 151, 316
Wool scouring 134, 182, 197, 209,
    223, 257, 258, 275, 316
  first-stage processing 316
  scoured wool 60, 198, 224, 257,
    258, 316
Wool types 48-51, 58, 59, 63, 311
  Australian wool 11, 327, 329
  coarse wool 48, 50, 53, 63, 70, 71,
    91, 92, 94, 105, 121, 122, 126,
    129, 144, 153, 156, 159, 163,
    177, 221, 252, 266, 267, 269,
    274
  fine and improved fine wool 11,
    49-52, 54, 59, 60, 62, 65, 85, 86,
    90, 94, 116, 125-127, 138, 139,
    144, 147, 151, 152, 154, 155,
    224, 251, 252, 291, 329, 330
  fine wool 2-4, 10, 27, 48-52, 54-
    63, 65, 67, 70, 85-87, 90-95,
    105, 117, 121-123, 125-129,
    136-145, 150-160, 162, 163,
    175, 176, 177, 178, 180,
    192-194, 196, 199, 200, 204,
    206, 208, 210, 215, 217-219,
    221, 222, 225, 238, 241-246,
    251-255, 258, 267, 270-272,
    274-276, 278, 280-282, 290,
    291-294, 310, 311, 326, 327,

329-331, 336
semi-fine and improved semi-fine wool 50, 57, 59, 60, 62, 63, 85, 86, 93, 126, 139, 152
semi-fine wool 50, 51, 55-63, 70, 85, 86, 90-93, 105, 126, 128, 129, 139, 145, 151, 156, 159, 162, 163, 252, 271, 274, 275, 280, 293, 310, 331

Xinjiang Academy of Agricultural Reclamation Sciences 152
Xinjiang Academy of Animal Sciences 148, 154
Xinjiang Agricultural Commission 148
Xinjiang Animal Husbandry Industrial and Commercial Company 157
Xinjiang Uygur Autonomous Region (XUAR) 3, 4, 10, 15, 16, 26, 28-34, 37, 39, 43, 52, 55-63, 65-68, 85, 93, 116, 118, 122, 135-164, 276, 285, 290, 293, 294, 303, 304, 307, 319, 325, 330, 331, 335

Yaks 36, 85, 122, 123, 250, 274